THIRD
EDITION

Data Mining
For Business Analytics

Concepts, Techniques, and Applications
with XLMiner®

GALIT SHMUELI
PETER C. BRUCE
NITIN R. PATEL

Library of Congress Cataloging-in-Publication Data applied for.

Names: Shmueli, Galit, 1971- author. | Patel, Nitin R. (Nitin Ratilal),
    author. | Bruce, Peter C., 1953- author.
Title: Data mining for business analytics : concepts, techniques, and
    applications in Microsoft Office Excel with XLMiner / Galit Shmueli, Nitin
    R. Patel, Peter C. Bruce.
Other titles: Data mining for business intelligence
Description: Third edition. | Hoboken, New Jersey : John Wiley & Sons, 2016.
    | Includes index. | Originally published as: Data mining for business
    intelligence, 2007.
Identifiers: LCCN 2015040496 (print) | LCCN 2015047680 (ebook) | ISBN
    9781118729274 (cloth) | ISBN 9781118729137 (Adobe PDF) | ISBN
    9781118729243 (ePub)
Subjects: LCSH: Business–Data processing. | Data mining. | Microsoft Excel
    (Computer file)
Classification: LCC HF5548.2 .S44843 2016 (print) | LCC HF5548.2 (ebook) |
    DDC 005.54–dc23
LC record available at http://lccn.loc.gov/2015040496
    ISBN: 9781118729274

Printed in the United States of America

10 9 8 7 6 5 4 3 2

XLMiner for Education
www.solver.com/xlminer

Your new textbook, *Data Mining for Business Analytics: Concepts, Techniques, and Applications with XLMiner®, Third Edition*, uses this software throughout. Here's how to get it for your course.

For Instructors: Setting Up the Course Code

To set up a course code for your course, please email Frontline Systems at academic@solver.com, or call **775-831-0300**, press 0, and ask for the Academic Coordinator. Course codes MUST be renewed each time you teach your course.

The course code is free, and it can usually be issued within 24 to 48 hours (often the same day). It will enable your students to download and install XLMiner for Education with a semester-long (140 day) license, and will enable Frontline Systems to assist students with installation, and provide technical support to you during the course.

Please give the course code, plus the instructions, to your students. If you're evaluating the book for adoption, you can use the course code yourself to download and install the software as described below.

For Students: Installing XLMiner for Education

1) To download and install XLMiner for Education from Frontline Systems, to work with Microsoft Excel for Windows, please visit:

www.solver.com/xlminer-wiley

2) **Fill out the registration form on this page**, supplying your name, school, email address (key information will be sent to this address), Course Code (obtain this from your instructor), and Textbook Code (enter **SDMBI3**).

3) On the download page, change 32-bit to 64-bit ONLY if you've confirmed that you have 64-bit Excel (see below). **Click the Download Now button**, and save the downloaded file (SolverSetup.exe or SolverSetup64.exe).

4) **Close any Excel windows you have open**.

5) **Run SolverSetup/SolverSetup64 to install the software.** When prompted, enter the installation password and the license activation code contained in the email sent to the address you entered on the form above.

If you have problems downloading or installing, please email support@solver.com or call **775-831-0300** and press 4 (tech support). Say that you have XLMiner for Education, and have your course code and textbook code available.

If you have problems setting up or solving your model, or interpreting the results, please ask your instructor for assistance. Frontline Systems cannot help you with homework problems.

- *If you purchase this textbook but you aren't enrolled in a course, call 775-831-0300 and press 0 for assistance with the software.*
- If you have a Mac, you'll need to install "dual-boot" or VM software, Microsoft Windows, and Office or Excel for Windows first. Excel for Mac will NOT work. For more information see www.solver.com/using-frontline-solvers-macintosh. Ask your instructor if you can use your browser and xlminer.com.
- For Excel 2007, always download SolverSetup. In Excel 2010, choose File > Help and look in the lower right. In Excel 2016 and Excel 2013, choose File > Account > About Excel and look at the top of the dialog. Download SolverSetup64 ONLY if you see "64-bit" displayed.

To our families
Boaz and Noa
Liz, Lisa, and Allison
Tehmi, Arjun, and in
memory of Aneesh

Contents

CHAPTER 20   **Text Mining** 450

PART VIII   CASES

CHAPTER 21   **Cases** 468

Foreword

Data is the new gold and mining this gold to create business value in today's context of a highly networked and digital society requires a skillset that we haven't traditionally delivered in business or statistics or even engineering programs on their own. For those businesses and organizations that feel overwhelmed by today's big data, the phrase *you ain't seen nothing yet* comes to mind. Yesterday's three major sources of big data—the 20+ years of investment in enterprise systems (ERP, CRM, SCM, ...), the three billion plus people on the online social grid, and the close to 5 billion people carrying increasingly sophisticated mobile devices—are going to be dwarfed by tomorrow's smarter physical ecosystems fueled by the Internet of Things (IoT) movement.

The idea that we can use sensors to connect physical objects such as homes, automobiles, roads, even garbage bins, and street lights, to digitally optimized systems of governance goes hand in glove with bigger data and the need for deeper analytical capabilities. We are not far away from a smart refrigerator sensing that you are short on, say, eggs, populating your grocery store's mobile app's shopping list, and arranging a Task Rabbit to do a grocery run for you. Or the refrigerator negotiating a deal with an Uber driver to deliver an evening meal to you. Nor are we far away from sensors embedded in roads and vehicles that can compute traffic congestion, track roadway wear and tear, record vehicle use, and factor these into dynamic usage based pricing, insurance rates, and even taxation. This brave new world is going to be fueled by analytics and the ability to harness data for competitive advantage.

Business Analytics is an emerging discipline that is going to help us ride this new wave. This new Business Analytics discipline requires individuals who are grounded in the fundamentals of business such that they know the right questions to ask, who have the ability to harness, store, and optimally process vast datasets from a variety of structured and unstructured sources, and who can then use an array of techniques from machine learning and statistics to uncover new insights for decision making. Such individuals are a rare commodity today, but their creation has been the focus of this book for close to a decade now. This book's forte is that it relies on explaining the core set of concepts required for today's business analytics professionals using real-world data-rich cases in a hands-on manner, without sacrificing academic rigor. I say this with the confidence of someone who was probably the first adopter of the zeroth edition of this book (Spring 2006 at the Indian School of Business).

What particularly pleases me about the third edition is the addition of chapters on social network analytics and text mining, as well as a more detailed focus on ensemble methods that are going to become increasingly useful in more complex application domains. Also completing the picture are treatments of collaborative filtering and recommendation engines. All in all, this edition represents a comprehensive treatment of business analytics that goes beyond business intelligence, the notion of viewing the existing data that you have in a richer manner to get insights. Instead, it provides a modern-day foundation for Business Analytics, the notion of linking the X's to the Y's of interest in a predictive sense.

I look forward to using it in multiple fora, from executive education to MBA classrooms to specialized Business Analytics programs. I trust you will too!

RAVI BAPNA

Carlson School of Management, University of Minnesota, 2016

Preface to the Third Edition

S ince the book's first appearance in early 2007, it has been used by numerous practitioners and in many courses, ranging from dedicated data mining classes to more general business analytics courses (including our own experience teaching this material both online and in person for more than 10 years). Following feedback from instructors teaching MBA, undergraduate, and executive courses, and from students, we revised some of the existing materials as well as added new material.

The first noticeable change is the title: we now use *business analytics* in place of *business intelligence*. This update reflects the change in terminology since the second edition: BI today refers mainly to reporting and data visualization ("what is happening now"), while BA has taken over the "advanced analytics," which include predictive analytics and data mining. In this new edition we therefore also updated these terms in the book, using them as is currently common.

We added the new Part VII on *Data Analytics*, covering two new topics: Social Network Analysis and Text Mining. The Data Analytics chapters expand data mining into the realm of new data structures: networks and text. As in the rest of the book, Excel-based tools are used to present these topics.

The new chapter on Social Network Analysis (Chapter 19) introduces metrics and graphs that help you understand connections among people and entities, and use that information to understand social networks, and to contribute additional depth to predictive models.

The new Text Mining chapter (Chapter 20) discusses how to process text, and convert it into a useful form for predictive modeling, where the modeling then follows the same paradigm as presented earlier in the book.

Another new chapter to the third edition is "Combining Methods: Ensembles and Uplift Modeling" (Chapter 13). This chapter, which is the last in Part IV on "Prediction and Classification Methods," introduces two important approaches. The first, ensembles, are the combination of multiple models for improving predictive power. Ensembles have routinely proved their usefulness in practical applications and in data mining contests. The second topic, Uplift Modeling, introduces an improved approach for measuring the impact of an intervention, and also introduces the application of analytics by governments. Similar to other chapters, the new chapters include real-world examples and end-of-chapter problems.

In addition to the new chapters, we extended coverage of methods in some of the chapters. The chapter on association rules is now expanded to recommendation algorithms, with an additional section on the popular collaborative filtering approach.

The Cases chapter includes two new cases: one on political persuasion and uplift modeling, and another on taxi cancellations. Both use real data.

An important update in the third edition is an extensive revision of each of the examples, boxes with special topics, and exercises that rely on XLMiner software. Since the second edition, XLMiner has undergone several upgrades that improved speed, functionality, presentation, and algorithmic implementation. The XLMiner screenshots in the third edition are based on using the latest XLMiner version (currently, V2015-R2).

Since the second edition's appearance, the landscape of the courses using the textbook has greatly expanded: whereas initially the book was used mainly in semester-long elective MBA-level courses, it is now used in a variety of courses in Business Analytics degree and certificate programs, ranging from undergraduate programs, to postgraduate and executive education programs. Courses in such programs also vary in their duration and coverage. In many schools, our book is used across multiple courses. The book is designed to continue supporting the general "Predictive Analytics" or "Data Mining" course as well as supporting a set of courses in dedicated business analytics programs.

A general "Business Analytics," "Predictive Analytics," or "Data Mining" course, common in MBA and undergraduate programs as a one-semester elective, would cover Parts I to III, and choose a subset of methods from Parts IV and V. Instructors can choose to use Cases as team assignments, class discussions, or projects. For a two-semester course, Part VI might be considered, and we recommend introducing the new Part VII (Data Analytics).

For a set of courses in a dedicated business analytics program, here are a few courses that have been using our book:

Predictive Analytics: Supervised Learning In a dedicated Business Analytics program, the topic of Predictive Analytics is typically instructed across a set of courses. The first course would cover Parts I to IV and instructors typically choose a subset of methods from Part IV according to the course length. We recommend including the new Chapter 3 in such a course, as well as the new "Part VII: Data Analytics."

Predictive Analytics: Unsupervised Learning This course introduces data exploration and visualization, dimension reduction, mining relationships, and clustering (Parts III and V). If this course follows the Predictive

Analytics: Supervised Learning course, then it is useful to examine examples and approaches that integrate unsupervised and supervised learning, such as the new part on "Data Analytics."

Forecasting Analytics A dedicated course on time series forecasting would rely on Part VI.

Advanced Analytics A course that integrates the learnings from Predictive Analytics (supervised and unsupervised learning). Such a course can focus on Part VII: Data Analytics, where social network analytics and text mining are introduced. Some instructors choose to use the Cases chapter in such a course.

In all courses, we strongly recommend including a project component, where data are either collected by students according to their interest or provided by the instructor (e.g., from the many data mining competition datasets available). From our experience and other instructors' experience, such projects enhance the learning and provide students with an excellent opportunity to understand the strengths of data mining and the challenges that arise in the process.

Important Note: A cloud based version of XLMiner is now available on the web that significantly expands the range of potential users, freeing them from the constraints of using Excel or Windows. The cloud version functions nearly identically to the Excel-based version illustrated in these pages, so can effectively be used with this book.

Preface to the First Edition

This book arose out of a data mining course at MIT's Sloan School of Management and was refined during its use in data mining courses at the University of Maryland's R. H. Smith School of Business and at statistics.com. Preparation for the course revealed that there are a number of excellent books on the business context of data mining, but their coverage of the statistical and machine-learning algorithms that underlie data mining is not sufficiently detailed to provide a practical guide if the instructor's goal is to equip students with the skills and tools to implement those algorithms. On the other hand, there are also a number of more technical books about data mining algorithms, but these are aimed at the statistical researcher or more advanced graduate student, and do not provide the case-oriented business focus that is successful in teaching business students.

Hence, this book is intended for the business student (and practitioner) of data mining techniques, and its goal is threefold:

1. To provide both a theoretical and a practical understanding of the key methods of classification, prediction, reduction, and exploration that are at the heart of data mining.
2. To provide a business decision-making context for these methods.
3. Using real business cases, to illustrate the application and interpretation of these methods.

The presentation of the cases in the book is structured so that the reader can follow along and implement the algorithms on his or her own with a very low learning hurdle.

Just as a natural science course without a lab component would seem incomplete, a data mining course without practical work with actual data is missing a key ingredient. The MIT data mining course that gave rise to this book followed an introductory quantitative course that relied on Excel—this made its practical work universally accessible. Using Excel for data mining seemed a natural progression. An important feature of this book is the use of Excel, an environment familiar to business analysts. All required data mining algorithms (plus illustrative datasets) are provided in an Excel add-in, XLMiner. Data for both the cases and exercises are available at www.dataminingbook.com.

Although the genesis for this book lay in the need for a case-oriented guide to teaching data mining, analysts and consultants who are considering the application of data mining techniques in contexts where they are not currently in use will also find this a useful, practical guide.

Acknowledgments

The authors thank the many people who assisted us in improving the first edition and improving it further in the second edition and now in the third edition. Anthony Babinec, who has been using earlier editions of this book for years in his data mining courses at Statistics.com, provided us with detailed and expert corrections. Similarly, Dan Toy and John Elder IV greeted our project with enthusiasm and provided detailed and useful comments on earlier drafts.

Boaz Shmueli and Raquelle Azran gave detailed editorial comments and suggestions on the first two editions; Bruce McCullough and Adam Hughes did the same for the first edition. Noa Shmueli provided careful proofs of the third edition. Ravi Bapna, who used an early draft in a data mining course at the Indian School of Business, has provided invaluable comments and helpful suggestions since the book's start. Useful comments and feedback have also come from the many instructors, too numerous to mention, who have used the book in their classes. Susan Palocsay, Scott Nestler, Margret Bjarnadottir, and Mia Stephens provided suggestions and feedback on the second edition and drafts of the third edition.

From the Smith School of Business at the University of Maryland, colleagues Shrivardhan Lele, Wolfgang Jank, and Paul Zantek provided practical advice and comments. We thank Robert Windle, and MBA students Timothy Roach, Pablo Macouzet, and Nathan Birckhead for invaluable datasets. We also thank MBA students Rob Whitener and Daniel Curtis for the heatmap and map charts. And we thank the many MBA students from the University of Maryland and the Indian School of Business for fruitful discussions and interesting data mining projects that have helped shape and improve the book.

This book would not have seen the light of day without the nurturing support of the faculty at the Sloan School of Management at MIT. Our special thanks to Dimitris Bertsimas, James Orlin, Robert Freund, Roy Welsch, Gordon Kaufmann, and Gabriel Bitran. As teaching assistants for the data mining course at Sloan, Adam Mersereau gave detailed comments on the notes and cases that were the genesis of this book, Romy Shioda helped with the preparation of several cases and exercises used here, and Mahesh Kumar helped with the material on clustering. We are grateful to the MBA students at Sloan for stimulating discussions in the class that led to refinement of the notes as well as XLMiner.

Chris Albright, Wayne Winston, and Uday Karmarkar gave us helpful advice on the use of XLMiner. Anand Bodapati provided both data and advice. Jake Hofman from Microsoft Research and Sharad Borle assisted with data access. Suresh Ankolekar and Mayank Shah helped develop several cases and provided valuable pedagogical comments. Vinni Bhandari helped write the Charles Book Club case.

We would like to thank Marvin Zelen, L. J. Wei, and Cyrus Mehta at Harvard, as well as Anil Gore at Pune University, for thought-provoking discussions on the relationship between statistics and data mining. Our thanks to Richard Larson of the Engineering Systems Division, MIT, for sparking many stimulating ideas on the role of data mining in modeling complex systems. They helped us develop a balanced philosophical perspective on the emerging field of data mining.

Our thanks to Ajay Sathe, who energetically shepherded XLMiner's development, and to his colleagues on the initial XLMiner team: Suresh Ankolekar, Poonam Baviskar, Kuber Deokar, Rupali Desai, Yogesh Gajjar, Ajit Ghanekar, Ayan Khare, Bharat Lande, Dipankar Mukhopadhyay, S. V. Sabnis, Usha Sathe, Anurag Srivastava, V. Subramaniam, Ramesh Raman, and Sanhita Yeolkar. We also thank Dan Fylstra, the founder of Frontline Systems, who has overseen the incorporation of XLMiner into the Solver suite. Frontline's programming team members Eissa Nematollahi and Oleg Shirokikh have continued the extension of XLMiner's capabilities.

Steve Quigley at Wiley showed confidence in this book from the beginning and helped us navigate through the publishing process with great speed. Curt Hinrichs' vision, tips, and encouragement helped bring this book to the starting gate. Jon Gurstelle, Allison McGinniss, Sari Friedman, and Katrina Maceda at Wiley, and Shikha Pahuja from Thomson Digital, were all helpful and responsive as we finalized this third edition. We are also grateful to Ashwini Kumthekar, Achala Sabane, Michael Shapard, Amy Hendrickson, and Heidi Sestrich who assisted with typesetting, figures, and indexing, and to Valerie Troiano who has shepherded many instructors through the use of XLMiner and early drafts of this text.

We also thank Catherine Plaisant at the University of Maryland's Human–Computer Interaction Lab, who helped out in a major way by contributing exercises and illustrations to the data visualization chapter, Marietta Tretter at Texas A&M for her helpful comments and thoughts on the time series chapters, and Stephen Few and Ben Shneiderman for feedback and suggestions on the data visualization chapter and overall design tips.

Gregory Piatetsky-Shapiro, founder of KDNuggets.com, has been generous with his time and counsel over the many years of this project. Ken Strasma, founder of the microtargeting firm HaystaqDNA and director of targeting for the 2004 Kerry campaign and the 2008 Obama campaign, provided the scenario and data for the section on uplift modeling. We also thank Jen Golbeck, director of the Social Intelligence Lab at the University of Maryland and author of *Analyzing the Social Web*, whose book inspired our presentation in the chapter on Social Network Analytics.

We also thank the many students who have used and commented on this text at Statistics.com and, for trouble-shooting and followup, the Statistics.com team, led by Kuber Deokar, Instructional Operations Supervisor, Dhanashree Vishwasrao, Assistant Teacher, and, with special gratitude for her hours of double-checking XLMiner, Shweta Jadhav.

## PART I

Preliminaries

## CHAPTER 1

Introduction

1.1 WHAT IS BUSINESS ANALYTICS?

Business analytics (BA) is the practice and art of bringing quantitative data to bear on decision making. The term means different things to different organizations.

Consider the role of analytics in helping newspapers survive the transition to a digital world. One tabloid newspaper with a working-class readership in Britain had launched a web version of the paper, and did tests on its home page to determine which images produced more hits: cats, dogs, or monkeys. This simple application, for this company, was considered analytics. By contrast, the *Washington Post* has a highly influential audience that is of interest to big defense contractors: it is perhaps the only newspaper where you routinely see advertisements for aircraft carriers. In the digital environment, the *Post* can track readers by time of day, location, and user subscription information. In this fashion, the display of the aircraft carrier advertisement in the online paper may be focused on a very small group of individuals—say, the members of the House and Senate Armed Services Committees who will be voting on the Pentagon's budget.

Business analytics, or more generically, *analytics*, includes a range of data analysis methods. Many powerful applications involve little more than counting, rule checking, and basic arithmetic. For some organizations, this is what is meant by analytics.

The next level of business analytics, now termed *business intelligence* (BI), refers to data visualization and reporting for understanding "what happened and what is happening." This is done by use of charts, tables, and dashboards to display, examine, and explore data. BI, which earlier consisted mainly of

Data Mining for Business Analytics: Concepts, Techniques, and Applications with XLMiner®, Third Edition.
Galit Shmueli, Peter C. Bruce, and Nitin R. Patel.
© 2016 John Wiley & Sons Inc. Published 2016 by John Wiley & Sons, Inc.

generating static reports, has evolved into more user-friendly and effective tools and practices, such as creating interactive dashboards that allow the user not only to access real-time data but also to directly interact with it. Effective dashboards are those that tie directly into company data, and give managers a tool to quickly see what might not readily be apparent in a large complex database. One such tool for industrial operations managers displays customer orders in a single two-dimensional display, using color and bubble size as added variables, showing customer name, type of product, size of order, and length of time to produce.

Business analytics now typically includes BI as well as sophisticated data analysis methods, such as statistical models and data mining algorithms used for exploring data, quantifying and explaining relationships between measurements, and predicting new records. Methods like regression models are used to describe and quantify "on average" relationships (e.g., between advertising and sales), to predict new records (e.g., whether a new patient will react positively to a medication), and to forecast future values (e.g., next week's web traffic).

Readers familiar with earlier editions of this book might have noticed that the book title changed from *Data Mining for Business Intelligence* to *Data Mining for business analytics* in this edition. The change reflects the more recent term BA, which overtook the earlier term BI to denote advanced analytics. Today, BI is used to refer to data visualization and reporting.

WHO USES PREDICTIVE ANALYTICS?

The widespread adoption of predictive analytics, coupled with the accelerating availability of data, has increased organizations' capabilities throughout the economy. A few examples:

Credit scoring: One long-established use of predictive modeling techniques for business prediction is credit scoring. A credit score is not some arbitrary judgment of credit-worthiness; it is based mainly on a predictive model that uses prior data to predict repayment behavior.

Future purchases: A more recent (and controversial) example is Target's use of predictive modeling to classify sales prospects as "pregnant" or "not-pregnant." Those classified as pregnant could then be sent sales promotions at an early stage of pregnancy, giving Target a head start on a significant purchase stream.

Tax evasion: The US Internal Revenue Service found it was 25 times more likely to find tax evasion when enforcement activity was based on predictive models, allowing agents to focus on the most likely tax cheats (Siegel, 2013).

The business analytics toolkit also includes statistical experiments, the most common of which is known to marketers as A-B testing. These are often used for pricing decisions:

- Orbitz, the travel site, found that it could price hotel options higher for Mac users than Windows users.
- Staples online store found it could charge more for staplers if a customer lived far from a Staples store.

Beware the organizational setting where analytics is a solution in search of a problem: A manager, knowing that business analytics and data mining are hot areas, decides that her organization must deploy them too, to capture that hidden value that must be lurking somewhere. Successful use of analytics and data mining requires both an understanding of the business context where value is to be captured, and an understanding of exactly what the data mining methods do.

1.2 WHAT IS DATA MINING?

In this book, data mining refers to business analytics methods that go beyond counts, descriptive techniques, reporting, and methods based on business rules. While we do introduce data visualization, which is commonly the first step into more advanced analytics, the book focuses mostly on the more advanced data analytics tools. Specifically, it includes statistical and machine-learning methods that inform decision making, often in automated fashion. Prediction is typically an important component, often at the individual level. Rather than "what is the relationship between advertising and sales," we might be interested in "what specific advertisement, or recommended product, should be shown to a given online shopper at this moment?" Or we might be interested in clustering customers into different "personas" that receive different marketing treatment, then assigning each new prospect to one of these personas.

The era of big data has accelerated the use of data mining. Data mining methods, with their power and automaticity, have the ability to cope with huge amounts of data and extract value.

1.3 DATA MINING AND RELATED TERMS

The field of analytics is growing rapidly, both in terms of the breadth of applications, and in terms of the number of organizations using advanced analytics. As a result there is considerable overlap and inconsistency of definitions.

The term *data mining* itself means different things to different people. To the general public, it may have a general, somewhat hazy and pejorative meaning of digging through vast stores of (often personal) data in search of something interesting. One major consulting firm has a "data mining department," but its responsibilities are in the area of studying and graphing past data in search of general trends. And, to confuse matters, their more advanced predictive models are the responsibility of an "advanced analytics department." Other terms that organizations use are *predictive analytics*, *predictive modeling*, and *machine learning*.

Data mining stands at the confluence of the fields of statistics and machine learning (also known as *artificial intelligence*). A variety of techniques for exploring data and building models have been around for a long time in the world of statistics: linear regression, logistic regression, discriminant analysis, and principal components analysis, for example. But the core tenets of classical statistics—computing is difficult and data are scarce—do not apply in data mining applications where both data and computing power are plentiful.

This gives rise to Daryl Pregibon's description of data mining as "statistics at scale and speed" (Pregibon, 1999). Another major difference between the fields of statistics and machine learning is the focus in statistics on inference from a sample to the population regarding an "average effect"—for example, "a $1 price increase will reduce average demand by 2 boxes." In contrast, the focus in machine learning is on predicting individual records—"the predicted demand for person *i* given a $1 price increase is 1 box, while for person *j* it is 3 boxes." The emphasis that classical statistics places on inference (determining whether a pattern or interesting result might have happened by chance in our sample) is absent from data mining.

In comparison to statistics, data mining deals with large datasets in an open-ended fashion, making it impossible to put the strict limits around the question being addressed that inference would require. As a result the general approach to data mining is vulnerable to the danger of *overfitting*, where a model is fit so closely to the available sample of data that it describes not merely structural characteristics of the data but random peculiarities as well. In engineering terms, the model is fitting the noise, not just the signal.

In this book, we use the term *machine learning* to refer to algorithms that learn directly from data, especially local patterns, often in layered or iterative fashion. In contrast, we use *statistical models* to refer to methods that apply global structure to the data. A simple example is a linear regression model (statistical) versus a *k*-nearest-neighbors algorithm (machine learning). A given record would be treated by linear regression in accord with an overall linear equation that applies to *all* the records. In *k*-nearest-neighbors, that record would be classified in accord with the values of a small number of nearby records.

Last, many practitioners, particularly those from the IT and computer science communities, use the term *machine learning* to refer to all the methods discussed in this book.

1.4 BIG DATA

Data mining and big data go hand in hand. *Big data* is a relative term—data today are big by reference to the past, and to the methods and devices available to deal with them. The challenge big data presents is often characterized by the four V's—volume, velocity, variety, and veracity. *Volume* refers to the amount of data. *Velocity* refers to the flow rate—the speed at which it is being generated and changed. *Variety* refers to the different types of data being generated (currency, dates, numbers, text, etc.). *Veracity* refers to the fact that data is being generated by organic distributed processes (e.g., millions of people signing up for services or free downloads) and not subject to the controls or quality checks that apply to data collected for a study.

Most large organizations face both the challenge and the opportunity of big data because most routine data processes now generate data that can be stored and, possibly, analyzed. The scale can be visualized by comparing the data in a traditional statistical analysis (e.g., 15 variables and 5000 records) to the Walmart database. If you consider the traditional statistical study to be the size of a period at the end of a sentence, then the Walmart database is the size of a football field. And that probably does not include other data associated with Walmart—social media data, for example, which comes in the form of unstructured text. If the analytical challenge is substantial, so can be the reward:

- OKCupid, the online dating site, uses statistical models with their data to predict what forms of message content are most likely to produce a response.
- Telenor, a Norwegian mobile phone service company, was able to reduce subscriber turnover 37% by using models to predict which customers were most likely to leave, and then lavishing attention on them.
- Allstate, the insurance company, tripled the accuracy of predicting injury liability in auto claims by incorporating more information about vehicle type.

The examples above are from Eric Siegel's book *Predictive Analytics* (2013, Wiley).

Some extremely valuable tasks were not even feasible before the era of big data. Consider web searches, the technology on which Google was built. In early days, a search for "Ricky Ricardo Little Red Riding Hood" would have yielded various links to the *I Love Lucy* TV show, other links to Ricardo's career as a band leader, and links to the children's story of Little Red Riding Hood. Only once the Google database had accumulated sufficient data (including records of what users clicked on) would the search yield, in the top position, links to the specific *I Love Lucy* episode in which Ricky enacts, in a comic mixture of Spanish and English, Little Red Riding Hood for his infant son.

1.5 DATA SCIENCE

The ubiquity, size, value, and importance of big data has given rise to a new profession: the *data scientist*. *Data science* is a mix of skills in the areas of statistics, machine learning, math, programming, business, and IT. The term itself is thus broader than the other concepts we discussed above, and it is a rare individual who combines deep skills in all the constituent areas. In their book *Analyzing the Analyzers* (Harris et al., 2013), the authors describe the skillsets of most data scientists as resembling a "T"—deep in one area (the vertical bar of the T), and shallower in other areas (the top of the T).

At a large data science conference session (Strata-Hadoop World, October 2014) most attendees felt that programming was an essential skill, though there was a sizable minority who felt otherwise. And, although big data is the motivating power behind the growth of data science, most data scientists do not actually spend most of their time working with terabyte-size or larger data.

Data of the terabyte or larger size would be involved at the deployment stage of a model. There are manifold challenges at that stage, most of them IT and programming issues related to data handling and tying together different components of a system. Much work must precede that phase. It is that earlier piloting and prototyping phase on which this book focuses—developing the statistical and machine learning models that will eventually be plugged into a deployed system. What methods do you use with what sorts of data and problems? How do the methods work? What are their requirements, their strengths, their weaknesses? How do you assess their performance?

1.6 WHY ARE THERE SO MANY DIFFERENT METHODS?

As can be seen in this book or any other resource on data mining, there are many different methods for prediction and classification. You might ask yourself why they coexist, and whether some are better than others. The answer is that each method has advantages and disadvantages. The usefulness of a method can depend on factors such as the size of the dataset, the types of patterns that

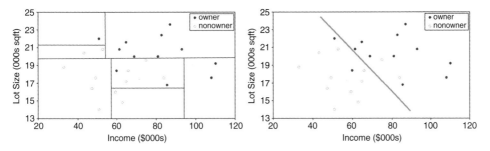

FIGURE 1.1   TWO METHODS FOR SEPARATING OWNERS FROM NONOWNERS

exist in the data, whether the data meet some underlying assumptions of the method, how noisy the data are, and the particular goal of the analysis. A small illustration is shown in Figure 1.1, where the goal is to find a combination of *household income level* and *household lot size* that separates owners (solid circles) from nonowners (hollow circles) of riding mowers. The first method (left panel) looks only for horizontal and vertical lines to separate owners from nonowners; the second method (right panel) looks for a single diagonal line.

Different methods can lead to different results, and their performance can vary. It is therefore customary in data mining to apply several different methods and select the one that appears most useful for the goal at hand.

1.7   TERMINOLOGY AND NOTATION

Because of the hybrid parentry of data mining, its practitioners often use multiple terms to refer to the same thing. For example, in the machine learning (artificial intelligence) field, the variable being predicted is the output variable or target variable. To a statistician, it is the dependent variable or the response. Here is a summary of terms used:

Algorithm   A specific procedure used to implement a particular data mining technique: classification tree, discriminant analysis, and the like.

Attribute   See **Predictor**.

Case   See **Observation**.

Confidence   A performance measure in association rules of the type "IF *A* and *B* are purchased, THEN *C* is also purchased." Confidence is the conditional probability that *C* will be purchased IF *A* and *B* are purchased.

Confidence   Also has a broader meaning in statistics (*confidence interval*), concerning the degree of error in an estimate that results from selecting one sample as opposed to another.

Dependent Variable   See **Response**.

Estimation   See **Prediction**.

Feature   See **Predictor**.

Holdout Data   (or **holdout set**) A sample of data not used in fitting a model, but instead used to assess the performance of that model. This book uses the terms *validation set* and *test set* instead of *holdout set*.

Input Variable   See **Predictor**.

Model   An algorithm as applied to a dataset, complete with its settings (many of the algorithms have parameters that the user can adjust).

Observation   The unit of analysis on which the measurements are taken (a customer, a transaction, etc.); also called *instance, sample, example, case, record, pattern,* or *row.* In spreadsheets, each row typically represents a record; each column, a variable. Note that the use of the term "sample" here is different from its usual meaning in statistics, where it refers to a collection of observations.

Outcome Variable   See **Response**.

Output Variable   See **Response**.

$P(A|B)$ The conditional probability of event A occurring given that event B has occurred. Read as "the probability that A will occur given that B has occurred."

Profile   A set of measurements on an observation (e.g., the height, weight, and age of a person).

Prediction   The prediction of the numerical value of a continuous output variable; also called *estimation*.

Predictor   A variable, usually denoted by X, used as an input into a predictive model. Also called a *feature, input variable, independent variable,* or from a database perspective, a *field*.

Record   See **Observation**.

Response   A variable, usually denoted by Y, which is the variable being predicted in supervised learning; also called *dependent variable, output variable, target variable,* or *outcome variable*.

Sample   In the statistical community, "sample" means a collection of observations. In the machine learning community, "sample" means a single observation.

Score   A predicted value or class. *Scoring new data* means using a model developed with training data to predict output values in new data.

Success Class   The class of interest in a binary outcome (e.g., *purchasers* in the outcome *purchase/no purchase*).

Supervised Learning   The process of providing an algorithm (logistic regression, regression tree, etc.) with records in which an output variable of interest is known and the algorithm "learns" how to predict this value with new records where the output is unknown.

Target   See **Response**.

Test Data   (or **test set**) The portion of the data used only at the end of the model building and selection process to assess how well the final model might perform on new data.

Training Data   (or **training set**) The portion of the data used to fit a model.

Unsupervised Learning   An analysis in which one attempts to learn patterns in the data other than predicting an output value of interest.

Validation Data   (or **validation set**) The portion of the data used to assess how well the model fits, to adjust models, and to select the best model from among those that have been tried.

Variable   Any measurement on the records, including both the input (X) variables and the output (Y) variable.

1.8   ROAD MAPS TO THIS BOOK

The book covers many of the widely used predictive and classification methods as well as other data mining tools. Figure 1.2 outlines data mining from a process perspective and where the topics in this book fit in. Chapter numbers are indicated beside the topic. Table 1.1 provides a different perspective: it organizes data mining procedures according to the type and structure of the data.

Order of Topics

The book is divided into five parts: Part I (Chapters 1–2) gives an overview of data mining and its components. Part II (Chapters 3–4) focuses on the early stages of data exploration and dimension reduction.

FIGURE 1.2 DATA MINING FROM A PROCESS PERSPECTIVE. NUMBERS IN PARENTHESES INDICATE CHAPTER NUMBERS

TABLE 1.1 ORGANIZATION OF DATA MINING METHODS IN THIS BOOK, ACCORDING TO THE NATURE OF THE DATA

	Supervised		Unsupervised
	Continuous Response	**Categorical Response**	**No Response**
Continuous Predictors	Linear regression (6) Neural nets (11) k-Nearest-neighbors (7) Ensembles (13)	Logistic regression (10) Neural nets (11) Discriminant analysis (12) k-Nearest-neighbors (7) Ensembles (13)	Principal components (4) Cluster analysis (15) Collaborative filtering (14)
Categorical Predictors	Linear regression (6) Neural nets (11) Regression trees (9) Ensembles (13)	Neural nets (11) Classification trees (9) Logistic regression (10) Naive Bayes (8) Ensembles (13)	Association rules (14) Collaborative filtering (14)

Note: Numbers in parentheses indicate chapter number.

Part III (Chapter 5) discusses performance evaluation. Although it contains only one chapter, we discuss a variety of topics, from predictive performance metrics to misclassification costs. The principles covered in this part are crucial for the proper evaluation and comparison of supervised learning methods.

Part IV includes eight chapters (Chapters 6–13), covering a variety of popular supervised learning methods (for classification and/or prediction). Within this part, the topics are generally organized according to the level of sophistication of the algorithms, their popularity, and ease of understanding. The final chapter introduces ensembles and combinations of methods.

Part V focuses on unsupervised mining of relationships. It presents association rules and collaborative filtering (Chapter 14) and cluster analysis (Chapter 15).

Part VI includes three chapters (Chapters 16–18), with the focus on forecasting time series. The first chapter covers general issues related to handling and understanding time series. The next two chapters present two popular forecasting approaches: regression-based forecasting and smoothing methods.

Part VII (Chapters 19–20) presents two broad data analytics topics: social network analysis and text mining. These methods apply data mining to specialized data structures: social networks and text.

Finally, Part VIII includes a set of cases.

Although the topics in the book can be covered in the order of the chapters, each chapter stands alone. We advise, however, that Parts I–III be read before proceeding to chapters in Parts IV–V. Similarly, Chapter 16 should precede other chapters in Part VI.

USING XLMINER SOFTWARE

To facilitate hands-on data mining experience, this book comes with access to XLMiner, a comprehensive data mining add-in for Excel. For those familiar with Excel, the use of an Excel add-in dramatically shortens the software learning curve. XLMiner will help you get started quickly on data mining and offers a variety of methods for analyzing data. The illustrations, exercises, and cases in this book are written in relation to this software. XLMiner has extensive coverage of statistical and data mining techniques for classification, prediction, mining associations and text, forecasting, and data exploration and reduction. It offers a variety of supervised data mining tools: neural nets, classification and regression trees, *k*-nearest-neighbor classification, naive Bayes, logistic regression, linear regression, and discriminant analysis, all for predictive modeling. It provides for automatic partitioning of data into training, validation, and test samples and for the deployment of the model to new data. It also offers unsupervised algorithms: association rules, principal components analysis, *k*-means clustering, and hierarchical clustering, as well as visualization tools and data-handling utilities. With its short learning curve, affordable price, and reliance on the familiar Excel platform, it is an ideal companion to a book on data mining for the business student.

Download To download the XLMiner setup program, visit www.solver.com/xlminer-data-mining and follow the instructions there.

Installation Close any Excel windows, then run the XLMiner setup program. Dialog boxes will guide you through the installation procedure. The final dialog box gives you an option to start Excel and open a "Getting Started" workbook. You'll also find XLMiner options under *Start > All Programs > Frontline Systems*.

Use XLMiner is loaded when you start Excel, and appears as an Excel Ribbon, as shown in Figure 1.3. By choosing the appropriate menu item, you can run any of XLMiner's procedures on the dataset that is open in your Excel worksheet.

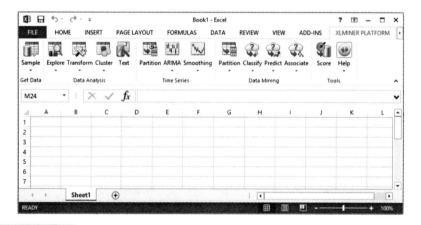

FIGURE 1.3   XLMINER SCREEN

## CHAPTER 2

Overview of the Data Mining Process

In this chapter we give an overview of the steps involved in data mining, starting from a clear goal definition and ending with model deployment. The general steps are shown schematically in Figure 2.1. We also discuss issues related to data collection, cleaning, and preprocessing. We introduce the notion of data partitioning, where methods are trained on a set of training data and then their performance is evaluated on a separate set of validation data, as well as explain how this practice helps avoid overfitting. Finally, we illustrate the steps of model building by applying them to data.

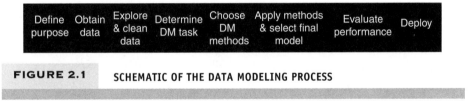

FIGURE 2.1 SCHEMATIC OF THE DATA MODELING PROCESS

2.1 INTRODUCTION

In Chapter 1 we saw some very general definitions of data mining. In this chapter we introduce the variety of methods sometimes referred to as *data mining*. The core of this book focuses on what has come to be called *predictive analytics,* the tasks of classification and prediction as well as pattern discovery, which have become key elements of a "business analytics" function in most large firms. These terms are described and illustrated below.

Not covered in this book to any great extent are two simpler database methods that are sometimes considered to be data mining techniques: (1) OLAP (online analytical processing) and (2) SQL (structured query language). OLAP and SQL searches on databases are descriptive in nature and are based on business rules set by the user (e.g., "find all credit card customers in a certain zip code with annual charges > \$20,000, who own their own home and who pay the entire amount of their monthly bill at least 95% of the time.") They do not involve statistical modeling or automated algorithmic methods, although SQL queries are often used to obtain the data in data mining.

2.2 Core Ideas in Data Mining

Classification

Classification is perhaps the most basic form of data analysis. The recipient of an offer can respond or not respond. An applicant for a loan can repay on time, repay late, or declare bankruptcy. A credit card transaction can be normal or fraudulent. A packet of data traveling on a network can be benign or threatening. A bus in a fleet can be available for service or unavailable. The victim of an illness can be recovered, still be ill, or be deceased.

A common task in data mining is to examine data where the classification is unknown or will occur in the future, with the goal of predicting what that classification is or will be. Similar data where the classification is known are used to develop rules, which are then applied to the data with the unknown classification.

Prediction

Prediction is similar to classification, except that we are trying to predict the value of a numerical variable (e.g., amount of purchase) rather than a class (e.g., purchaser or nonpurchaser). Of course, in classification we are trying to predict a class, but the term *prediction* in this book refers to the prediction of the value of a continuous variable. (Sometimes in the data mining literature, the terms *estimation* and *regression* are used to refer to the prediction of the value of a continuous variable, and *prediction* may be used for both continuous and categorical data.)

Association Rules and Recommendation Systems

Large databases of customer transactions lend themselves naturally to the analysis of associations among items purchased, or "what goes with what." *Association rules*, or *affinity analysis*, is designed to find such general associations patterns between items in large databases. The rules can then be used in a variety of ways. For example, grocery stores can use such information for product placement. They can use the rules for weekly promotional offers or for bundling products. Association rules derived from a hospital database on patients' symptoms during consecutive hospitalizations can help find "which symptom is followed by what other symptom" to help predict future symptoms for returning patients.

Online recommendation systems, such as those used on Amazon.com and Netflix.com, use *collaborative filtering*, a method that uses individual users' preferences and tastes given their historic purchase, rating, browsing, or any other measurable behavior indicative of preference, as well as other users' history. In contrast to *association rules* that generate rules general to an entire population, collaborative filtering generates "what goes with what" at the individual user level. Hence collaborative filtering is used in many recommendation systems that aim to deliver personalized recommendations to users with a wide range of preferences.

Predictive Analytics

Classification, prediction, and, to some extent, association rules and collaborative filtering constitute the analytical methods employed in *predictive analytics*. The term predictive analytics is sometimes used to also include data pattern identification methods such as clustering.

Data Reduction and Dimension Reduction

The performance of data mining algorithms is often improved when the number of variables is limited, and when large numbers of records can be grouped into homogeneous groups. For example, rather than dealing with thousands of product types, an analyst might wish to group them into a smaller number of groups and build separate models for each group. Or a marketer might want to classify customers into different "personas," and must therefore group customers into homogeneous groups to define the personas. This process of consolidating a large number of records (or cases) into a smaller set is termed *data reduction*. Methods for reducing the number of cases are often called *clustering*.

Reducing the number of variables is typically called *dimension reduction*. Dimension reduction is a common initial step before deploying supervised learning methods, intended to improve predictive power, manageability, and interpretability.

Data Exploration and Visualization

One of the earliest stages of engaging with a dataset is exploring it. Exploration is aimed at understanding the global landscape of the data, and detecting unusual values. Exploration is used for data cleaning and manipulation as well as for visual discovery and "hypothesis generation."

Methods for exploring data include looking at various data aggregations and summaries, both numerically and graphically. This includes looking at each variable separately as well as looking at relationships between variables. The purpose is to discover patterns and exceptions. Exploration by creating charts and dashboards is called *data visualization* or *visual analytics*. For numerical variables, we use histograms and boxplots to learn about the distribution of their values, to detect outliers (extreme observations), and to find other information that is relevant to the analysis task. Similarly, for categorical variables we use bar charts. We can also look at scatter plots of pairs of numerical variables to learn about possible relationships, the type of relationship, and again, to detect outliers. Visualization can be greatly enhanced by adding features such as color and interactive navigation.

Supervised and Unsupervised Learning

A fundamental distinction among data mining techniques is between supervised and unsupervised methods. *Supervised learning algorithms* are those used in classification and prediction. We must have data available in which the value of the outcome of interest (e.g., purchase or no purchase) is known. These *training data* are the data from which the classification or prediction algorithm "learns," or is "trained," about the relationship between predictor variables and the outcome variable. Once the algorithm has learned from the training data, it is then applied to another sample of data (the *validation data*) where the outcome is known, to see how well it does in comparison to other models. If many different models are being tried out, it is prudent to save a third sample, which also includes known outcomes (the *test data*) to use with the model finally selected to predict how well it will do. The model can then be used to classify or predict the outcome of interest in new cases where the outcome is unknown.

Simple linear regression is an example of a supervised learning algorithm (although rarely called that in the introductory statistics course where you probably first encountered it). The Y variable is the (known) outcome variable and the X variable is a predictor variable. A regression line is drawn to minimize the sum of squared deviations between the actual Y values and the values predicted by this line. The regression line can now be used to predict Y values for new values of X for which we do not know the Y value.

Unsupervised learning algorithms are those used where there is no outcome variable to predict or classify. Hence there is no "learning" from cases where such an outcome variable is known. Association rules, dimension reduction methods, and clustering techniques are all unsupervised learning methods.

Supervised and unsupervised methods are sometimes used in conjunction. For example, unsupervised clustering methods are used to separate loan applicants into several risk-level groups. Then supervised algorithms are applied separately to each risk-level group for predicting loan default propensity.

SUPERVISED LEARNING REQUIRES GOOD SUPERVISION

In some cases, the value of the target variable is known because it is an inherent component of the data. Web logs will show whether a person clicked on a link or didn't click. Bank records will show whether a loan was paid on time or not. In other cases, the value of the known target must be supplied by a human labeling process to accumulate enough data to train a model. Email must be labeled as spam or legitimate, documents in legal discovery must be labeled as relevant or irrelevant. In either case, the data mining algorithm can be led astray if the quality of the supervision is poor.

Gene Weingarten reported in the January 5, 2014, *Washington Post* magazine how the strange phrase "defiantly recommend" is making its way into English via auto-correction. "Defiantly" is closer to the common misspelling *definatly* than is *definitely*, so Google.com, in the early days, offered it as a correction when users typed the misspelled word "definatly." In the ideal supervised learning model, humans guide the auto-correction process by rejecting *defiantly* and substituting *definitely*. Google's algorithm would then learn that this is the best first-choice correction of "definatly." The problem was that too many people were lazy, just accepting the first correction that Google presented. All these acceptances then cemented "defiantly" as the proper correction.

2.3 THE STEPS IN DATA MINING

This book focuses on understanding and using data mining algorithms (steps 4 to 7 below). However, some of the most serious errors in analytics projects result from a poor understanding of the problem—an understanding that must be developed before we get into the details of algorithms to be used. Here is a list of steps to be taken in a typical data mining effort:

1. *Develop an understanding of the purpose of the data mining project.* How will the stakeholder use the results? Who will be affected by the results? Will the analysis be a one-shot effort or an ongoing procedure?

2. *Obtain the dataset to be used in the analysis.* This often involves random sampling from a large database to capture records to be used in an analysis. It may also involve pulling together data from different databases or sources. The databases could be internal (e.g., past purchases made by customers) or external (credit ratings). While data mining deals with very large databases, usually the analysis to be done requires only thousands or tens of thousands of records.

3. *Explore, clean, and preprocess the data.* This step involves verifying that the data are in reasonable condition. How should missing data be handled? Are the values in a reasonable range, given what you would expect for each variable? Are there obvious outliers? The data are reviewed graphically: for example, a matrix of scatterplots showing the relationship of each variable with every other variable. We also need to ensure consistency in the definitions of fields, units of measurement, time periods, and so on. In this step, new variables are also typically created from existing ones. For example, "duration" can be computed from start and end dates.

4. *Reduce the data dimension, if necessary.* Dimension reduction can involve operations such as eliminating unneeded variables, transforming variables (e.g., turning "money spent" into "spent > \$100" vs. "spent ≤ \$100"), and creating new variables (e.g., a variable that records whether at least one of several products was purchased). Make sure that you know what each variable means and whether it is sensible to include it in the model.

5. *Determine the data mining task.* (classification, prediction, clustering, etc.). This involves translating the general question or problem of step 1 into a more specific data mining question.

6. *Partition the data (for supervised tasks).* If the task is supervised (classification or prediction), randomly partition the dataset into three parts: training, validation, and test datasets.

7. *Choose the data mining techniques to be used.* (regression, neural nets, hierarchical clustering, etc.).

8. *Use algorithms to perform the task.* This is typically an iterative process—trying multiple variants, and often using multiple variants of the same algorithm (choosing different variables or settings within the algorithm). Where appropriate, feedback from the algorithm's performance on validation data is used to refine the settings.

9. *Interpret the results of the algorithms.* This involves making a choice as to the best algorithm to deploy and, where possible, testing the final choice on the test data to get an idea as to how well it will perform. (Recall that each algorithm may also be tested on the validation data for tuning purposes; in this way the validation data become a part of the fitting process and

are likely to underestimate the error in the deployment of the model that is finally chosen.)

10. *Deploy the model.* This step involves integrating the model into operational systems and running it on real records to produce decisions or actions. For example, the model might be applied to a purchased list of possible customers, and the action might be "include in the mailing if the predicted amount of purchase is > $10." A key step here is "scoring" the new records, or using the chosen model to predict the target value for each new record.

The foregoing steps encompass the steps in SEMMA, a methodology developed by the software company SAS:

Sample   Take a sample from the dataset; partition into training, validation, and test datasets.

Explore   Examine the dataset statistically and graphically.

Modify   Transform the variables and impute missing values.

Model   Fit predictive models (e.g., regression tree, neural network).

Assess   Compare models using a validation dataset.

IBM SPSS Modeler (previously SPSS-Clementine) has a similar methodology, termed CRISP-DM (CRoss-Industry Standard Process for Data Mining). All these frameworks include the same main steps involved in predictive modeling.

2.4   PRELIMINARY STEPS

Organization of Datasets

Datasets are nearly always constructed and displayed so that variables are in columns and records are in rows. In the example shown in Section 2.6 (home values in West Roxbury, Boston, in 2014), 14 variables are recorded for over 5000 homes. The spreadsheet is organized so that each row represents a home—the first home's assessed value was $344,200, its tax was $4430, its size was 9965 ft^2, it was built in 1880, and so on. In supervised learning situations, one of these variables will be the outcome variable, typically listed in the first or last column (in this case it is TOTAL VALUE, in the first column).

Sampling from a Database

Quite often, we want to perform our data mining analysis on less than the total number of records that are available. Data mining algorithms will have varying limitations on what they can handle in terms of the numbers of records and variables, limitations that may be specific to computing power and capacity as well as software limitations. Even within those limits, many algorithms will execute faster with smaller samples.

Accurate models can often be built with as few as several hundred or thousand records. Hence we will want to sample a subset of records for model building.

Oversampling Rare Events in Classification Tasks

If the event we are interested in classifying is rare, such as customers purchasing a product in response to a mailing, or fraudulent credit card transactions, sampling a random subset of records may yield so few events (e.g., purchases) that we have little information on them. We would end up with lots of data on nonpurchasers and nonfraudulent transactions but little on which to base a model that distinguishes purchasers from nonpurchasers or fraudulent from nonfraudulent. In such cases, we would want our sampling procedure to overweight the rare class (purchasers or frauds) relative to the majority class (nonpurchasers, nonfrauds) so that our sample would end up with a healthy complement of purchasers or frauds.

Assuring an adequate number of responder or "success" cases to train the model is just part of the picture. A more important factor is the costs of misclassification. Whenever the response rate is extremely low, we are likely to attach more importance to identifying a responder than to identifying a nonresponder. In direct-response advertising (whether by traditional mail, email or web advertising), we may encounter only one or two responders for every hundred records—the value of finding such a customer far outweighs the costs of reaching him or her. In trying to identify fraudulent transactions, or customers unlikely to repay debt, the costs of failing to find the fraud or the nonpaying customer are likely to exceed the cost of more detailed review of a legitimate transaction or customer.

If the costs of failing to locate responders are comparable to the costs of misidentifying responders as nonresponders, our models would usually achieve highest overall accuracy if they identified everyone (or almost everyone, if it is easy to identify a few responders without catching many nonresponders) as a nonresponder. In such a case, the misclassification rate is very low—equal to the rate of responders—but the model is of no value.

More generally, we want to train our model with the asymmetric costs in mind so that the algorithm will catch the more valuable responders, probably at the cost of "catching" and misclassifying more nonresponders as responders than would be the case if we assume equal costs. This subject is discussed in detail in Chapter 5.

Preprocessing and Cleaning the Data

Types of Variables   There are several ways of classifying variables. Variables can be numerical or text (character/string). They can be continuous (able to assume any real numerical value, usually in a given range), integer (taking only integer values), or categorical (assuming one of a limited number of values). Categorical variables can be either coded as numerical (1, 2, 3) or text (payments current, payments not current, bankrupt). Categorical variables can also be unordered (called *nominal variables*) with categories such as North America, Europe, and Asia; or they can be ordered (called *ordinal variables*) with categories such as high value, low value, and nil value.

Continuous variables can be handled by most data mining routines. In XLMiner, all supervised routines take continuous predictor variables, with the exception of the naive Bayes classifier, which deals exclusively with categorical predictor variables. The machine learning roots of data mining grew out of problems with categorical outcomes; the roots of statistics lie in the analysis of continuous variables. Sometimes, it is desirable to convert continuous variables to categorical variables. This is done most typically in the case of outcome variables, where the numerical variable is mapped to a decision (e.g., credit scores above a certain level mean "grant credit," a medical test result above a certain level means "start treatment"). XLMiner has a facility for this type of conversion.

Handling Categorical Variables   Categorical variables can also be handled by most data mining routines, but often require special handling. If the categorical variable is ordered (age group, degree of creditworthiness, etc.), we can sometimes code the categories numerically (1, 2, 3,...) and treat the variable as if it were a continuous variable. The smaller the number of categories, and the less they represent equal increments of value, the more problematic this approach becomes, but it often works well enough.

Nominal categorical variables, however, often cannot be used as is. In many cases they must be decomposed into a series of binary variables, called *dummy variables*. For example, a single categorical variable that can have possible values of "student," "unemployed," "employed," or "retired" would be split into four separate dummy variables:

Student—Yes/No
Unemployed—Yes/No
Employed—Yes/No
Retired—Yes/No

In many cases, only three of the dummy variables need to be used; if the values of three are known, the fourth is also known. For example, given that these four values are the only possible ones, we can know that if a person is neither student, unemployed, nor employed, he or she must be retired. In some routines (e.g., linear regression and logistic regression), you should not use all four variables—the redundant information will cause the algorithm to fail. XLMiner has a utility to convert categorical variables to binary dummies.

Variable Selection   More is not necessarily better when it comes to selecting variables for a model. Other things being equal, parsimony, or compactness, is a desirable feature in a model. For one thing, the more variables we include, the greater the number of records we will need to assess relationships among the variables. Fifteen records may suffice to give us a rough idea of the relationship between Y and a single predictor variable X. If we now want information about the relationship between Y and 15 predictor variables $X_1, \ldots, X_{15}$, 15 records will not be enough (each estimated relationship would have an average of only one record's worth of information, making the estimate very unreliable). In addition, models based on many variables are often less robust, as they require the collection of more data in the future, are subject to more data quality and availability issues, and require more data cleaning and preprocessing.

How Many Variables and How Much Data?   Statisticians give us procedures to learn with some precision how many records we would need to achieve a given degree of reliability with a given dataset and a given model. These are called "power calculations" and are intended to assure that an average population effect will be estimated with sufficient precision from a sample. Data miners' needs are usually different because the focus is not on identifying an average effect but rather on predicting individual records. This purpose typically requires larger samples than those used for statistical inference. A good rule of thumb is to have 10 records for every predictor variable. Another, used by Delmaster and Hancock (2001, p. 68) for classification procedures, is to have at least $6 \times m \times p$ records, where m is the number of outcome classes and p is the number of variables.

In general, compactness or parsimony is a desirable feature in a data mining model. Even when we start with a small number of variables, we often end up with many more after creating new variables (e.g., converting a categorical variable into a set of dummy variables). Data visualization and dimension reduction methods help reduce the number of variables so that redundancies are avoided.

Even when we have an ample supply of data, there are good reasons to pay close attention to the variables that are included in a model. Someone with domain knowledge (i.e., knowledge of the business process and the data) should be consulted, as knowledge of what the variables represent is typically critical for building a good model and avoiding errors.

For example, suppose we're trying to predict the total purchase amount spent by customers, and we have a few predictor columns that are coded $X_1, X_2, X_3, \ldots$, where we don't know what those codes mean. We might find that X_1 is an excellent predictor of the total amount spent. However, if we discover that X_1 is the amount spent on shipping, calculated as a percentage of the purchase amount, then obviously a model that uses shipping amount cannot be used to predict purchase amount, since the shipping amount is not known until the transaction is completed. Another example is if we are trying to predict loan default at the time a customer applies for a loan. If our dataset includes only information on approved loan applications, we will not have information about what distinguishes defaulters from nondefaulters among denied applicants. A model based on approved loans alone can therefore not be used to predict defaulting behavior at the time of loan application but rather only once a loan is approved.

Outliers   The more data we are dealing with, the greater the chance of encountering erroneous values resulting from measurement error, data-entry error, or the like. If the erroneous value is in the same range as the rest of the data, it may be harmless. If it is well outside the range of the rest of the data (e.g., a misplaced decimal), it may have a substantial effect on some of the data mining procedures we plan to use.

Values that lie far away from the bulk of the data are called *outliers*. The term *far away* is deliberately left vague because what is or is not called an outlier is basically an arbitrary decision. Analysts use rules of thumb such as "anything over 3 standard deviations away from the mean is an outlier," but no statistical rule can tell us whether such an outlier is the result of an error. In this statistical sense, an outlier is not necessarily an invalid data point, it is just a distant one.

The purpose of identifying outliers is usually to call attention to values that need further review. We might come up with an explanation looking at the data—in the case of a misplaced decimal, this is likely. We might have no explanation but know that the value is wrong—a temperature of 178° F for a sick person. Or, we might conclude that the value is within the realm of possibility and leave it alone. All these are judgments best made by someone with *domain knowledge*, knowledge of the particular application being considered: direct mail, mortgage finance, and so on, as opposed to technical knowledge of statistical or data mining procedures. Statistical procedures can do little beyond identifying the record as something that needs review.

If manual review is feasible, some outliers may be identified and corrected. In any case, if the number of records with outliers is very small, they might be treated as missing data. How do we inspect for outliers? One technique in Excel is to sort the records by the first column, then review the data for very large or very small values in that column. Then repeat for each successive column. Another option is to examine the minimum and maximum values of each column using Excel's min and max functions. For a more automated approach that considers each record as a unit, clustering techniques could be used to identify clusters of one or a few records that are distant from others. Those records could then be examined.

Missing Values   Typically, some records will contain missing values. If the number of records with missing values is small, those records might be omitted. However, if we have a large number of variables, even a small proportion of missing values can affect a lot of records. Even with only 30 variables, if only 5% of the values are missing (spread randomly and independently among cases and variables), almost 80% of the records would have to be omitted from the analysis. (The chance that a given record would escape having a missing value is $0.95^{30} = 0.215$.)

An alternative to omitting records with missing values is to replace the missing value with an imputed value, based on the other values for that variable across all records. For example, if among 30 variables, household income is missing for a particular record, we might substitute the mean household income across all records. Doing so does not, of course, add any information about how household income affects the outcome variable. It merely allows us to proceed with the analysis and not lose the information contained in this record for the other 29 variables. Note that using such a technique will understate the variability in a dataset. However, we can assess variability and the performance of our data mining technique, using the validation data, and therefore this need not present a major problem. XLMiner offers the ability to replace missing values using fairly simple substitutes (e.g. mean, median). More sophisticated procedures do exist—for example, using linear regression, based on other variables, to fill in the missing values. These methods have been elaborated mainly for analysis of medical and scientific studies, where each patient or subject record comes at great expense. In data mining, where data are typically plentiful, simpler methods usually suffice.

Some datasets contain variables that have a very large number of missing values. In other words, a measurement is missing for a large number of records. In that case, dropping records with missing values will lead to a large loss of data. Imputing the missing values might also be useless, as the imputations are based on a small number of existing records. An alternative is to examine the importance of the predictor. If it is not very crucial, it can be dropped. If it is important, perhaps a proxy variable with fewer missing values can be used instead. When such a predictor is deemed central, the best solution is to invest in obtaining the missing data.

Significant time may be required to deal with missing data, as not all situations are susceptible to automated solutions. In a messy dataset, for example, a "0" might mean two things: (1) the value is missing, or (2) the value is actually zero. In the credit industry, a "0" in the "past due" variable might mean a customer who is fully paid up, or a customer with no credit history at all—two very different situations. Human judgment may be required for individual cases or to determine a special rule to deal with the situation.

Normalizing (Standardizing) and Rescaling Data   Some algorithms require that the data be normalized before the algorithm can be implemented effectively. To normalize a variable, we subtract the mean from each value and then divide by the standard deviation. This operation is also sometimes called *standardizing*. In effect, we are expressing each value as the "number of standard deviations away from the mean," also called a *z-score*.

Normalizing is one way to bring all variables to the same scale. Another popular approach is rescaling each variable to a [0,1] scale. This is done by subtracting the minimum value and then dividing by the range. Subtracting the minimum shifts the variable origin to zero. Dividing by the range shrinks or expands the data to the range [0,1].

To consider why normalizing or scaling to [0,1] might be necessary, consider the case of clustering. Clustering typically involves calculating a distance measure that reflects how far each record is from a cluster center or from other records. With multiple variables, different units will be used: days, dollars, counts, and so on. If the dollars are in the thousands and everything else is in the tens, the dollar variable will come to dominate the distance measure. Moreover, changing units from, say, days to hours or months could alter the outcome completely.

Data mining software, including XLMiner, typically have an option to normalize the data in those algorithms where it may be required. It is an option rather than an automatic feature of such algorithms because there are situations where we want each variable to contribute to the distance measure in proportion to its original scale.

2.5 PREDICTIVE POWER AND OVERFITTING

In supervised learning, a key question presents itself: How well will our prediction or classification model perform when we apply it to new data? We are particularly interested in comparing the performance of different models so that we can choose the model that we think will do the best when it is implemented in practice. A key concept is to make sure that our chosen model generalizes beyond the dataset that we have at hand. To assure generalization, we use the concept of *data partitioning* and try to avoid *overfitting*. These two important concepts are described next.

Creation and Use of Data Partitions

At first glance, we might think it is best to choose the model that did the best job of classifying or predicting the outcome variable of interest with the data at hand. However, when we use the same data both to develop the model and to assess its performance, we introduce an "optimism" bias. This is because when we choose the model that works best with the data, this model's superior performance comes from two sources:

- A superior model
- Chance aspects of the data that happen to match the chosen model better than they match other models

The latter is a particularly serious problem with techniques (e.g., trees and neural nets) that do not impose linear or other structure on the data, and thus end up overfitting it.

To address the overfitting problem, we simply divide (partition) our data and develop our model using only one of the partitions. After we have a model, we try it out on another partition and see how it performs, which we can measure in several ways. In a classification model, we can count the proportion of held-back records that were misclassified. In a prediction model, we can measure the residuals (prediction errors) between the predicted values and the actual values. This evaluation approach in effect mimics the deployment scenario, where our model is applied to data that it hasn't "seen."

We typically deal with two or three partitions: a training set, a validation set, and sometimes an additional test set. Partitioning the data into training, validation, and test sets is done either randomly according to predetermined proportions or by specifying which records go into which partition according to some relevant variable (e.g., in time series forecasting, the data are partitioned according to their chronological order). In most cases the partitioning should be done randomly to minimize the chance of getting a biased partition. It is also possible (although cumbersome) to divide the data into more than three partitions by successive partitioning (e.g., divide the initial data into three partitions, then take one of those partitions and partition it further).

Training Partition   The training partition, typically the largest partition, contains the data used to build the various models we are examining. The same training partition is generally used to develop multiple models.

Validation Partition   The validation partition (sometimes called the *test partition*) is used to assess the predictive performance of each model so that you can compare models and choose the best one. In some algorithms (e.g., classification and regression trees, k-nearest-neighbors), the validation partition may be used in an automated fashion to tune and improve the model.

Test Partition   The test partition (sometimes called the *holdout* or *evaluation partition*) is used to assess the performance of the chosen model with new data.

Why have both a validation and a test partition? When we use the validation data to assess multiple models and then choose the model that performs best with the validation data, we again encounter another (lesser) facet of the overfitting problem—chance aspects of the validation data that happen to match the chosen model better than they match other models. In other words, by using the validation data to choose one of several models, the performance of the chosen model on the validation data will be overly optimistic.

The random features of the validation data that enhance the apparent performance of the chosen model will probably not be present in new data to which the model is applied. Therefore we may have overestimated the accuracy of our model. The more models we test, the more likely it is that one of them will be particularly effective in modeling the noise in the validation data. Applying the model to the test data, which it has not seen before, will provide an unbiased estimate of how well the model will perform with new data. Figure 2.2 shows the three data partitions and their use in the data mining process. When we are concerned mainly with finding the best model and less with exactly how well it will do, we might use only training and validation partitions.

Note that with some algorithms, such as nearest-neighbor algorithms, records in the validation and test partitions, and in new data, are compared

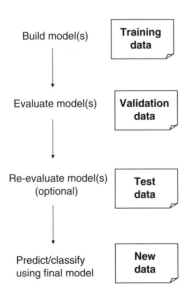

FIGURE 2.2   THREE DATA PARTITIONS AND THEIR ROLE IN THE DATA MINING PROCESS

to records in the training data to find the nearest neighbor(s). As k-nearest-neighbors is implemented in XLMiner and as discussed in this book, the use of two partitions is an essential part of the classification or prediction process, not merely a way to improve or assess it. Nonetheless, we can still interpret the error in the validation data in the same way that we would interpret error from any other model.

> XLMiner has a facility for partitioning a dataset randomly or according to a user-specified variable. For user-specified partitioning, create a new variable that contains the value "t" (training), "v" (validation), or "s" (test), according to the designation of that record. Partitioning is done either through the Partition menu or directly from within the menus of the supervised algorithms ("partition-on-the-fly"). For more details see www.solver.com/partition-data.

Overfitting

The more variables we include in a model, the greater the risk of overfitting the particular data used for modeling. What is overfitting?

In Table 2.1 we show hypothetical data about advertising expenditures in one time period and sales in a subsequent time period. A scatter plot of the data is shown in Figure 2.3. We could connect up these points with a smooth but complicated function, one that interpolates all these data points perfectly and leaves no error (residuals). This can be seen in Figure 2.4. However, we can see that such a curve is unlikely to be accurate, or even useful, in predicting future sales on the basis of advertising expenditures. For instance, it is hard to believe that increasing expenditures from $400 to $500 will actually decrease revenue.

TABLE 2.1

Advertising	Sales
239	514
364	789
602	550
644	1386
770	1394
789	1440
911	1354

A basic purpose of building a model is to represent relationships among variables in such a way that this representation will do a good job of predicting future outcome values on the basis of future predictor values. Of course, we

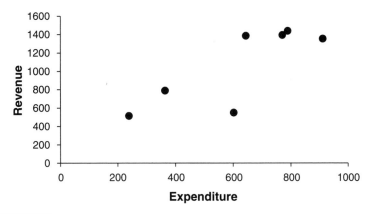

FIGURE 2.3   **SCATTER PLOT FOR ADVERTISING AND SALES DATA**

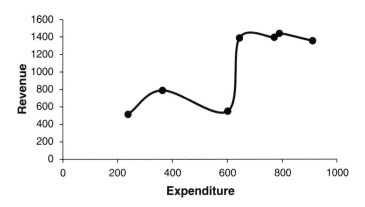

FIGURE 2.4   **OVERFITTING: THIS FUNCTION FITS THE DATA WITH NO ERROR**

want the model to do a good job of describing the data we have, but we are more interested in its performance with future data.

In the example above, a simple straight line might do a better job than the complex function in terms of predicting future sales on the basis of advertising. Instead, we devised a complex function that fit the data perfectly, and in doing so, we overreached. We ended up modeling some variation in the data that is nothing more than chance variation. We mistreated the noise in the data as if it were a signal.

Similarly, we can add predictors to a model to sharpen its performance with the data at hand. Consider a database of 100 individuals, 50 of whom have contributed to a charitable cause. Information about income, family size, and zip code might do a fair job of predicting whether or not someone is a contributor. If we keep adding additional predictors, we can improve the performance of the model with the data at hand and reduce the misclassification error to a negligible level. However, this low error rate is misleading because it probably includes spurious effects that are specific to the 100 individuals but not beyond that sample.

For example, one of the variables might be height. We have no basis in theory to suppose that tall people might contribute more or less to charity, but if there are several tall people in our sample and they just happened to contribute heavily to charity, our model might include a term for height—the taller you are, the more you will contribute. Of course, when the model is applied to additional data, it is likely that this will not turn out to be a good predictor.

If the dataset is not much larger than the number of predictor variables, it is very likely that a spurious relationship like this will creep into the model. Continuing with our charity example, with a small sample just a few of whom are tall, whatever the contribution level of tall people may be, the algorithm is tempted to attribute it to their being tall. If the dataset is very large relative to the number of predictors, this is less likely to occur. In such a case, each predictor must help predict the outcome for a large number of cases, so the job it does is much less dependent on just a few cases that might be flukes.

Somewhat surprisingly, even if we know for a fact that a higher degree curve is the appropriate model, if the model-fitting dataset is not large enough, a lower degree function (i.e., not as likely to fit the noise) is likely to perform better. Overfitting can also result from the application of many different models, from which the best performing model is selected.

2.6 BUILDING A PREDICTIVE MODEL WITH XLMINER

Let us go through the steps typical to many data mining tasks using a familiar procedure: multiple linear regression. This will help you understand the overall process before we begin tackling new algorithms. We illustrate the procedure using XLMiner.

Predicting Home Values in the West Roxbury Neighborhood

The Internet has revolutionized the real estate industry. Realtors now list houses and their prices on the web, and estimates of house and condominium prices have become widely available, even for units not on the market. Zillow (www.zillow.com) is the most popular online real estate information site,[1] and in 2014 they purchased their major rival, Trulia. By 2015 Zillow had become the dominant platform for checking house prices and, as such, the dominant online advertising venue for realtors. What used to be a comfortable 6% commission structure for realtors, affording them a handsome surplus (and an oversupply of realtors), was being rapidly eroded by an increasing need to pay for advertising

---

[1] "Zestimates may not be as right as you'd like" *Washington Post* Feb. 7, 2015, p. T10, by K. Harney.

on Zillow. (This, in fact, is the key to Zillow's business model—redirecting the 6% commission away from realtors and to itself.)

Zillow gets much of the data for its "Zestimates" of home values directly from publicly available city housing data, used to estimate property values for tax assessment. A competitor seeking to get into the market would likely take the same approach. So might realtors seeking to develop an alternative to Zillow.

A simple approach would be a naive, model-less method—just use the assessed values as determined by the city. Those values, however, do not necessarily include all properties, and they might not include changes warranted by remodeling, additions, and the like. Moreover, the assessment methods used by cities may not be transparent or always reflect true market values. However, the city property data can be used as a starting point to build a model, to which additional data (e.g., that collected by large realtors) can be added later.

Let's look at how Boston property assessment data, available from the city of Boston, might be used to predict home values. The data in *West Roxbury.xls* includes information on single family owner-occupied homes in West Roxbury, a neighborhood in southwest Boston, in 2014. The data include values for various predictor variables, and for a target-assessed home value ("total value"). This dataset has 14 variables, and a description of each variable is given in Table 2.2 (the full data dictionary provided by the City of Boston is at http://goo.gl/QBRlYF; we have modified a few variable names). The dataset includes 5802 homes. A sample of the data is shown in Figure 2.3.

As we saw earlier, below the header row, each row in the data represents a home. For example, the first home was assessed at a total value of $344.2 thousand (TOTAL VALUE). Its tax bill was $4330. It has a lot size of 9965 square feet (ft^2), was built in year 1880, has two floors, 6 rooms, and so on.

TABLE 2.2   DESCRIPTION OF VARIABLES IN WEST ROXBURY (BOSTON) HOME VALUE DATASET

TOTAL VALUE	Total assessed value for property, in thousands of USD
TAX	Tax bill amount based on total assessed value multiplied by the tax rate, in USD
LOT SQ FT	Total lot size of parcel in square feet
YR BUILT	Year property was built
GROSS AREA	Gross floor area
LIVING AREA	Total living area for residential properties (ft^2)
FLOORS	Number of floors
ROOMS	Total number of rooms
BEDROOMS	Total number of bedrooms
FULL BATH	Total number of full baths
HALF BATH	Total number of half baths
KITCHEN	Total number of kitchens
FIREPLACE	Total number of fireplaces
REMODEL	When house was remodeled (Recent/Old/None)

Modeling Process

We now describe in detail the various model stages using the West Roxbury home values example.

1. *Determine the purpose.* Let's assume that the purpose of our data mining project is to predict the value of homes in West Roxbury.

2. *Obtain the data.* We will use the 2014 West Roxbury housing data. The dataset in question is small enough that we do not need to sample from it—we can use it in its entirety.

3. *Explore, clean, and preprocess the data.* Let's look first at the description of the variables, also known as the "data dictionary," to be sure that we understand them all. These descriptions are available on the "description" worksheet in the Excel file and in Table 2.3. The variable names and descriptions in this dataset all seem fairly straightforward, but this is not always the case. Often variable names are cryptic and their descriptions are unclear or missing.

TABLE 2.3		FIRST 10 RECORDS IN THE WEST ROXBURY HOME VALUES DATASET											
TOTAL VALUE	TAX	LOT SQFT	YR BUILT	GROSS AREA	LIVING AREA	FLOORS	ROOMS	BED ROOMS	FULL BATH	HALF BATH	KIT CHEN	FIRE PLACE	REMODEL
344.2	4330	9965	1880	2436	1352	2	6	3	1	1	1	0	None
412.6	5190	6590	1945	3108	1976	2	10	4	2	1	1	0	Recent
330.1	4152	7500	1890	2294	1371	2	8	4	1	1	1	0	None
498.6	6272	13773	1957	5032	2608	1	9	5	1	1	1	1	None
331.5	4170	5000	1910	2370	1438	2	7	3	2	0	1	0	None
337.4	4244	5142	1950	2124	1060	1	6	3	1	0	1	1	Old
359.4	4521	5000	1954	3220	1916	2	7	3	1	1	1	0	None
320.4	4030	10000	1950	2208	1200	1	6	3	1	0	1	0	None
333.5	4195	6835	1958	2582	1092	1	5	3	1	0	1	1	Recent
409.4	5150	5093	1900	4818	2992	2	8	4	2	0	1	0	None

It is useful to pause and think about what the variables mean and whether they should be included in the model. Consider the variable TAX. At first glance, we consider that the tax on a home is usually a function of its assessed value, so there is some circularity in the model—we want to predict a home's value using TAX as a predictor, yet TAX itself is determined by a home's value. TAX might be a very good predictor of home value in a numerical sense, but would it be useful if we wanted to apply our model to homes whose assessed value might not be known? For this reason, we will exclude TAX from the analysis.

TABLE 2.4	OUTLIER IN WEST ROXBURY DATA
FLOORS	**ROOMS**
15	8
2	10
1.5	6
1	6

It is also useful to check for outliers that might be errors. For example, suppose that the column FLOORS (number of floors) looked like the one in Table 2.4, after sorting the data in descending order based on floors. We can tell right away that the 15 is in error—it is unlikely that a home has 15 floors. All other values are between 1 and 2. Probably, the decimal was misplaced and the value should be 1.5.

Last, we create dummy variables for categorical variables. Here we have one categorical variable: REMODEL, which has three categories.

4. *Reduce the data dimension.* Our dataset has been prepared for presentation with fairly low dimension—it has only 13 variables, and the single categorical variable considered has only three categories (and hence adds two dummy variables). If we had many more variables, at this stage we might want to apply a variable reduction technique such as condensing multiple categories into a smaller number, or applying principal components analysis to consolidate multiple similar numerical variables (e.g., LIVING AREA, ROOMS, BEDROOMS, BATH, HALF BATH) into a smaller number of variables.

5. *Determine the data mining task.* The specific task is to predict the value of TOTAL VALUE using the predictor variables. This is a supervised prediction task. For simplicity, we excluded several additional variables present in the original dataset, which have many categories (BLDG TYPE, ROOF TYPE, and EXT FIN). We therefore use all the numerical variables (except TAX) and the dummies created for the remaining categorical variables.

6. *Partition the data (for supervised tasks).* In XLMiner, select *Partition* from the *Data Mining* menus, and the dialog box shown in Figure 2.5 appears. Here we specify the data range to be partitioned and the variables to be included in the partitioned dataset. The partitioning can be handled in one of two ways:

   a. The dataset can have a partition variable that governs the division into training and validation partitions ("t" = training, "v" = validation).

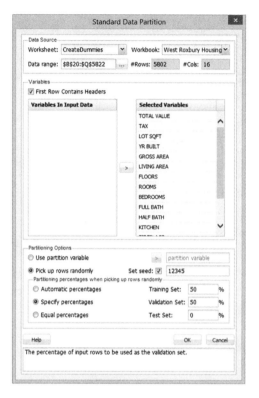

FIGURE 2.5 PARTITIONING THE DATA. THE DEFAULT IN XLMINER PARTITIONS THE DATA INTO 60% TRAINING DATA, 40% VALIDATION DATA, AND 0% TEST DATA. HERE A PARTITION OF 50% TRAINING AND 50% VALIDATION IS USED.

b. The partitioning can be done randomly. If the partitioning is done randomly, we have the option of specifying a seed for randomization (which has the advantage of letting us duplicate the same random partition later, should we need to). In this example, a seed of 12345 is used.

In this case we divide the data into two partitions: training and validation. The training partition is used to build the model, and the validation partition is used to see how well the model does when applied to new data. We need to specify the percent of the data used in each partition. *Note:* Although not used in our example, a test partition might also be used.

7. *Choose the technique.* In this case, it is multiple linear regression. Having divided the data into training and validation partitions, we can use XLMiner to build a multiple linear regression model with the training data. We want to predict the value of a house in West Roxbury on the basis of all the other predictors (except TAX).

8. *Use the algorithm to perform the task.* In XLMiner, we select *Multiple Linear Regression* from the *Prediction* menu. The first dialog box is shown in Figure 2.6. The variable TOTAL VALUE is selected as the output (dependent) variable. All the other variables, except TAX and one of the REMODEL dummy variables, are selected as input (predictor) variables. We ask XLMiner to show us the fitted values on the training data as well as the predicted values (scores) on the validation data, as shown in Figure 2.7. XLMiner produces standard regression output, but for now we defer that as well as the more advanced options displayed above (see Chapter 6 or the user documentation for XLMiner for more information.) Rather, we review the predictions themselves. Figure 2.8 shows the predicted values for the first few records in the training data along with the actual values and the residuals (prediction errors). Note that the predicted values would often be called the *fitted values*, since they are for the records to which the model was fit. The results for the validation data are shown in Figure 2.9. The prediction errors for the training and validation data are compared in Figure 2.10.

   Prediction error can be measured in several ways. Three measures produced by XLMiner are shown in Figure 2.10. On the right is the *average error*, simply the average of the residuals (errors). In both cases, it is quite small relative to the units of TOTAL VALUE, indicating that, on balance, predictions average about right—our predictions are "unbiased." Of course, this simply means that the positive and negative errors balance out. It tells us nothing about how large these errors are.

   The *total sum of squared errors* on the left adds up the squared errors, so whether an error is positive or negative, it contributes just the same. However, this sum does not yield information about the size of the typical error.

   The *RMS error* (root-mean-squared error) is perhaps the most useful term of all. It takes the square root of the average squared error, so it gives an idea of the typical error (whether positive or negative) in the same scale as that used for the original outcome variable. As we might expect, the RMS error for the validation data (45.2 thousand dollars), which the model is seeing for the first time in making these predictions, is larger than for the training data (40.9 thousand dollars), which were used in training the model.

9. *Interpret the results.* At this stage we would typically try other prediction algorithms (e.g., regression trees) and see how they do error-wise. We might also try different "settings" on the various models (e.g., we could use the *best subsets* option in multiple linear regression to choose a reduced set of variables that might perform better with the validation data). After choosing the best model—typically, the model with the lowest error on the validation data while also recognizing that "simpler is better"—we

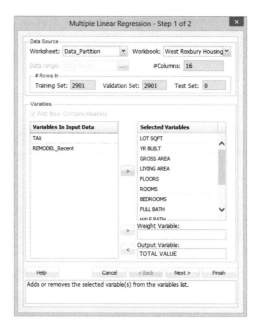

FIGURE 2.6   **USING XLMINER FOR MULTIPLE LINEAR REGRESSION**

FIGURE 2.7   **SPECIFYING THE OUTPUT**

XLMiner : Multiple Linear Regression—Prediction of Training Data

Predicted Value	Actual Value	Residual	Confidence Intervals		Prediction Intervals		LOT SQFT	YR BUILT	GROSS AREA	LIVING AREA	FLOORS	ROOMS	BEDROOMS	FULL BATH	HALF BATH	KITCHEN	FIREPLACE	REMODEL _None	REMODEL _Old
			Lower	Upper	Lower	Upper													
378.2729	344.2	−34.073	372.3562	384.1896	297.6552	458.8907	9965	1880	2436	1352	2	6	3	1	1	1	0	1	0
342.9992	331.5	−11.499	338.1995	347.7988	262.4557	423.5426	5000	1910	2370	1438	2	7	3	2	0	1	0	1	0
316.3789	320.4	4.02106	311.4061	321.3518	235.825	396.9329	####	1950	2208	1200	1	6	3	1	0	1	0	1	0
362.3197	313	−49.32	358.617	366.0225	281.8342	442.8053	5000	1960	2624	1485	1.5	6	3	2	0	1	1	1	0
384.3895	344.5	−39.89	380.6254	388.1536	303.9011	464.8779	6768	1958	2844	1460	1.5	6	3	2	0	1	1	1	0
347.9685	326.2	−21.769	345.0256	350.9115	267.5144	428.4227	5000	1954	2536	1272	1.5	6	3	1	1	1	1	1	0
272.2333	298.2	25.9667	266.8726	277.5939	191.6544	352.8121	5000	1940	2129	864	1	7	3	2	0	1	0	1	0
361.2712	344.9	−16.371	355.8066	366.7358	280.6854	441.857	####	1950	2099	1445	1	7	3	1	1	1	1	1	0
378.3477	348	−30.348	373.234	383.4615	297.7849	458.9105	9001	1875	2840	1632	2	7	3	1	0	1	0	1	0
276.3738	317.5	41.1262	271.3227	281.4249	195.815	356.9326	4450	1920	1400	1232	2	7	3	1	0	1	0	1	0

FIGURE 2.8   PREDICTIONS FOR A SAMPLE OF TRAINING DATA

XLMiner : Multiple Linear Regression—Prediction of Validation Data

Predicted Value	Actual Value	Residual	Confidence Intervals		Prediction Intervals		LOT SQFT	YR BUILT	GROSS AREA	LIVING AREA	FLOORS	ROOMS	BEDROOMS	FULL BATH	HALF BATH	KITCHEN	FIREPLACE	REMODEL _None	REMODEL _Old
			Lower	Upper	Lower	Upper													
385.435	390.3	4.865	382.0741	388.796	304.9645	465.9056	5500	1930	2986	1710	2	7	3	1	1	1	0	1	0
395.9625	409.5	13.537	390.2393	401.6857	315.3588	476.5663	8114	1958	2705	1479	1	6	3	1	1	1	1	0	0
319.6694	321	1.3306	315.7557	323.5831	239.1739	400.1649	4794	1954	2361	1261	1.5	6	2	1	1	1	0	1	0
411.0741	365.4	−45.67	406.6657	415.4824	330.553	491.5951	5107	1960	2642	1634	2	7	3	1	1	1	2	1	0
634.523	593.5	−41.02	628.2185	640.8275	553.8759	715.1701	8580	1905	6075	3382	2	10	5	2	1	1	1	1	0
441.9512	384.4	−57.55	434.7909	449.1115	361.2326	522.6697	7950	1954	4012	1975	1.5	9	4	1	1	1	0	0	1
391.4235	355	−36.42	387.3284	395.5187	310.919	471.9281	9331	1950	2522	1428	1.5	6	2	1	1	1	1	1	0
319.2621	345.4	26.138	315.2613	323.2629	238.7623	399.7619	6000	1954	2302	1103	1	5	2	1	1	1	1	1	0
433.9011	476.3	42.399	429.0982	438.704	353.3574	514.4447	5500	1925	2936	1836	2	8	4	1	1	1	1	0	0
310.1917	302.7	−7.492	305.8799	314.5035	229.6758	390.7075	4140	1925	2895	1295	1	7	3	1	0	1	1	1	0

FIGURE 2.9   PREDICTIONS FOR A SAMPLE OF VALIDATION DATA

Training Data Scoring—Summary Report

Total Sum of Squared Errors	RMS Error	Average Error
4854025	40.90507	6.98463E-13

Validation Data Scoring—Summary Report

Total Sum of Squared Errors	RMS Error	Average Error
5927610	45.20286	0.427140769

FIGURE 2.10   ERROR RATES FOR TRAINING (TOP) AND VALIDATION (BOTTOM) DATA (ERROR FIGURES ARE IN THOUSANDS OF $)

	A	B	C	D	E	F	G	H	I	J	K	L	M	N	O
1	TAX	LOT SQFT	YR BUILT	GROSS AREA	LIVING AREA	FLOORS	ROOMS	BEDR OOMS	FULL BATH	HALF BATH	KITC HEN	FIREP LACE	REMODEL _None	REMODEL _Old	REMODEL _Recent
2	3850	6877	1963	2240	1808	1	6	3	1	1	1	0	1	0	0
3	5386	5965	1963	2998	1637	1.5	8	3	3	0	1	1	0	1	0
4	4608	5662	1961	2805	1750	2	7	4	2	0	1	1	0	0	1

FIGURE 2.11   **WORKSHEET WITH THREE RECORDS TO BE SCORED**

use that model to predict the output variable in fresh data. These steps are covered in more detail in the analysis of cases.

10. *Deploy the model.* After the best model is chosen, it is applied to new data to predict TOTAL VALUE for homes where this value is unknown. This was, of course, the original purpose. Predicting the output value for new records is called *scoring*. For predictive tasks, scoring produces predicted numerical values. For classification tasks, scoring produces classes and/or propensities. In XLMiner, we can score new records using one of the models we developed. To do that, we must first create a worksheet or file with the records to be predicted. For these records, we must include all the predictor values. Figure 2.11 shows an example of a worksheet with three homes to be scored using our linear regression model. Note that all the required predictor columns are present, and the output column is absent.

Figure 2.12 shows the Score dialog box. We chose "match by name" to match the predictor columns in our model with the new records' worksheet. The result is shown in Figure 2.13, where the predictions are in the first column. Note: In XLMiner, scoring new data can also be done directly from a specific prediction or classification[2] method dialog box ("New Data Scoring," typically in the last step). In our example, scoring can be done in step 2 in the multiple linear regression dialog shown in Figure 2.7.

XLMiner has a facility for drawing a sample from an external database. The sample can be drawn at random or it can be stratified. It also has a facility to score data in the external database using the model that was obtained from the training data.

[2] Note: In some versions of XLMiner, propensities for new records are produced only if "New Data Scoring" is selected in the classification method dialog, they are not available in the option to score stored models.

FIGURE 2.12   SCORE DIALOG BOX. NEW RECORDS COLUMN NAMES ARE MATCHED WITH PREDICTOR COLUMN NAMES IN MODEL.

Multiple Linear Regression Prediction: Score Data

Model Workbook	West Roxbury Housing Ch2 MLR.xlsx
Model Worksheet	MLR_Stored
Data Workbook	West Roxbury Housing Ch2 MLR.xlsx
Data Worksheet	RecordsToScore
Data Range	$A$1:$O$4

Predicted	LOT SQFT	YR BUILT	GROSS AREA	LIVING AREA	FLOORS	ROOMS	BEDR OOMS	FULL BATH	HALF BATH	KITCHEN	FIREPLACE	REMODEL _None	REMODEL _Old
337.7738	6877	1963	2240	1808	1	6	3	1	1	1	0	1	0
412.7901	5965	1963	2998	1637	1.5	8	3	3	0	1	1	0	1
433.8581	5662	1961	2805	1750	2	7	4	2	0	1	1	0	0

FIGURE 2.13   THREE RECORDS SCORED USING LINEAR REGRESSION MODEL

2.7   USING EXCEL FOR DATA MINING

An important aspect of this modeling process is that the heavy-duty analysis does not necessarily require a huge number of records. The dataset to be analyzed might have millions of records, but in applying multiple linear regression or applying a classification tree, the use of a sample of 20,000 is likely to yield as accurate an answer as that obtained when using the entire dataset. The principle involved is the same as the principle behind polling: If sampled judiciously, 2000 voters can give an estimate of the entire population's opinion within one or two percentage points. (See "How Many Variables and How Much Data" in Section 2.4 for further discussion.)

Therefore, in most cases, the number of records required in each partition (training, validation, and test) can be accommodated within the rows allowed by Excel. Of course, we need to get those records into Excel, and for this purpose the standard version of XLMiner provides an interface for random sampling of records from an external database.

Similarly, we may need to apply the results of our analysis to a large database, and for this purpose the standard version of XLMiner has a facility for storing models and scoring them to an external database. For example, XLMiner would write an additional column (variable) to the database consisting of the predicted purchase amount for each record.

2.8 Automating Data Mining Solutions

In most supervised data mining applications, the goal is not a static, one-time analysis of a particular dataset. Rather, we want to develop a model that can be used on an ongoing basis to predict or classify new records. Our initial analysis will be in prototype mode, while we explore and define the problem and test different models. We will follow all the steps outlined earlier in this chapter.

At the end of that process, we will typically want our chosen model to be deployed in automated fashion. For example, the US Internal Revenue Service (IRS) receives several hundred million tax returns per year—it does not want to have to pull each tax return out into an Excel sheet or other environment separate from its main database to determine the predicted probability that the return is fraudulent. Rather, it would prefer that determination to be made as part of the normal tax filing environment and process. Music streaming services, such as Pandora or Spotify, need to determine "recommendations" for next songs quickly for each of millions of users; there is no time to extract the data for manual analysis.

In practice, this is done by building the chosen algorithm into the computational setting in which the rest of the process lies. A tax return is entered directly into the IRS system by a tax preparer, a predictive algorithm is immediately applied to the new data in the IRS system, and a predicted classification is decided by the algorithm. Business rules would then determine what happens with that classification. In the IRS case, the rule might be "if no predicted fraud, continue routine processing; if fraud is predicted, alert an examiner for possible audit."

This flow of the tax return from data entry, into the IRS system, through a predictive algorithm, then back out to a human user is an example of a "data pipeline." The different components of the system communicate with one another via application programming interfaces (APIs) that establish locally valid rules for transmitting data and associated communications. An API for a data mining algorithm would establish the required elements for a predictive algorithm to work—the exact predictor variables, their order, data formats, and so on. It would also establish the requirements for communicating the results of the algorithm. Algorithms to be used in an automated data pipeline will need to be compliant with the rules of the APIs where they operate.

Finally, once the computational environment is set and functioning, the data miner's work is not done. The environment in which a model operates is typically dynamic, and predictive models often have a short shelf life—one leading consultant finds they rarely continue to function effectively for more than a year. So, even in a fully deployed state, models must be periodically checked and re-evaluated. Once performance flags, it is time to return to prototype mode and see if a new model can be developed.

In this book, our focus will be on the prototyping phase—all the steps that go into properly defining the model and developing and selecting a model. You should be aware, though, that most of the actual work of implementing a data mining model lies in the automated deployment phase. Much of this work is not in the analytic domain; rather, it lies in the domains of databases and computer science, to assure that detailed nuts and bolts of an automated dataflow all work properly.

DATA MINING SOFTWARE TOOLS: THE STATE OF THE MARKET

by Herb Edelstein

The data mining market has changed in some important ways since the last edition of this book. The most significant trends have been the increasing volume of information available and the growing use of the cloud for data storage and analytics. Data mining and analysis have evolved to meet these new demands.

The term "Big Data" reflects the surge in the amount and types of data collected. There is still an enormous amount of transactional data, data warehousing data, scientific data, and clickstream data. However, adding to the massive storage requirements of traditional data is the influx of information from unstructured sources (e.g., customer service calls and images), social media, and more recently the Internet of Things, which produces a flood of sensor data. The number of organizations collecting such data has greatly increased as virtually every business is expanding in these areas.

Rapid technological change has been an important factor. The price of data storage has dropped precipitously. At this writing, hard disk storage has fallen to about $50 per terabyte and solid state drives are about $200 per terabyte. Concomitant with the decrease in storage costs has been a dramatic increase in bandwidth at ever lower costs. This has enabled the spread of cloud-based computing.

Cloud-based computing refers to using remote platforms for storage, data management, and now analysis. Because scaling up the hardware and software infrastructure for Big Data is so complex, many organizations are entrusting their data to outside vendors. The largest cloud players (Amazon, Google, IBM, and Microsoft) are each reporting annual revenues in excess of US$5 billion, according to *Forbes* magazine.

Managing this much data is a challenging task. While traditional relational DBMSs—such as Oracle, Microsofts SQL Server, IBMs DB2, and SAPs Adaptive Server Enterprise (formerly Sybase)—are still among the leading data management tools, open source DBMSs such as Oracles MySQL are becoming increasingly popular. In addition, nonrelational data storage is making headway in storing extremely large amounts of data. For example, Hadoop, an open source tool for managing large distributed database architectures, has become an important player in the cloud database space. However, Hadoop is a tool for the application developer community rather than end users. Consequently many of the data mining analytics vendors such as SAS have built interfaces to Hadoop.

All the major database management system vendors offer data mining capabilities, usually integrated into their DBMS. Leading products include Microsoft SQL Server Analysis Services, Oracle Data Mining, and Teradata Warehouse Miner. The target user for embedded data mining is a database professional. Not surprisingly, these products take advantage of database functionality, including using the DBMS to transform variables, storing models in the database, and extending the data access language to include model-building and scoring the database. A few products also supply a separate graphical interface for building data mining models. Where the DBMS has parallel processing capabilities, embedded data mining tools will generally take advantage of it, resulting in greater performance. As with the data mining suites described below, these tools offer an assortment of algorithms. Not only does IBM have embedded analytics in DB2, but following its acquisition of SPSS, IBM has incorporated Clementine and SPSS into IBM Modeler.

There are still a large number of stand-alone data mining tools based on a single algorithm or on a collection of algorithms called a suite. Target users include both statisticians and business intelligence analysts. The leading suites include SAS Enterprise Miner, SAS JMP, IBM Modeler, Salford Systems SPM, Statistica, XLMiner, and RapidMiner. Suites are characterized by providing a wide range of functionality, frequently accessed via a graphical user interface designed to enhance model-building productivity. A popular approach for many of these GUIs is to provide a workflow interface in which the data mining steps and analysis are linked together.

Many suites have outstanding visualization tools and links to statistical packages that extend the range of tasks they can perform. They provide interactive data transformation tools as well as a procedural scripting language for more complex data transformations. The suite vendors are working to link their tools more closely to underlying DBMSs; for example, data transformations might be handled by the DBMS. Data mining models can be exported to be incorporated into the DBMS through generating SQL, procedural language code (e.g., C++, or Java), or a standardized data mining model language called Predictive Model Markup Language (PMML).

In contrast to the general-purpose suites, application-specific tools are intended for particular analytic applications such as credit scoring, customer retention, and product marketing. Their focus may be further sharpened to address the needs of specialized markets such as mortgage lending or financial services. The target user is an analyst with expertise in the application domain. Therefore the interfaces, the algorithms, and even the terminology are customized for that particular industry, application, or customer. While less flexible than general-purpose tools, they offer the advantage of already incorporating domain knowledge into the product design, and can provide very good solutions with less effort. Data mining companies including SAS, IBM, and RapidMiner offer vertical market tools, as do industry specialists such as Fair Isaac. Other companies, such as Domo, are focusing on creating dashboards with analytics and visualizations for business intelligence.

Another technological shift has occurred with the spread of open source model building tools and open core tools. A somewhat simplified view of open source software is that the source code for the tool is available at no charge to the community of users and can be modified or enhanced by them. These enhancements are submitted to the originator or copyright holder, who can add them to the base package. Open core is a more recent approach in which a core set of functionality remains open and free, but there are proprietary extensions that are not free.

The most important open source statistical analysis software is R. R is descended from a Bell Labs program called S, which was commercialized as S+. Many data mining algorithms have been added to R, along with a plethora of statistics, data management tools, and visualization tools. Because it is essentially a programming language, R has enormous flexibility but a steeper learning curve than many of the GUI-based tools. Although there are some GUIs for R, the overwhelming majority of use is through programming.

Some vendors, as well as the open source community, are adding statistical and data mining tools to Python, a popular programming language that is generally easier to use than C++ or Java, and faster than R.

As mentioned above, the cloud-computing vendors have moved into the data mining/predictive analytics business by offering AaaS (Analytics as a Service) and pricing their products on a transaction basis. These products are oriented more toward application developers than business intelligence analysts. A big part of the attraction of mining data in the cloud is the ability to store and manage enormous amounts of data without requiring the expense and complexity of building an in-house capability. This can also enable a more rapid implementation of large distributed multi-user applications. Cloud based data can be used with non-cloud-based analytics if the vendors analytics do not meet the users needs.

Amazon has added Amazon Machine Learning to its Amazon Web Services (AWS), taking advantage of predictive modeling tools developed for Amazons internal use. AWS supports both relational databases and Hadoop data management. Models cannot be exported, because they are intended to be applied to data stored on the Amazon cloud.

Google is very active in cloud analytics with its BigQuery and Prediction API. BigQuery allows the use of Google infrastructure to access large amounts of data using a SQL-like interface. The Prediction API can be accessed from a variety of languages including R and Python. It uses a variety of machine learning algorithms and automatically selects the best results. Unfortunately, this is not a transparent process. Furthermore, as with Amazon, models cannot be exported.

Microsoft is an active player in cloud analytics with its Azure Machine Learning Studio and Stream Analytics. Azure works with Hadoop clusters as well as with traditional relational databases. Azure ML offers a broad range of algorithms such as boosted trees and support vector machines as well as supporting R scripts and Python. Azure ML also supports a workflow interface making it more suitable for the nonprogrammer data scientist. The real-time analytics component is designed to allow streaming data from a variety of sources to be analyzed on the fly. XLMiner's cloud version is based on Microsoft Azure. Microsoft also acquired Revolution Analytics, a major player in the R analytics business, with a view to integrating Revolution's "R Enterprise" with SQL Server and Azure ML. R Enterprise includes extensions to R that eliminate memory limitations and take advantage of parallel processing.

One drawback of the cloud-based analytics tools is a relative lack of transparency and user control over the algorithms and their parameters. In some cases, the service will simply select a single model that is a black box to the user. Another drawback is that for the most part cloud-based tools are aimed at more sophisticated data scientists who are systems savvy.

Data science is playing a central role in enabling many organizations to optimize everything from production to marketing. New storage options and analytical tools promise even greater capabilities. The key is to select technology that's appropriate for an organization's unique goals and constraints. As always, human judgment is the most important component of a data mining solution.

This book's focus is on a comprehensive understanding of the different techniques and algorithms used in data mining, and less on the data management requirements of real-time deployment of data mining models. XLMiner's short learning curve and integration with Excel makes it ideal for this purpose, and for exploration, prototyping, and piloting of solutions.

Herb Edelstein is president of Two Crows Consulting (www.twocrows.com), a leading data mining consulting firm near Washington, DC. He is an internationally recognized expert in data mining and data warehousing, a widely published author on these topics, and a popular speaker. © 2015 Herb Edelstein.

PROBLEMS ▬▬▬▬▬▬▬▬▬▬▬▬▬▬▬▬▬▬▬▬▬▬▬▬▬▬▬▬▬ ■

2.1 Assuming that data mining techniques are to be used in the following cases, identify whether the task required is supervised or unsupervised learning.

   a. Deciding whether to issue a loan to an applicant based on demographic and financial data (with reference to a database of similar data on prior customers).

   b. In an online bookstore, making recommendations to customers concerning additional items to buy based on the buying patterns in prior transactions.

   c. Identifying a network data packet as dangerous (virus, hacker attack) based on comparison to other packets whose threat status is known.

   d. Identifying segments of similar customers.

   e. Predicting whether a company will go bankrupt based on comparing its financial data to those of similar bankrupt and nonbankrupt firms.

   f. Estimating the repair time required for an aircraft based on a trouble ticket.

   g. Automated sorting of mail by zip code scanning.

   h. Printing of custom discount coupons at the conclusion of a grocery store checkout based on what you just bought and what others have bought previously.

2.2 Describe the difference in roles assumed by the validation partition and the test partition.

2.3 Consider the sample from a database of credit applicants in Table 2.5. Comment on the likelihood that it was sampled randomly, and whether it is likely to be a useful sample.

TABLE 2.5   **SAMPLE FROM A DATABASE OF CREDIT APPLICATIONS**

OBS	CHEK ACCT	DURATION	HISTORY	NEW CAR	USED CAR	FURNITURE	RADIO TV	EDUC	RETRAIN	AMOUNT	SAVE ACCT	RESPONSE
1	0	6	4	0	0	0	1	0	0	1169	4	1
8	1	36	2	0	1	0	0	0	0	6948	0	1
16	0	24	2	0	0	0	1	0	0	1282	1	0
24	1	12	4	0	1	0	0	0	0	1804	1	1
32	0	24	2	0	0	1	0	0	0	4020	0	1
40	1	9	2	0	0	0	1	0	0	458	0	1
48	0	6	2	0	1	0	0	0	0	1352	2	1
56	3	6	1	1	0	0	0	0	0	783	4	1
64	1	48	0	0	0	0	0	0	1	14421	0	0
72	3	7	4	0	0	0	1	0	0	730	4	1
80	1	30	2	0	0	1	0	0	0	3832	0	1
88	1	36	2	0	0	0	0	1	0	12612	1	0
96	1	54	0	0	0	0	0	0	1	15945	0	0
104	1	9	4	0	0	1	0	0	0	1919	0	1
112	2	15	2	0	0	0	0	1	0	392	0	1

2.4 Consider the sample from a bank database shown in Table 2.6; it was selected randomly from a larger database to be the training set. *Personal Loan* indicates whether a solicitation for a personal loan was accepted and is the response variable. A campaign is planned for a similar solicitation in the future, and the bank is looking for a model that will identify likely responders. Examine the data carefully and indicate what your next step would be.

TABLE 2.6   SAMPLE FROM A BANK DATABASE

OBS	AGE	EXPERIENCE	INCOME	ZIP	FAMILY	CC AVG	EDUC	MORTGAGE	PERSONAL LOAN	SECURITIES ACCT
1	25	1	49	91107	4	1.6	1	0	0	1
4	35	9	100	94112	1	2.7	2	0	0	0
5	35	8	45	91330	4	1.0	2	0	0	0
9	35	10	81	90089	3	0.6	2	104	0	0
10	34	9	180	93023	1	8.9	3	0	1	0
12	29	5	45	90277	3	0.1	2	0	0	0
17	38	14	130	95010	4	4.7	3	134	1	0
18	42	18	81	94305	4	2.4	1	0	0	0
21	56	31	25	94015	4	0.9	2	111	0	0
26	43	19	29	94305	3	0.5	1	97	0	0
29	56	30	48	94539	1	2.2	3	0	0	0
30	38	13	119	94104	1	3.3	2	0	1	0
35	31	5	50	94035	4	1.8	3	0	0	0
36	48	24	81	92647	3	0.7	1	0	0	0
37	59	35	121	94720	1	2.9	1	0	0	0
38	51	25	71	95814	1	1.4	3	198	0	0
39	42	18	141	94114	3	5.0	3	0	1	1
41	57	32	84	92672	3	1.6	3	0	0	1

2.5 Using the concept of overfitting, explain why when a model is fit to training data, zero error with those data is not necessarily good.

2.6 In fitting a model to classify prospects as purchasers or nonpurchasers, a certain company drew the training data from internal data that include demographic and purchase information. Future data to be classified will be lists purchased from other sources, with demographic (but not purchase) data included. It was found that "refund issued" was a useful predictor in the training data. Why is this not an appropriate variable to include in the model?

2.7 A dataset has 1000 records and 50 variables with 5% of the values missing, spread randomly throughout the records and variables. An analyst decides to remove records with missing values. About how many records would you expect to be removed?

2.8 Normalize the data in Table 2.7, showing calculations.

TABLE 2.7

Age	Income ($)
25	49,000
56	156,000
65	99,000
32	192,000
41	39,000
49	57,000

2.9 Statistical distance between records can be measured in several ways. Consider Euclidean distance, measured as the square root of the sum of the squared differences. For the first two records in Table 2.7, it is

$$\sqrt{(25 - 56)^2 + (49,000 - 156,000)^2}.$$

Can normalizing the data change which two records are farthest from each other in terms of Euclidean distance?

2.10 Two models are applied to a dataset that has been partitioned. Model A is considerably more accurate than model B on the training data, but slightly less accurate than model B on the validation data. Which model are you more likely to consider for final deployment?

2.11 The dataset ToyotaCorolla.xls contains data on used cars for sale during the late summer of 2004 in The Netherlands. It has 1436 records containing details on 38 attributes, including Price, Age, Kilometers, HP, and other specifications.

a. Explore the data using the data visualization capabilities of XLMiner. Which of the pairs among the variables seem to be correlated?

b. We plan to analyze the data using various data mining techniques described in future chapters. Prepare the data for use as follows:

i. The dataset has two categorical attributes, Fuel Type and Metallic. Describe how you would convert these to binary variables. Confirm this using XLMiner's utility to transform categorical data into dummies.

ii. Prepare the dataset (as factored into dummies) for data mining techniques of supervised learning by creating partitions using XLMiner's data partitioning utility. Select all the variables and use default values for the random seed and partitioning percentages for training (50%), validation (30%), and test (20%) sets. Describe the roles that these partitions will play in modeling.

## PART II

Data Exploration and Dimension Reduction

## CHAPTER 3

Data Visualization

In this chapter we describe a set of plots that can be used to explore the multi-dimensional nature of a dataset. We present basic plots (bar charts, line graphs, and scatter plots), distribution plots (boxplots and histograms), and different enhancements that expand the capabilities of these plots to let the user visualize more information. We focus on how the different visualizations and operations can support data mining tasks, from supervised tasks (prediction, classification, and time series forecasting) to unsupervised tasks, and provide a few guidelines on specific visualizations to use with each data mining task. We also describe the advantages of interactive visualization over static plots. The chapter concludes with a presentation of specialized plots suitable for data with special structure (hierarchical, network, and geographical).

3.1 USES OF DATA VISUALIZATION[1]

The popular saying "a picture is worth a thousand words" refers to the ability to condense diffused verbal information into a compact and quickly understood graphical image. In the case of numbers, data visualization and numerical summarization provide us with both a powerful tool to explore data and an effective way to present results.

Where do visualization techniques fit into the data mining process, as described so far? They are primarily used in the preprocessing portion of the data mining process. Visualization supports data cleaning by finding incorrect values

---

[1]This and subsequent sections in this chapter © 2016 Statistics.com and Galit Shmueli. Used by permission.

Data Mining for Business Analytics: Concepts, Techniques, and Applications with XLMiner®, Third Edition.
Galit Shmueli, Peter C. Bruce, and Nitin R. Patel.

(e.g., patients whose age is 999 or −1), missing values, duplicate rows, columns with all the same value, and the like. Visualization techniques are also useful for variable derivation and selection: they can help determine which variables to include in the analysis and which might be redundant. They can also help with determining appropriate bin sizes, should binning of numerical variables be needed (e.g., a numerical outcome variable might need to be converted to a binary variable, as was done in the Boston housing data, if a yes/no decision is required). They can also play a role in combining categories as part of the data reduction process. Finally, if the data have yet to be collected and collection is expensive (as with the Pandora project at its outset; see Chapter 7), visualization methods can help determine, using a sample, which variables and metrics are useful.

In this chapter we focus on the use of graphical presentations for the purpose of *data exploration*, particularly with relation to predictive analytics. Although our focus is not on visualization for the purpose of data reporting, this chapter offers ideas as to the effectiveness of various graphical displays for the purpose of data presentation. These offer a wealth of useful presentations beyond tabular summaries and basic bar charts, which are currently the most popular form of data presentation in the business environment. For an excellent discussion of using graphs to report business data, see Few (2004). In terms of reporting data mining results graphically, we describe common graphical displays elsewhere in the book, some of which are technique specific [e.g., dendrograms for hierarchical clustering (Chapter 15), network charts for social network analysis (Chapter 19), and tree charts for classification and regression trees (Chapter 9)] while others are more general [e.g., receiver operating characteristic (ROC) curves and lift charts for classification (Chapter 5) and profile plots and heatmaps for clustering (Chapter 15)].

Data exploration is a mandatory initial step whether or not more formal analysis follows. Graphical exploration can support free-form exploration for the purpose of understanding the data structure, cleaning the data (e.g., identifying unexpected gaps or "illegal" values), identifying outliers, discovering initial patterns (e.g., correlations among variables and surprising clusters), and generating interesting questions. Graphical exploration can also be more focused, geared toward specific questions of interest. In the data mining context a combination is needed: free-form exploration performed with the purpose of supporting a specific goal.

Graphical exploration can range from generating very basic plots to using operations such as filtering and zooming interactively to explore a set of interconnected visualizations that include advanced features such as color and multiple-panels. This chapter is not meant to be an exhaustive guidebook on visualization techniques, but instead to discuss main principles and features that support data exploration in a data mining context. We start by describing varying levels of sophistication in terms of visualization, and show the advantages of different features and operations. Our discussion is from the perspective of how visualization supports the subsequent data mining goal. In particular, we distinguish between supervised and unsupervised learning; within supervised learning, we also further distinguish between classification (categorical Y) and prediction (numerical Y).

3.2 DATA EXAMPLES

To illustrate data visualization, we use two datasets that also appear in other chapters in the book. This allows the reader to compare some of the basic Excel plots used in other chapters to the improved plots, and easily see the merit of advanced visualization.

Example 1: Boston Housing Data

The Boston housing data contain information on census tracts in Boston[2] for which several measurements are taken (e.g., crime rate, pupil/teacher ratio). It has 14 variables. A description of each variable is given in Table 3.1 and a sample of the first nine records is shown in Table 3.2. In addition to the original 13 variables, the dataset contains the variable CAT.MEDV, which was created by categorizing median value (MEDV) into two categories, high and low.

TABLE 3.1	DESCRIPTION OF VARIABLES IN BOSTON HOUSING DATASET
CRIM	Crime rate
ZN	Percentage of residential land zoned for lots over 25,000 ft^2
INDUS	Percentage of land occupied by nonretail business
CHAS	Does tract bound Charles River (=1 if tract bounds river, =0 otherwise)
NOX	Nitric oxide concentration (parts per 10 million)
RM	Average number of rooms per dwelling
AGE	Percentage of owner-occupied units built prior to 1940
DIS	Weighted distances to five Boston employment centers
RAD	Index of accessibility to radial highways
TAX	Full-value property tax rate per $10,000
PTRATIO	Pupil-to-teacher ratio by town
LSTAT	Percentage of lower status of the population
MEDV	Median value of owner-occupied homes in $1000s
CAT.MEDV	Is median value of owner-occupied homes in tract above $30000 (CAT.MEDV=1) or not (CAT.MEDV=0)

---

[2] The Boston Housing dataset was originally published by Harrison and Rubinfeld in "Hedonic prices and the demand for clean air," *Journal of Environmental Economics and Management*, vol. 5, pp. 81–102, 1978.

We consider three possible tasks:

1. A supervised predictive task, where the outcome variable of interest is the median value of a home in the tract (MEDV).
2. A supervised classification task, where the outcome variable of interest is the binary variable CAT.MEDV that indicates whether the home value is above or below $30,000.
3. An unsupervised task, where the goal is to cluster census tracts.

(MEDV and CAT.MEDV are not used together in any of the three cases).

TABLE 3.2   FIRST NINE RECORDS IN THE BOSTON HOUSING DATA

CRIM	ZN	INDUS	CHAS	NOX	RM	AGE	DIS	RAD	TAX	PTRATIO	LSTAT	MEDV	CAT. MEDV
0.00632	18.0	2.31	0	0.538	6.575	65.2	4.0900	1	296	15.3	4.98	24.0	0
0.02731	0.0	7.07	0	0.469	6.421	78.9	4.9671	2	242	17.8	9.14	21.6	0
0.02729	0.0	7.07	0	0.469	7.185	61.1	4.9671	2	242	17.8	4.03	34.7	1
0.03237	0.0	2.18	0	0.458	6.998	45.8	6.0622	3	222	18.7	2.94	33.4	1
0.06905	0.0	2.18	0	0.458	7.147	54.2	6.0622	3	222	18.7	5.33	36.2	1
0.02985	0.0	2.18	0	0.458	6.430	58.7	6.0622	3	222	18.7	5.21	28.7	0
0.08829	12.5	7.87	0	0.524	6.012	66.6	5.5605	5	311	15.2	12.43	22.9	0
0.14455	12.5	7.87	0	0.524	6.172	96.1	5.9505	5	311	15.2	19.15	27.1	0
0.21124	12.5	7.87	0	0.524	5.631	100.0	6.0821	5	311	15.2	29.93	16.5	0

Example 2: Ridership on Amtrak Trains

Amtrak, a US railway company, routinely collects data on ridership. Here we focus on forecasting future ridership using the series of monthly ridership between January 1991 and March 2004. The data and their source are described in Chapter 16. Hence our task here is (numerical) time series forecasting.

3.3 BASIC CHARTS: BAR CHARTS, LINE GRAPHS, AND SCATTER PLOTS

The three most effective basic plots are bar charts, line graphs, and scatter plots. These plots are easy to create in Microsoft Excel and are the plots commonly used today in the business world, in both data exploration and presentation (unfortunately, pie charts are also popular, although usually ineffective visualizations). Basic charts support data exploration by displaying one or two columns of data (variables) at a time. This is useful in the early stages of getting familiar with the data structure, the amount and types of variables, the volume and type of missing values, and so on.

The nature of the data mining task and domain knowledge about the data will affect the use of basic charts in terms of the amount of time and effort allocated to different variables. In supervised learning, there will be more focus on the outcome variable. In scatter plots, the outcome variable is typically associated with the *y*-axis. In unsupervised learning (for the purpose of data reduction or clustering), basic plots that convey relationships (e.g., scatter plots) are preferred.

The top left panel in Figure 3.1 displays a line chart for the time series of monthly railway passengers on Amtrak. Line graphs are used primarily for showing time series. The choice of time frame to plot, as well as the temporal scale, should depend on the horizon of the forecasting task and on the nature of the data.

Bar charts are useful for comparing a single statistic (e.g., average, count, percentage) across groups. The height of the bar (or length, in a horizontal display) represents the value of the statistic, and different bars correspond to different groups. Two examples are shown in the bottom panels in Figure 3.1. The left panel shows a bar chart for a numerical variable (MEDV) and the right panel shows a bar chart for a categorical variable (CAT.MEDV). In each,

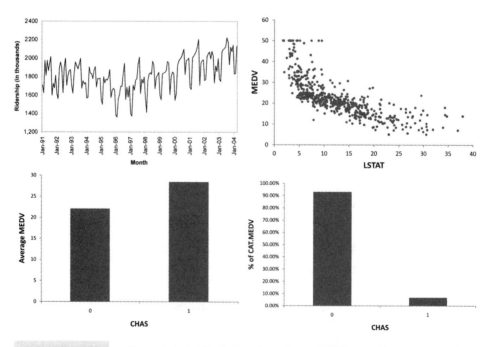

FIGURE 3.1   BASIC PLOTS: LINE GRAPH (TOP LEFT), SCATTER PLOT (TOP RIGHT), BAR CHART FOR NUMERICAL VARIABLE (BOTTOM LEFT), AND BAR CHART FOR CATEGORICAL VARIABLE (BOTTOM RIGHT). CHARTS PRODUCED IN EXCEL.

separate bars are used to denote homes in Boston that are near the Charles River vs. those that are not (thereby comparing the two categories of CHAS). The chart with the numerical output MEDV (bottom left) uses the average MEDV on the y-axis. This supports the predictive task: the numerical outcome is on the y-axis and the x-axis is used for a potential categorical predictor.[3] (Note that the x-axis on a bar chart must be used only for categorical variables because the order of bars in a bar chart should be interchangeable.) For the classification task, CAT.MEDV is on the y-axis (bottom right), but its aggregation is a percentage (the alternative would be a count). This graph shows us that the vast majority (over 90%) of the tracts do not border the Charles river (CHAS = 0). Note that the labeling of the y-axis can be confusing in this case: the value of CAT.MEDV plays no role and the y-axis is simply a percentage of all records.

The top right panel in Figure 3.1 displays a scatter plot of MEDV vs. LSTAT. This is an important plot in the prediction task. Note that the output MEDV is again on the y-axis (and LSTAT on the x-axis is a potential predictor). Because both variables in a basic scatter plot must be numerical, it cannot be used to display the relation between CAT.MEDV and potential predictors for the classification task (but we can enhance it to do so; see Section 3.4). For unsupervised learning, this particular scatter plot helps us study the association between two numerical variables in terms of information overlap as well as identifying clusters of observations.

All three basic plots highlight global information such as the overall level of ridership or MEDV, as well as changes over time (line chart), differences between subgroups (bar chart), and relationships between numerical variables (scatter plot).

Distribution Plots: Boxplots and Histograms

Before moving on to more sophisticated visualizations that enable multidimensional investigation, we note two important plots that are usually not considered "basic charts" but are very useful in statistical and data mining contexts. The *boxplot* and the *histogram* are two plots that display the entire distribution of a numerical variable. Although averages are very popular and useful summary statistics, there is usually much to be gained by looking at additional statistics such as the median and standard deviation of a variable, and even more so by examining the entire distribution. Whereas bar charts can only use a single aggregation, boxplots and histograms display the entire distribution of a numerical variable. Boxplots are also effective for comparing subgroups by generating

---

[3] We refer here to a bar chart with vertical bars. The same principles apply if using a bar chart with horizontal lines, except that the x-axis is now associated with the numerical variable and the y-axis with the categorical variable.

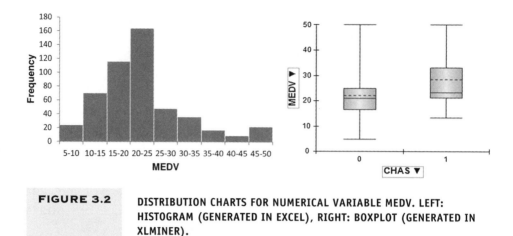

FIGURE 3.2 DISTRIBUTION CHARTS FOR NUMERICAL VARIABLE MEDV. LEFT: HISTOGRAM (GENERATED IN EXCEL), RIGHT: BOXPLOT (GENERATED IN XLMINER).

side-by-side boxplots, or for looking at distributions over time by creating a series of boxplots.

Distribution plots are useful in supervised learning for determining potential data mining methods and variable transformations. For example, skewed numerical variables might warrant transformation (e.g., moving to a logarithmic scale) if used in methods that assume normality (e.g., linear regression, discriminant analysis).

A histogram represents the frequencies of all x values with a series of vertical connected bars. For example, in the top left panel of Figure 3.2, there are about 20 tracts where the median value (MEDV) is between $7,500 and $12,500.

A boxplot represents the variable being plotted on the y-axis (although the plot can potentially be turned in a 90 degrees angle, so that the boxes are parallel to the x-axis). In the top right panel of Figure 3.2 there are two boxplots (called a side-by-side boxplot). The box encloses 50% of the data—for example, in the right-hand box half of the tracts have median values (MEDV) between $20,000 and $33,000. The horizontal line inside the box represents the median (50th percentile). The top and bottom of the box represent the 75th and 25th percentiles, respectively. Lines extending above and below the box cover the rest of the data range; outliers may be depicted as points or circles. Sometimes the average is marked by a + (or similar) sign, as in the top right panel of Figure 3.2. Comparing the average and the median helps in assessing how skewed the data are. Boxplots are often arranged in a series with a different plot for each of the various values of a second variable, shown on the x-axis.

Because histograms and boxplots are geared toward numerical variables, their basic form is useful for prediction tasks. Boxplots can also support unsupervised learning by displaying relationships between a numerical variable (y-axis) and a categorical variable (x-axis). To illustrate these points, the top panel in Figure 3.2

shows a histogram of MEDV, revealing a skewed distribution. Transforming the output variable to log(MEDV) would likely improve results of a linear regression predictor.

The right panel in Figure 3.2 shows side-by-side boxplots comparing the distribution of MEDV for homes that border the Charles River (1) or not (0), similar to Figure 3.1. We see that not only is the average MEDV for river-bounding homes higher than the non–river-bounding homes, the entire distribution is higher (median, quartiles, min, and max). We can also see that all river-bounding homes have MEDV above $10 thousand, unlike non–river-bounding homes. This information is useful for identifying the potential importance of this predictor (CHAS), and for choosing data mining methods that can capture the nonoverlapping area between the two distributions (e.g., trees). Boxplots and histograms applied to numerical variables can also provide directions for deriving new variables; for example, they can indicate how to bin a numerical variable (e.g., binning a numerical outcome in order to use a naive Bayes classifier, or in the Boston Housing example, choosing the cutoff to convert MEDV to CAT.MEDV).

Finally, side-by-side boxplots are useful in classification tasks for evaluating the potential of numerical predictors. This is done by using the *x*-axis for the categorical outcome and the *y*-axis for a numerical predictor. An example is shown in Figure 3.3, where we can see the effects of four numerical predictors on CAT.MEDV. The pairs that are most separated (e.g., PTRATIO and INDUS) indicate potentially useful predictors.

FIGURE 3.3   SIDE-BY-SIDE BOXPLOTS FOR EXPLORING THE CAT.MEDV OUTPUT VARIABLE BY DIFFERENT NUMERICAL PREDICTORS. IN A SIDE-BY-SIDE BOXPLOT, ONE AXIS IS USED FOR A CATEGORICAL VARIABLE, AND THE OTHER FOR A NUMERICAL VARIABLE. PLOTTING A CATEGORICAL OUTCOME VARIABLE AND A NUMERICAL PREDICTOR COMPARES THE PREDICTOR'S DISTRIBUTION ACROSS THE OUTCOME CATEGORIES. PLOTTING A NUMERICAL OUTCOME VARIABLE AND A CATEGORICAL PREDICTOR DISPLAYS THE DISTRIBUTION OF THE OUTCOME VARIABLE ACROSS DIFFERENT LEVELS OF THE PREDICTOR.

Boxplots and histograms are not readily available in Microsoft Excel (although they can be constructed through a manual process). They are available in a wide range of statistical software packages. In XLMiner they can be generated through the *Charts* menu (we note the current limitation of five categories for side-by-side boxplots).

The main weakness of basic charts and distribution plots, in their basic form (i.e., using position in relation to the axes to encode values), is that they can only display two variables and therefore cannot reveal high-dimensional information. Each of the basic charts has two dimensions, where each dimension is dedicated to a single variable. In data mining, the data are usually multivariate by nature, and the analytics are designed to capture and measure multivariate information. Visual exploration should therefore also incorporate this important aspect. In the next section we describe how to extend basic charts (and distribution charts) to multidimensional data visualization by adding features, employing manipulations, and incorporating interactivity. We then present several specialized charts that are geared toward displaying special data structures (Section 3.5).

Heatmaps: Visualizing Correlations and Missing Values

A *heatmap* is a graphical display of numerical data where color is used to denote values. In a data mining context, heatmaps are especially useful for two purposes: for visualizing correlation tables and for visualizing missing values in the data. In both cases the information is conveyed in a two-dimensional table. A correlation table for p variables has p rows and p columns. A data table contains p columns (variables) and n rows (observations). If the number of rows is huge, then a subset can be used. In both cases it is much easier and faster to scan the color-coding rather than the values. Note that heatmaps are useful when examining a large number of values, but they are not a replacement for more precise graphical display, such as bar charts, because color differences cannot be perceived accurately.

An example of a correlation table heatmap is shown in Figure 3.4, showing all the pairwise correlations between 13 variables (MEDV and 12 predictors).

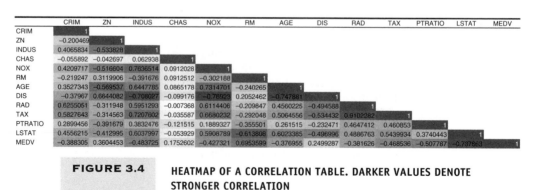

	CRIM	ZN	INDUS	CHAS	NOX	RM	AGE	DIS	RAD	TAX	PTRATIO	LSTAT	MEDV
CRIM	1												
ZN	−0.200469	1											
INDUS	0.4065834	−0.533828	1										
CHAS	−0.055892	−0.042697	0.062938	1									
NOX	0.4209717	−0.516604	0.7636514	0.0912028	1								
RM	−0.219247	0.3119906	−0.391676	0.0912512	−0.302188	1							
AGE	0.3527343	−0.569537	0.6447785	0.0865178	0.7314701	−0.240265	1						
DIS	−0.37967	0.6644082	−0.708027	−0.099176	−0.76923	0.2052462	−0.747881	1					
RAD	0.6255051	−0.311948	0.5951293	−0.007368	0.6114406	−0.209847	0.4560225	−0.494588	1				
TAX	0.5827643	−0.314563	0.7207602	−0.035587	0.6680232	−0.292048	0.5064556	−0.534432	0.9102282	1			
PTRATIO	0.2899456	−0.391679	0.3832476	−0.121515	0.1889327	−0.355501	0.261515	−0.232471	0.4647412	0.460853	1		
LSTAT	0.4556215	−0.412995	0.6037997	−0.053929	0.5908789	−0.613808	0.6023385	−0.496996	0.4886763	0.5439934	0.3740443	1	
MEDV	−0.388305	0.3604453	−0.483725	0.1752602	−0.427321	0.6953599	−0.376955	0.2499287	−0.381626	−0.468536	−0.507787	−0.737663	1

FIGURE 3.4   HEATMAP OF A CORRELATION TABLE. DARKER VALUES DENOTE STRONGER CORRELATION

Darker shades correspond to stronger (positive or negative) correlation. It is easy to quickly spot the high and low correlations. This heatmap was produced using Excel's *Conditional Formatting*.

In a missing-value heatmap, rows correspond to records and columns to variables. We use a binary coding of the original dataset where 1 denotes a missing value and 0 otherwise. This new binary table is then colored such that only missing-value cells (with value 1) are colored. Figure 3.5 shows an example of a missing-value heatmap for a dataset with over 1000 columns. The data include economic, social, political and "well-being" information on different countries around the world (each row is a country). The variables were merged from multiple sources, and for each source information was not always available on every country. The missing-data heatmap helps visualize the level and amount of "missingness" in the merged data file. Some patterns of "missingness" easily emerge: variables that are missing for nearly all observations, as well as clusters of rows (countries) that are missing many values. Variables with little missingness are also visible. This information can then be used for determining how to handle the missingness (e.g., dropping some variables, dropping some records, imputing, or via other techniques).

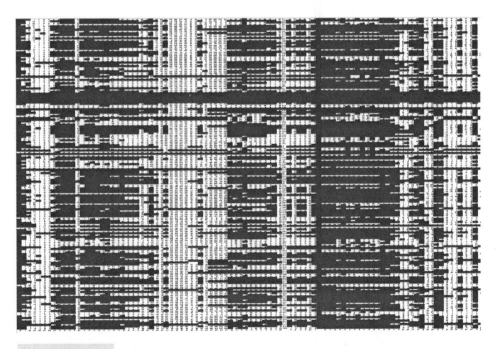

FIGURE 3.5   HEATMAP OF MISSING VALUES IN A DATASET. BLACK DENOTES MISSING VALUE

3.4 MULTIDIMENSIONAL VISUALIZATION

Basic plots can convey richer information with features such as color, size, and multiple panels, and by enabling operations such as rescaling, aggregation, and interactivity. These additions allow us to look at more than one or two variables at a time. The beauty of these additions is their effectiveness in displaying complex information in an easily understandable way. Effective features are based on understanding how visual perception works (see Few, 2009, for a discussion). The purpose is to make the information more understandable, not just represent the data in higher dimensions (e.g., as three-dimensional plots that are usually ineffective visualizations).

Adding Variables: Color, Size, Shape, Multiple Panels, and Animation

In order to include more variables in a plot, we must consider the type of variable to include. To represent additional categorical information, the best way is to use hue, shape, or multiple panels. For additional numerical information we can use color intensity or size. Temporal information can be added via animation.

Incorporating additional categorical and/or numerical variables into the basic (and distribution) plots means that we can now use all of them for both prediction and classification tasks! For example, we mentioned earlier that a basic scatter plot cannot be used for studying the relationship between a categorical outcome and predictors (in the context of classification). However, a very effective plot for classification is a scatter plot of two numerical predictors color-coded by the categorical outcome variable. An example is shown in the top panel of Figure 3.6, with color denoting CAT.MEDV.

In the context of prediction, color-coding supports the exploration of the conditional relationship between the numerical outcome (on the y-axis) and a numerical predictor. Color-coded scatter plots then help assess the need for creating interaction terms (e.g., is the relationship between MEDV and LSTAT different for homes near the river compared to homes away from the river?).

Color can also be used to include further categorical variables into a bar chart, as long as the number of categories is small. When the number of categories is large, a better alternative is to use multiple panels. Creating multiple panels (also called "trellising") is done by splitting the observations according to a categorical variable, and creating a separate plot (of the same type) for each category. An example is shown in the right-hand panel of Figure 3.6, where a bar chart of average MEDV by RAD is broken down into two panels by CHAS. We see that the average MEDV for different highway accessibility levels (RAD) behaves differently for homes near the river (lower panel) compared to homes away from the river (upper panel). This is especially salient for RAD = 1. We also see that there are no near-river homes in RAD levels 2, 6, and 7. Such information might lead us to create an interaction term between RAD and CHAS, and to

FIGURE 3.6

ADDING CATEGORICAL VARIABLES BY COLOR-CODING AND MULTIPLE PANELS. LEFT: SCATTER PLOT OF TWO NUMERICAL PREDICTORS, COLOR-CODED BY THE CATEGORICAL OUTCOME CAT.MEDV. RIGHT: BAR CHART OF MEDV BY TWO CATEGORICAL PREDICTORS (CHAS AND RAD), USING MULTIPLE PANELS FOR CHAS.

consider condensing some of the bins in RAD. All these explorations are useful for prediction and classification.

A special plot that uses scatter plots with multiple panels is the *scatter plot matrix*. In it, all pairwise scatter plots are shown in a single display. The panels in a matrix scatter plot are organized in a special way, such that each column corresponds to a variable and each row corresponds to a variable, thereby the intersections create all the possible pairwise scatter plots. The scatter plot matrix plot is useful in unsupervised learning for studying the associations between numerical variables, detecting outliers, and identifying clusters. For supervised learning, it can be used for examining pairwise relationships (and their nature) between predictors to support variable transformations and variable selection (see Correlation Analysis in Chapter 4). For prediction, it can also be used to depict the relationship of the outcome with the numerical predictors.

An example of a scatter plot matrix is shown in Figure 3.7, with MEDV and three predictors. Variable name indicates the *y*-axis variable. For example, the plots in the bottom row all have MEDV on the *y*-axis (which allows studying the individual outcome–predictor relations). We can see different types of relationships from the different shapes (e.g., an exponential relationship between MEDV and LSTAT and a highly skewed relationship between CRIM and IN-DUS), which can indicate needed transformations. Note that the plots above and to the right of the diagonal are mirror images of those below and to the left.

Once hue is used, further categorical variables can be added via shape and multiple panels. However, one must proceed cautiously in adding multiple variables, as the display can become overcluttered and then visual perception is lost.

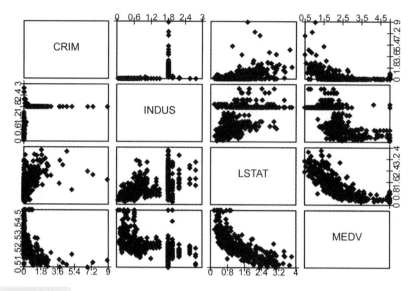

FIGURE 3.7   SCATTER PLOT MATRIX FOR MEDV AND THREE NUMERICAL PREDICTORS

Adding a numerical variable via size is useful especially in scatter plots (thereby creating "bubble plots") because, in a scatter plot, points represent individual observations. In plots that aggregate across observations (e.g., boxplots, histograms, bar charts), size and hue are not normally incorporated.

Finally, adding a temporal dimension to a plot to show how the information changes over time can be achieved via animation. A famous example is Rosling's animated scatter plots showing how world demographics changed over the years (www.gapminder.org). However, while animations of this type work for "statistical storytelling," they are not very effective for data exploration.

Manipulations: Rescaling, Aggregation and Hierarchies, Zooming, Filtering

Most of the time spent in data mining projects is spent in preprocessing. Typically, considerable effort is expended in getting all the data in a format that can actually be used in the data mining software. Additional time is spent processing the data in ways that improve the performance of the data mining procedures. This preprocessing step in data mining includes variable transformation and derivation of new variables to help models perform more effectively. Transformations include changing the numeric scale of a variable, binning numerical variables, condensing categories in categorical variables, and so on. The following manipulations support the preprocessing step as well as the choice of adequate data mining methods. They do so by revealing patterns and their nature.

FIGURE 3.8 RESCALING CAN ENHANCE PLOTS AND REVEAL PATTERNS. LEFT: ORIGINAL SCALE. RIGHT: LOGARITHMIC SCALE.

Rescaling Changing the scale in a display can enhance the plot and illuminate relationships. For example, in Figure 3.8 we see the effect of changing both axes of the scatter plot (top) and the y-axis of a boxplot (bottom) to logarithmic (log) scale. Whereas the original plots (left) are hard to understand, the patterns become visible in log scale (right). In the histograms the nature of the relationship between MEDV and CRIM is hard to determine in the original scale because too many of the points are "crowded" near the y-axis. The rescaling removes this crowding and allows a better view of the linear relationship between the two log-scaled variables (indicating a log-log relationship). In the boxplot displays, the crowding toward the x-axis in the original units does not allow us to compare the two box sizes, their locations, lower outliers, and most of the distribution information. Rescaling removes the "crowding to the x-axis" effect, thereby allowing a comparison of the two boxplots.

Aggregation and Hierarchies Another useful manipulation of scaling is changing the level of aggregation. For a temporal scale, we can aggregate by different granularity (e.g., monthly, daily, hourly) or even by a "seasonal" factor of interest such as month-of-year or day-of-week. A popular aggregation for time series is a moving average, where the average of neighboring values

FIGURE 3.9 TIME SERIES LINE GRAPHS USING DIFFERENT AGGREGATIONS (RIGHT PANELS), ADDING CURVES (TOP LEFT PANEL), AND ZOOMING IN (BOTTOM LEFT PANEL). GENERATED IN EXCEL.

within a given window size is plotted. Moving average plots enhance global trend visualization (see Chapter 16).

Nontemporal variables can be aggregated if some meaningful hierarchy exists: geographical (tracts within a zip code in the Boston Housing example), organizational (people within departments within units), and so on. Figure 3.9 illustrates two types of aggregation for the railway ridership time series. The original monthly series is shown in the top left panel. Seasonal aggregation (by month-of-year) is shown in the top right panel, where it is easy to see the peak in ridership in July–Aug and the dip in Jan–Feb. The bottom right panel shows temporal aggregation, where the series is now displayed in yearly aggregates. This plot reveals the global long-term trend in ridership and the generally increasing trend from 1996 on.

Examining different scales, aggregations, or hierarchies supports both supervised and unsupervised tasks in that it can reveal patterns and relationships at various levels, and can suggest new sets of variables with which to work.

Zooming and Panning The ability to zoom in and out of certain areas of the data on a plot is important for revealing patterns and outliers. We are often interested in more detail on areas of dense information or of special interest. Panning refers to the operation of moving the zoom window to other areas (popular in mapping applications such as Google Maps). An example of zooming is shown in the bottom left panel of Figure 3.9, where the ridership series is zoomed in to the first two years of the series.

Zooming and panning support supervised and unsupervised methods by detecting areas of different behavior, which may lead to creating new interaction terms, new variables, or even separate models for data subsets. In addition, zooming and panning can help choose between methods that assume global behavior (e.g., regression models) and data-driven methods (e.g., exponential smoothing forecasters and k-nearest-neighbors classifiers), and indicate the level of global/local behavior (as manifested by parameters such as k in k-nearest-neighbors, the size of a tree, or the smoothing parameters in exponential smoothing).

Filtering   Filtering means removing some of the observations from the plot. The purpose of filtering is to focus the attention on certain data while eliminating "noise" created by other data. Filtering supports supervised and unsupervised learning in a similar way to zooming and panning: it assists in identifying different or unusual local behavior.

Reference: Trend Line and Labels

Trend lines and using in-plot labels also help to detect patterns and outliers. Trend lines serve as a reference and allow us to more easily assess the shape of a pattern. Although linearity is easy to visually perceive, more elaborate relationships such as exponential and polynomial trends are harder to assess by eye. Trend lines are useful in line graphs as well as in scatter plots. An example is shown in the top left panel of Figure 3.9, where a polynomial curve is overlaid on the original line graph (see also Chapter 16).

In displays that are not overcrowded, the use of in-plot labels can be useful for better exploration of outliers and clusters. An example is shown in Figure 3.10. The figure shows different utilities on a scatter plot that compares fuel cost with total sales. We might be interested in clustering the data, and using clustering algorithms to identify clusters that differ markedly with respect to fuel cost and sales. The scatter plot with the labels helps visualize clusters and their members: for example, Nevada and Puget are part of a clear cluster with low fuel costs and high sales. For more on clustering and on this example, see Chapter 15.

Scaling up to Large Datasets

When the number of observations (rows) is large, plots that display each individual observation (e.g., scatter plots) can become ineffective. Aside from using aggregated charts such as boxplots, some alternatives are:

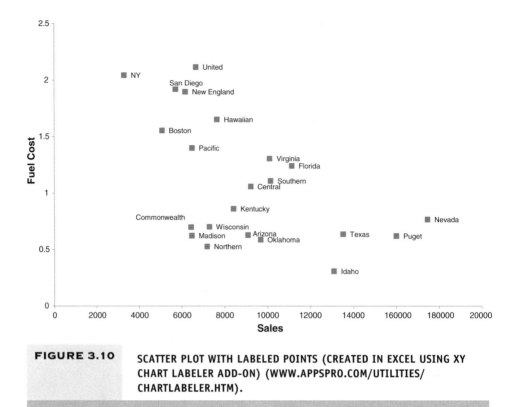

FIGURE 3.10   **SCATTER PLOT WITH LABELED POINTS (CREATED IN EXCEL USING XY CHART LABELER ADD-ON) (WWW.APPSPRO.COM/UTILITIES/ CHARTLABELER.HTM).**

1. Sampling—drawing a random sample and using it for plotting (XLMiner has a sampling utility)
2. Reducing marker size
3. Using more transparent marker colors and removing fill
4. Breaking down the data into subsets (e.g., by creating multiple panels)
5. Using aggregation (e.g., bubble plots where size corresponds to number of observations in a certain range)
6. Using jittering (slightly moving each marker by adding a small amount of noise)

An example of the advantage of plotting a sample over the large dataset is shown in Chapter 12 (Figure 12.2), comparing a scatter plot of a large dataset to that of a sample. Figure 3.11 illustrates an improved plot of the full dataset obtained by using smaller markers, jittering to uncover overlaid points, and more transparent colors. We can see that larger areas of the plot are dominated by the gray class, the black class is mainly on the right, while there is a lot of overlap in the top right area.

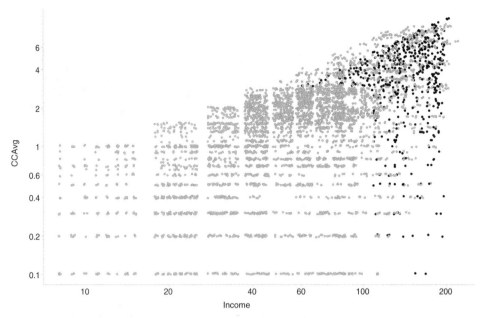

FIGURE 3.11 SCATTER PLOT OF LARGE DATASET WITH REDUCED MARKER SIZE, JITTERING, AND MORE TRANSPARENT COLORING. PRODUCED USING SPOTFIRE.

Multivariate Plot: Parallel Coordinates Plot

Another approach toward presenting multidimensional information in a two-dimensional plot is via specialized plots such as the *parallel coordinates plot*. In this plot a vertical axis is drawn for each variable. Then each observation is represented by drawing a line that connects its values on the different axes, thereby creating a "multivariate profile." An example is shown in Figure 3.12 for the Boston Housing data. In this display separate panels are used for the two values of CAT.MEDV, in order to compare the profiles of homes in the two classes (for a classification task). We see that the more expensive homes (bottom panel) consistently have low CRIM, low LSAT, and high RM compared to cheaper homes (top panel), which are more mixed on CRIM, and LSAT, and have a medium level of RM. This observation gives indication of useful predictors and suggests possible binning for some numerical predictors.

Parallel coordinate plots are also useful in unsupervised tasks. They can reveal clusters, outliers, and information overlap across variables. A useful manipulation is to re-order the columns to better reveal observation clusterings. Parallel coordinate plots are available in XLMiner.

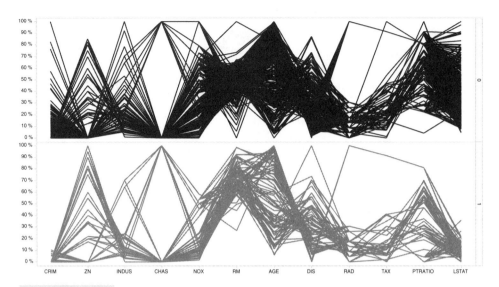

FIGURE 3.12 PARALLEL COORDINATES PLOT FOR BOSTON HOUSING DATA. EACH OF THE VARIABLES (SHOWN ON THE HORIZONTAL AXIS) IS SCALED TO 0–100%. PANELS ARE USED TO DISTINGUISH CAT.MEDV (TOP PANEL = HOMES BELOW $30,000). PRODUCED USING SPOTFIRE.

Interactive Visualization

Similar to the interactive nature of the data mining process, interactivity is key to enhancing our ability to gain information from graphical visualization. In the words of Stephen Few (Few, 2009), an expert in data visualization,

> We can only learn so much when staring at a static visualization such as a printed graph. . . . If we can't interact with the data . . . , we hit the wall.

By interactive visualization we mean an interface that supports the following principles:

1. Making changes to a chart is *easy, rapid, and reversible*.
2. Multiple concurrent charts and tables can be easily combined and displayed on a single screen.
3. A set of visualizations can be linked, so that operations in one display are reflected in the other displays.

Let us consider a few examples where we contrast a static plot generator (e.g., Excel) with an interactive visualization interface.

Histogram Re-binning Consider the situation where we need to bin a numerical variable, and plan to use a histogram for that process. A static histogram would require re-plotting for each new binning choice (in Excel it might even require creating the new bins manually). If the user generates multiple plots,

FIGURE 3.13   MULTIPLE INTER-LINKED PLOTS IN A SINGLE VIEW (USING SPOTFIRE). NOTE THE MARKED OBSERVATION IN THE TOP LEFT PANEL, WHICH IS ALSO HIGHLIGHTED IN ALL OTHER PLOTS.

then the screen becomes cluttered. If the same plot is recreated, then it is hard to compare different binning choices. In contrast, an interactive visualization would provide an easy way to change bin width interactively (e.g., see the slider below the histogram in Figure 3.13), and then the histogram would automatically and rapidly replot as the user changes the bin width.

Aggregation and Zooming   Consider a time series forecasting task, given a long series of data. Temporal aggregation at multiple levels is needed for determining short- and long-term patterns. Zooming and panning are used to identify unusual periods. A static plotting software requires the user to create new data columns for each temporal aggregation (e.g., aggregate daily data to obtain weekly aggregates). Zooming and panning in Excel requires manually changing the min and max values on the axis scale of interest (thereby losing the ability to quickly move between different areas without creating multiple charts). An interactive visualization would provide immediate temporal hierarchies which the user can easily switch between. Zooming would be enabled as a slider near the axis (e.g., see the sliders on the top left panel in Figure 3.13), thereby allowing direct manipulation and rapid reaction.

Combining Multiple Linked Plots That Fit in a Single Screen   To support a classification task, multiple plots are created of the outcome variable vs. potential categorical and numerical predictors. These can include side-by-side boxplots, color-coded scatter plots, multipanel bar charts, and so on. The user wants to detect possible multidimensional relationships (and identify possible outliers) by selecting a certain subset of the data (e.g., a single category of some variable) and locating the observations on the other plots. In a static interface, the user would have to manually organize the plots of interest and re-size them in order to fit within a single screen. A static interface would usually not support inter-plot linkage, and even if it did, the entire set of plots would have to be re-generated each time that a selection is made. In contrast, an interactive visualization would provide an easy way to automatically organize and re-size the set of plots to fit within a screen. Linking the set of plots would be easy, and in response to the users selection on one plot, the appropriate selection would be automatically highlighted in the other plots (e.g., see Figure 3.13).

In earlier sections we used plots to illustrate the advantages of visualizations, because "a picture is worth a thousand words." The advantages of an interactive visualization are even harder to convey in words. As Ben Shneiderman, a well-known researcher in information visualization and interfaces, puts it:

> A picture is worth a thousand words. An interface is worth a thousand pictures.

Interactive Visualization Software   Some added features such as color, shape, and size are often available in software that produce static plots, while others (multiple panels, hierarchies, labels) are only available in more advanced visualization tools. Even when a feature is available (e.g., color), the ease of applying it to a plot can widely vary. For example, incorporating color into an Excel scatter plot is a daunting task.[4] Plot manipulation possibilities (e.g., zooming, filtering, and aggregation) and ease of implementation are also quite limited in standard "static plot" software.

Although we do not intend to provide a market survey of interactive visualization tools, we mention a few prominent packages. Spotfire (http://spotfire.tibco.com) and Tableau (www.tableausoftware.com) are two designated data visualization tools (several of the plots in this chapter were created using Spotfire). They both provide a high level of interactivity, can support large data sets, and produce high-quality plots that are also easy to export. JMP by SAS (www.jmp.com) is a "statistical discovery" software that also has strong interactive visualization capabilities. All three offer free trial versions. Finally, Watson Analytics by IBM (www.ibm.com/analytics/watson-analytics/) allows uploading your data and visualizing it via different interactive visualizations.

---

[4] see www.bzst.com/2009/08/creating-color-coded-scatterplots-in.html

Programming environments like R and Python, which have become popular for statistical analysis, data mining, and presentation graphics, are less suitable for interactive visualization, where a sophisticated and highly engineered user interface is required.

3.5   SPECIALIZED VISUALIZATIONS

In this section we mention a few specialized visualizations that are able to capture data structures beyond the standard time series and cross-sectional structures—special types of relationships that are usually hard to capture with ordinary plots. In particular, we address hierarchical data, network data, and geographical data—three types of data that are becoming more available.

Visualizing Networked Data

Network analysis techniques were spawned by the explosion of social and product network data. Examples of social networks are networks of sellers and buyers on eBay and networks of people on Facebook. An example of a product network is the network of products on Amazon (linked through the recommendation system). Network data visualization is available in various network-specialized software, and also in general-purpose software.

A network diagram consists of actors and relations between them. "Nodes" are the actors (e.g., people in a social network or products in a product network), and represented by circles. "Edges" are the relations between nodes, and are represented by lines connecting nodes. For example, in a social network such Facebook, we can construct a list of users (nodes) and all the pairwise relations (edges) between users who are "Friends." Alternatively, we can define edges as a posting that one user posts on another user's Facebook page. In this setup we might have more than a single edge between two nodes. Networks can also have nodes of multiple types. A common structure is networks with two types of nodes. An example of a two-type node network is shown in Figure 3.14, where we see a set of transactions between a network of sellers and buyers on the online auction site www.eBay.com (the data are for auctions selling Swarovski beads, and took place during a period of several months; from Jank and Yahav, 2010). The circles on the left side represent sellers and on the right side buyers. Circle size represents the number of transactions that the node (seller or buyer) was involved in within this network. Line width represents the number of auctions that the bidder–seller pair interacted in (in this case we use arrows to denote the directional relationship from buyer to seller). We can see that this marketplace is dominated by three or four high-volume sellers. We can also see that many

Network of eBay Sellers and Buyers

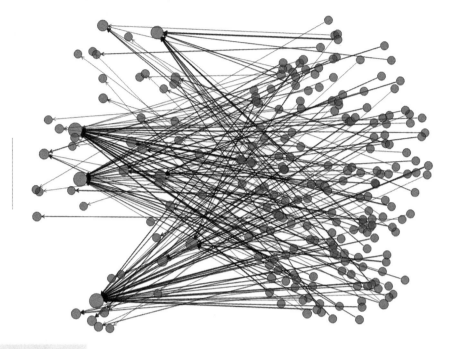

FIGURE 3.14

NETWORK GRAPH OF EBAY SELLERS (LEFT SIDE) AND BUYERS (RIGHT SIDE) OF SWAROVSKI BEADS. CIRCLE SIZE REPRESENTS THE NODE'S NUMBER OF TRANSACTIONS. LINE WIDTH REPRESENTS THE NUMBER OF TRANSACTIONS BETWEEN THAT PAIR OF SELLER–BUYER. PRODUCED USING SPOTFIRE.

buyers interact with a single seller. The market structures for many individual products can be reviewed quickly in this way. Network providers can use the information, for example, to identify possible partnerships to explore with sellers.

Figure 3.14 was produced using Spotfire's network visualization. An Excel-based tool is NodeXL (http://nodexl.codeplex.com), which is a template for Excel that allows entering a network edge list. The graph's appearance can be customized, and various interactive features are available such as zooming, scaling, and panning the graph, dynamically filtering nodes and edges, altering the graph's layout, finding clusters of related nodes, and calculating graph metrics. Networks can be imported from and exported to a variety of data formats, and built-in connections for getting networks from Twitter, Flickr, and your local email are provided (see Chapter 19 for details and examples).

Network graphs can be potentially useful in the context of association rules (see Chapter 14). For example, consider mining a dataset of consumers' grocery purchases to learn which items are purchased together ("what goes with what"). A network can be constructed with items as nodes and edges connecting items that were purchased together. After a set of rules is generated by the data mining algorithm (which often contains an excessive number of rules, many of which are unimportant), the network graph can help visualize different rules for the purpose of choosing the interesting ones. For example, a popular "beer and diapers" combination would appear in the network graph as a pair of nodes with very high connectivity. An item that is almost always purchased regardless of other items (e.g., milk) would appear as a very large node with high connectivity to all other nodes.

Visualizing Hierarchical Data: Treemaps

We discussed hierarchical data and the exploration of data at different hierarchy levels in the context of plot manipulations. *Treemaps* are useful visualizations specialized for exploring large data sets that are hierarchically structured (tree-structured). They allow exploration of various dimensions of the data while maintaining the hierarchical nature of the data. An example is shown in Figure 3.15, which displays a large set of auctions from eBay.com,[5] hierarchically

Treemap of eBay Auctions

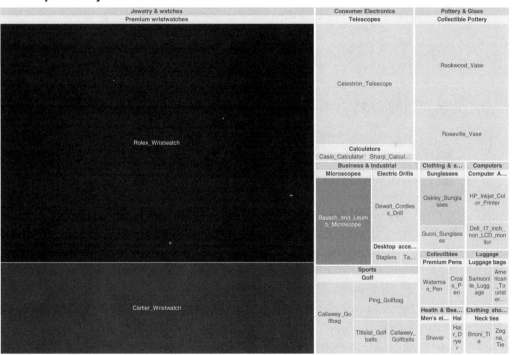

FIGURE 3.15   TREEMAP SHOWING NEARLY 11,000 eBAY AUCTIONS, ORGANIZED BY ITEM CATEGORY, SUBCATEGORY, AND BRAND. RECTANGLE SIZE REPRESENTS AVERAGE CLOSING PRICE (REFLECTING ITEM VALUE). SHADE REPRESENTS % OF SELLERS WITH NEGATIVE FEEDBACK (DARKER = HIGHER %).

---

[5] We thank Sharad Borle for sharing this dataset.

ordered by item category, sub-category, and brand. The levels in the hierarchy of the treemap are visualized as rectangles containing sub-rectangles. Categorical variables can be included in the display by using hue. Numerical variables can be included via rectangle size and color intensity (ordering of the rectangles is sometimes used to reinforce size). In the example in Figure 3.15 size is used to represent the average closing price (which reflects item value), and color intensity represents the percentage of sellers with negative feedback (a negative seller feedback indicates buyer dissatisfaction in past transactions and often indicative of fraudulent seller behavior). Consider the task of classifying ongoing auctions in terms of a fraudulent outcome. From the treemap we see that the highest proportion of sellers with negative ratings (black) is concentrated in expensive item auctions (Rolex and Cartier wristwatches).

Ideally, treemaps should be explored interactively, zooming to different levels of the hierarchy. One example of an interactive online application of treemaps is currently available at www.drasticdata.nl/drastictreemap.htm. One of their examples displays player-level data from the 2014 World Cup, aggregated to team level. The user can choose to explore players and team data.

Excel 2016 is planned to include treemaps in its in-built charts.

Visualizing Geographical Data: Map Charts

Many datasets used for data mining now include geographical information. Zip codes are one example of a categorical variable with many categories, where it is not straightforward to create meaningful variables for analysis. Plotting the data on a geographic map can often reveal patterns that are harder to identify otherwise. A map chart uses a geographical map as its background, and then color, hue, and other features are used to include categorical or numerical variables. Besides specialized mapping software, maps are now becoming part of general-purpose software, and Google Maps provides APIs (application programming interfaces) that allow organizations to overlay their data on a Google map. While Google Maps is readily available, the resulting map charts (e.g., Figure 3.16) are somewhat inferior in terms of effectiveness compared to map charts in dedicated interactive visualization software.

Figure 3.17 shows two world map charts (created with Spotfire), comparing countries' "well-being" (according to a 2006 Gallup survey) in the top map to gross domestic product (GDP) in the bottom map. Darker shade means higher value (white areas are missing data).

FIGURE 3.16 MAP CHART OF STUDENTS (LIGHT PINS) AND INSTRUCTORS'
(DARK PINS) LOCATIONS ON A GOOGLE MAP (FROM STATISTICS.COM)

Well-Being Score

GDP

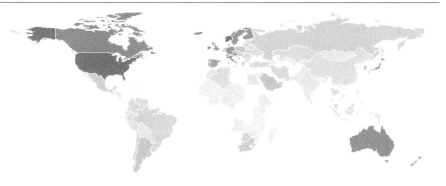

FIGURE 3.17 WORLD MAPS COMPARING "WELL-BEING" (TOP) TO GDP (BOTTOM).
SHADING BY AVERAGE "GLOBAL WELL-BEING" SCORE (TOP) OR GDP
(BOTTOM) OF COUNTRY. DARKER CORRESPONDS TO HIGHER SCORE OR
LEVEL. DATA FROM VEENHOVEN'S WORLD DATABASE OF HAPPINESS.

3.6 SUMMARY: MAJOR VISUALIZATIONS AND OPERATIONS, BY DATA MINING GOAL

Prediction

- Plot outcome on the y-axis of boxplots, bar charts, scatter plots.
- Study relation of outcome to categorical predictors via side-by-side boxplots, bar charts, and multiple panels.
- Study relation of outcome to numerical predictors via scatter plots.
- Use distribution plots (boxplot, histogram) for determining needed transformations of the outcome variable (and/or numerical predictors).
- Examine scatter plots with added color/panels/size to determine the need for interaction terms.
- Use various aggregation levels and zooming to determine areas of the data with different behavior, and to evaluate the level of global vs. local patterns.

Classification

- Study relation of outcome to categorical predictors using bar charts with the outcome on the y-axis.
- Study relation of outcome to pairs of numerical predictors via color-coded scatter plots (color denotes the outcome).
- Study relation of outcome to numerical predictors via side-by-side boxplots: plot boxplots of a numerical variable by outcome. Create similar displays for each numerical predictor. The most separable boxes indicate potentially useful predictors.
- Use color to represent the outcome variable on a parallel coordinate plot.
- Use distribution plots (boxplot, histogram) for determining needed transformations of numerical predictor variables.
- Examine scatter plots with added color/panels/size to determine the need for interaction terms.
- Use various aggregation levels and zooming to determine areas of the data with different behavior, and to evaluate the level of global vs. local patterns.

Time Series Forecasting

- Create line graphs at different temporal aggregations to determine types of patterns.

- Use zooming and panning to examine various shorter periods of the series to determine areas of the data with different behavior.
- Use various aggregation levels to identify global and local patterns.
- Identify missing values in the series (that will require handling).
- Overlay trend lines of different types to determine adequate modeling choices.

Unsupervised Learning

- Create scatter plot matrices to identify pairwise relationships and clustering of observations.
- Use heatmaps to examine the correlation table.
- Use various aggregation levels and zooming to determine areas of the data with different behavior.
- Generate a parallel coordinate plot to identify clusters of observations.

PROBLEMS

3.1 **Shipments of Household Appliances: Line Graphs.** The file ApplianceShipments.xls contains the series of quarterly shipments (in millions of dollars) of US household appliances between 1985 and 1989.

    a. Create a well-formatted time plot of the data using Excel.

    b. Does there appear to be a quarterly pattern? For a closer view of the patterns, zoom in to the range of 3500–5000 on the y-axis.

    c. Using Excel, create one chart with four separate lines, one line for each of Q1, Q2, Q3, and Q4. In Excel this can be achieved by sorting the data by Q1, Q2, Q3, Q4 (alphabetical sorting will work), and then plotting them as separate series on the line graph. Zoom in to the range of 3500–5000 on the y-axis. Does there appear to be a difference between quarters?

    d. Using Excel, create a line graph of the series at a yearly aggregated level (i.e., the total shipments in each year).

    e. Re-create the plots above using an interactive visualization tool. Be sure to enter the quarter information in a format that is recognized by the software as a date.

    f. Compare the two processes of generating the line graphs in terms of effort as well as the quality of the resulting plots. What are the advantages of each?

3.2 **Sales of Riding Mowers: Scatter Plots.** A company that manufactures riding mowers wants to identify the best sales prospects for an intensive sales campaign. In particular, the manufacturer is interested in classifying households as prospective owners or nonowners on the basis of Income (in $1000s) and Lot Size (in 1000 ft^2). The marketing expert looked at a random sample of 24 households, given in the file RidingMowers.xls.

    a. Using Excel, create a scatter plot of Lot Size vs. Income, color coded by the outcome variable owner/nonowner. Make sure to obtain a well-formatted plot (remove excessive background and gridlines; create legible labels and a legend, etc.). **Hint:** First sort the data by the outcome variable, and then plot the data for each category as separate series.

    b. Create the same plot, this time using an interactive visualization tool.

    c. Compare the two processes of generating the plot in terms of effort as well as the quality of the resulting plots. What are the advantages of each? Explain.

3.3 **Laptop Sales at a London Computer Chain: Bar Charts and Boxplots.** The file LaptopSalesJanuary2008.xls contains data for all sales of laptops at a computer chain in London in January 2008. This is a subset of the full dataset that includes data for the entire year.

    a. Create a bar chart, showing the average retail price by store. Which store has the highest average? Which has the lowest?

    b. To better compare retail prices across stores, create side-by-side boxplots of retail price by store. Now compare the prices in the two stores from (a). Does there seem to be a difference between their price distributions?

3.4 **Laptop Sales at a London Computer Chain: Interactive Visualization.** *The next exercises are designed for using an interactive visualization tool. The file LaptopSales.txt is a comma-separated file with nearly 300,000 rows. ENBIS (the European Network for Business and Industrial Statistics) provided these data as part of a contest organized in the fall of 2009.*

Scenario: Imagine that you are a new analyst for a company called Acell (a company selling laptops). You have been provided with data about products and sales. You need to help the company with their business goal of planning a product strategy and pricing policies that will maximize Acell's projected revenues in 2009. Using an interactive visualization tool, answer the following questions.

a. **Price Questions:**

    i. At what price are the laptops actually selling?

    ii. Does price change with time? (**Hint:** Make sure that the date column is recognized as such. The software should then enable different temporal aggregation choices, e.g., plotting the data by weekly or monthly aggregates, or even by day of week.)

    iii. Are prices consistent across retail outlets?

    iv. How does price change with configuration?

b. **Location Questions:**

    i. Where are the stores and customers located?

    ii. Which stores are selling the most?

    iii. How far would customers travel to buy a laptop?

        ○ *Hint 1:* You should be able to aggregate the data; for example, plot the sum or average of the prices.

        ○ *Hint 2:* Use the coordinated highlighting between multiple visualizations in the same page; for example, select a store in one view to see the matching customers in another visualization.

        ○ *Hint 3:* Explore the use of filters to see differences. Make sure to filter in the zoomed-out view; for example, try to use a "store location" slider as an alternative way to dynamically compare store locations. It might be more useful to spot outlier patterns if there were 50 store locations to compare.

    iv. Try an alternative way of looking at how far customers traveled. Do this by creating a new data column that computes the distance between customer and store.

c. **Revenue Questions:**

    i. How do the sales volume in each store relate to Acell's revenues?

    ii. How does this relationship depend on the configuration?

d. **Configuration Questions:**

    i. What are the details of each configuration? How does this relate to price?

    ii. Do all stores sell all configurations?

## CHAPTER 4

Dimension Reduction

In this chapter we describe the important step of dimension reduction. The dimension of a dataset, which is the number of variables, must be reduced for the data mining algorithms to operate efficiently. This process is part of the pilot/prototype phase of data mining and is done before deploying a model. We present and discuss several dimension reduction approaches: (1) incorporating domain knowledge to remove or combine categories, (2) using data summaries to detect information overlap between variables (and remove or combine redundant variables or categories), (3) using data conversion techniques such as converting categorical variables into numerical variables, and (4) employing automated reduction techniques such as principal components analysis (PCA), where a new set of variables (which are weighted averages of the original variables) is created. These new variables are uncorrelated, and a small subset of them usually contains most of their combined information (hence we can reduce dimension by using only a subset of the new variables). Finally, we mention data mining methods such as regression models and classification and regression trees, which can be used for removing redundant variables and for combining "similar" categories of categorical variables.

4.1 INTRODUCTION

In data mining one often encounters situations where there are a large number of variables in the database. Even when the initial number of variables is small, this set quickly expands in the data preparation step, where new derived variables

Data Mining for Business Analytics: Concepts, Techniques, and Applications with XLMiner®, Third Edition.
Galit Shmueli, Peter C. Bruce, and Nitin R. Patel.
© 2016 John Wiley & Sons Inc. Published 2016 by John Wiley & Sons, Inc.

are created (e.g., dummies for categorical variables and new forms of existing variables). In such situations it is likely that subsets of variables are highly correlated with each other. Including highly correlated variables in a classification or prediction model, or including variables that are unrelated to the outcome of interest, can lead to overfitting, and accuracy and reliability can suffer. A large number of variables also poses computational problems for some supervised as well as unsupervised algorithms (aside from questions of correlation). In model deployment, superfluous variables can increase costs due to the collection and processing of these variables. The *dimensionality* of a model is the number of independent or input variables used by the model.

4.2 CURSE OF DIMENSIONALITY

The *curse of dimensionality* is the affliction caused by adding variables to multivariate data models. As variables are added, the data space becomes increasingly sparse, and classification and prediction models fail because the available data are insufficient to provide a useful model across so many variables. An important consideration is the fact that the difficulties posed by adding a variable increase exponentially with the addition of each variable. One way to think of this intuitively is to consider the location of an object on a chessboard. It has two dimensions and 64 squares or choices. If you expand the chessboard to a cube, you increase the dimensions by 50%—from 2 dimensions to 3 dimensions. However, the location options increase by 800%, to 512 ($8 \times 8 \times 8$). In statistical distance terms, the proliferation of variables means that nothing is close to anything else anymore—too much noise has been added and patterns and structure are no longer discernible. The problem is particularly acute in Big Data applications, including genomics, for example, where an analysis might have to deal with values for thousands of different genes. One of the key steps in data mining, therefore, is finding ways to reduce dimensionality with minimal sacrifice of accuracy. In the artificial intelligence literature, dimension reduction is often referred to as *factor selection* or *feature extraction*.

4.3 PRACTICAL CONSIDERATIONS

Although data mining prefers automated methods over domain knowledge, it is important at the first step of data exploration to make sure that the variables measured are reasonable for the task at hand. The integration of expert knowledge through a discussion with the data provider (or user) will probably lead to better results. Practical considerations include: Which variables are most important for the task at hand, and which are most likely to be useless? Which variables are likely to contain much error? Which variables will be available for measurement (and what will it cost to measure them) in the future if the analysis is repeated? Which variables can actually be measured before the outcome occurs? For example, if we want to predict the closing price of an ongoing online auction, we cannot use the number of bids as a predictor because this will not be known until the auction closes.

Example 1: House Prices in Boston

We return to the Boston housing example introduced in Chapter 3. For each neighborhood, a number of variables are given, such as the crime rate, the student/teacher ratio, and the median value of a housing unit in the neighborhood. A description of all 14 variables is given in Table 4.1. The first nine records of the data are shown in Table 4.2. The first row represents the first neighborhood, which had an average per capita crime rate of 0.006, 18% of the residential land zoned for lots over 25,000 ft^2, 2.31% of the land devoted to nonretail business, no border on the Charles River, and so on.

TABLE 4.1	DESCRIPTION OF VARIABLES IN THE BOSTON HOUSINGDATASET
CRIM	Crime rate
ZN	Percentage of residential land zoned for lots over 25,000 ft^2
INDUS	Percentage of land occupied by nonretail business
CHAS	Does tract bound Charles River (= 1 if tract bounds river, = 0 otherwise)
NOX	Nitric oxide concentration (parts per 10 million)
RM	Average number of rooms per dwelling
AGE	Percentage of owner-occupied units built prior to 1940
DIS	Weighted distances to five Boston employment centers
RAD	Index of accessibility to radial highways
TAX	Full-value property tax rate per $10,000
PTRATIO	Pupil-to-teacher ratio by town
LSTAT	Percentage of lower status of the population
MEDV	Median value of owner-occupied homes in $1000s
CAT.MEDV	Is median value of owner-occupied homes in tract above $30,000 (CAT.MEDV = 1) or not (CAT.MEDV = 0)

4.4 DATA SUMMARIES

As we have seen in the chapter on data visualization, an important initial step of data exploration is getting familiar with the data and their characteristics through summaries and graphs. The importance of this step cannot be overstated. The better you understand the data, the better will be the results from the modeling or mining process.

TABLE 4.2 FIRST NINE RECORDS IN THE BOSTON HOUSING DATA

CRIM	ZN	INDUS	CHAS	NOX	RM	AGE	DIS	RAD	TAX	PTRATIO	LSTAT	MEDV	CAT.MEDV
0.00632	18.0	2.31	0	0.538	6.575	65.2	4.0900	1	296	15.3	4.98	24.0	0
0.02731	0.0	7.07	0	0.469	6.421	78.9	4.9671	2	242	17.8	9.14	21.6	0
0.02729	0.0	7.07	0	0.469	7.185	61.1	4.9671	2	242	17.8	4.03	34.7	1
0.03237	0.0	2.18	0	0.458	6.998	45.8	6.0622	3	222	18.7	2.94	33.4	1
0.06905	0.0	2.18	0	0.458	7.147	54.2	6.0622	3	222	18.7	5.33	36.2	1
0.02985	0.0	2.18	0	0.458	6.430	58.7	6.0622	3	222	18.7	5.21	28.7	0
0.08829	12.5	7.87	0	0.524	6.012	66.6	5.5605	5	311	15.2	12.43	22.9	0
0.14455	12.5	7.87	0	0.524	6.172	96.1	5.9505	5	311	15.2	19.15	27.1	0
0.21124	12.5	7.87	0	0.524	5.631	100.0	6.0821	5	311	15.2	29.93	16.5	0

Numerical summaries and graphs of the data are very helpful for data reduction. The information that they convey can assist in combining categories of a categorical variable, in choosing variables to remove, in assessing the level of information overlap between variables, and more. Before discussing such strategies for reducing the dimension of a data set, let us consider useful summaries and tools.

Summary Statistics

Excel has several functions and facilities that assist in summarizing data. The functions *average, stdev, min, max, median,* and *count* are very helpful for learning about the characteristics of each variable. First, they give us information about the scale and type of values that the variable takes. The min and max functions can be used to detect extreme values that might be errors. The average and median give a sense of the central values of that variable, and a large deviation between the two also indicates skew. The standard deviation gives a sense of how dispersed the data are (relative to the mean). Other functions, such as *countblank,* which gives the number of empty cells, can tell us about missing values. It is also possible to use Excel's *Descriptive Statistics* facility in the *Data>Data Analysis* menu. This will generate a set of 13 summary statistics for each of the variables.

Figure 4.1 shows six summary statistics for the Boston housing example. We immediately see that the different variables have very different ranges of values. We will soon see how variation in scale across variables can distort analyses if not treated properly. Another observation that can be made is that the average of the first variable, CRIM (as well as several others), is much larger than the median, indicating right skew. None of the variables have empty cells. There also do not appear to be indications of extreme values that might result from typing errors. Next, we summarize relationships between two or more variables. For numerical variables, we can compute pairwise correlations (using

	Average	Median	Min	Max	Std	Count	Countblank
CRIM	3.61	0.26	0.01	88.98	8.60	506	0
ZN	11.36	0.00	0.00	100.00	23.32	506	0
INDUS	11.14	9.69	0.46	27.74	6.86	506	0
CHAS	0.07	0.00	0.00	1.00	0.25	506	0
NOX	0.55	0.54	0.39	0.87	0.12	506	0
RM	6.28	6.21	3.56	8.78	0.70	506	0
AGE	68.57	77.50	2.90	100.00	28.15	506	0
DIS	3.80	3.21	1.13	12.13	2.11	506	0
RAD	9.55	5.00	1.00	24.00	8.71	506	0
TAX	408.24	330.00	187.00	711.00	168.54	506	0
PTRATIO	18.46	19.05	12.60	22.00	2.16	506	0
LSTAT	12.65	11.36	1.73	37.97	7.14	506	0
MEDV	22.53	21.20	5.00	50.00	9.20	506	0

FIGURE 4.1   SUMMARY STATISTICS FOR THE BOSTON HOUSING DATA

	CRIM	ZN	INDUS	CHAS	NOX	RM	AGE	DIS	RAD	TAX	PTRATIO	LSTAT	MEDV
CRIM	1.00												
ZN	−0.20	1.00											
INDUS	0.41	−0.53	1.00										
CHAS	−0.06	−0.04	0.06	1.00									
NOX	0.42	−0.52	0.76	0.09	1.00								
RM	−0.22	0.31	−0.39	0.09	−0.30	1.00							
AGE	0.35	−0.57	0.64	0.09	0.73	−0.24	1.00						
DIS	−0.38	0.66	−0.71	−0.10	−0.77	0.21	−0.75	1.00					
RAD	0.63	−0.31	0.60	−0.01	0.61	−0.21	0.46	−0.49	1.00				
TAX	0.58	−0.31	0.72	−0.04	0.67	−0.29	0.51	−0.53	0.91	1.00			
PTRATIO	0.29	−0.39	0.38	−0.12	0.19	−0.36	0.26	−0.23	0.46	0.46	1.00		
LSTAT	0.46	−0.41	0.60	−0.05	0.59	−0.61	0.60	−0.50	0.49	0.54	0.37	1.00	
MEDV	−0.39	0.36	−0.48	0.18	−0.43	0.70	−0.38	0.25	−0.38	−0.47	−0.51	−0.74	1.00

FIGURE 4.2   CORRELATION TABLE FOR BOSTON HOUSING DATA, GENERATED USING EXCEL'S DATA ANALYSIS MENU

the Excel function *correl*). We can also obtain a complete matrix of correlations between each pair of variables in the data using Excel's *Correlation* facility in the *Data>Data Analysis* menu. Figure 4.2 shows the correlation matrix for a subset of the Boston housing variables. We see that most correlations are low and that many are negative. Recall also the visual display of a correlation matrix via a heatmap (see Figure 3.4 in Chapter 3 for the heatmap corresponding to this correlation table). We will return to the importance of the correlation matrix soon, in the context of correlation analysis.

Pivot Tables

Another very useful tool is Excel's *pivot tables*, in the *Insert* menu. These are interactive tables that can combine information from multiple variables and compute a range of summary statistics (count, average, percentage, etc.). A simple example is the average MEDV for neighborhoods that bound the Charles River vs. those that do not (the variable CHAS is chosen as the column area). First, we get a count of neighborhoods bordering the river. This is shown in the top panel of Figure 4.3. The Excel pivot table was obtained by selecting CHAS as a "row labels" field and MEDV as a "values" field (using the "count" summary). It appears that the majority of neighborhoods (471 of 506) do not bound the river. By double-clicking on a certain cell, the complete data for records in that cell are shown on a new worksheet. For instance, double-clicking on the cell containing 471 will display the complete records of neighborhoods that do not bound the river. Pivot tables can be used for multiple variables. For categorical variables we obtain a breakdown of the records by the combination of categories. For instance, the bottom panel of Figure 4.3 shows the average MEDV by CHAS (row) and RM (column). Note that the numerical variable RM (the average number of rooms per dwelling in the neighborhood) is grouped into bins of 3–4, 5–6, and so on. Excel's bin notation means that the lower number is included, and all values up to but not including the higher number. So 3–4 includes only 3 rooms and not 4. Note also the empty cells, denoting that there are no neighborhoods in the dataset with those combinations (e.g., bounding the river and having on average 3 rooms). There are many more possibilities and options for using Excel's pivot tables. We leave it to the reader to explore these using Excel's documentation.

Count of MEDV	
CHAS	Total
0	471
1	35
Grand Total	506

Average of MEDV	CHAS		
RM	0	1	Grand Total
3-4	25.3		25.3
4-5	16.023077		16.02307692
5-6	17.133333	22.21818182	17.48734177
6-7	21.76917	25.91875	22.01598513
7-8	35.964444	44.06666667	36.91764706
8-9	45.7	35.95	44.2
Grand Total	22.093843	28.44	22.53280632

FIGURE 4.3   PIVOT TABLES FOR THE BOSTON HOUSING DATA

In classification tasks, where the goal is to find predictor variables that do a good job of distinguishing between two classes, a good exploratory step is to produce summaries for each class. This can assist in detecting useful predictors that display some separation between the two classes. Data summaries are useful for almost any data mining task and are therefore an important preliminary step for cleaning and understanding the data before carrying out further analyses.

4.5 CORRELATION ANALYSIS

In datasets with a large number of variables (which are likely to serve as predictors), there is usually much overlap in the information covered by the set of variables. One simple way to find redundancies is to look at a correlation matrix. This shows all the pairwise correlations between variables. Pairs that have a very strong (positive or negative) correlation contain a lot of overlap in information and are good candidates for data reduction by removing one of the variables. Removing variables that are strongly correlated to others is useful for avoiding multicollinearity problems that can arise in various models. (*Multicollinearity* is the presence of two or more predictors sharing the same linear relationship with the outcome variable.)

Correlation analysis is also a good method for detecting duplications of variables in the data. Sometimes the same variable appears accidentally more than once in the dataset (under a different name) because the dataset was merged from multiple sources, the same phenomenon is measured in different units, and so on. Using correlation table heatmaps, as shown in Chapter 3, can make the task of identifying strong correlations easier.

4.6   REDUCING THE NUMBER OF CATEGORIES IN CATEGORICAL VARIABLES

When a categorical variable has many categories, and this variable is destined to be a predictor, many data mining methods will require converting it into many dummy variables. In particular, a variable with m categories will be transformed into either m or $m - 1$ dummy variables (depending on the method). This means that even if we have very few original categorical variables, they can greatly inflate the dimension of the dataset. One way to handle this is to reduce the number of categories by combining close or similar categories. Combining categories requires incorporating expert knowledge and common sense. Pivot tables are useful for this task. With pivot tables we can examine the sizes of the various categories and how the response behaves at each category. Generally, categories that contain very few observations are good candidates for combining with other categories. Use only the categories that are most relevant to the analysis, and label the rest as "other." In classification tasks (with a categorical output), a pivot table broken down by the output classes can help identify categories that do not separate the classes. Those categories too are candidates for inclusion in the "other" category. An example is shown in Figure 4.4, where the distribution of output variable CAT.MEDV is broken down by ZN (treated here as a categorical variable). We can see that the distribution of CAT.MEDV

FIGURE 4.4   DISTRIBUTION OF CAT.MEDV (BLACK DENOTES CAT.MEDV=0) BY ZN. SIMILAR BARS INDICATE LOW SEPARATION BETWEEN CLASSES, AND CAN BE COMBINED

is identical for ZN = 17.5, 90, 95, and 100 (where all neighborhoods have CAT.MEDV = 1). These four categories can then be combined into a single category. Similarly categories ZN = 12.5, 25, 28, 30 and 70 can be combined. Further combination is also possible based on similar bars.

In a time series context where we might have a categorical variable denoting season (e.g., month or hour of day) that will serve as a predictor, reducing categories can be done by examining the time series plot and identifying similar periods. For example, the time plot in Figure 4.5 shows the quarterly revenues of Toys "R" Us between 1992 and 1995. Only quarter 4 periods appear different, and therefore we can combine quarters 1–3 into a single category.

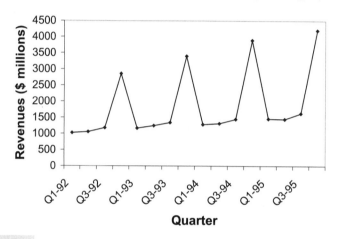

FIGURE 4.5   QUARTERLY REVENUES OF TOYS "R" US, 1992–1995

4.7   CONVERTING A CATEGORICAL VARIABLE TO A NUMERICAL VARIABLE

Sometimes the categories in a categorical variable represent intervals. Common examples are age group or income bracket. If the interval values are known (e.g., category 2 is the age interval 20–30), we can replace the categorical value ("2" in the example) with the mid-interval value (here "25"). The result will be a numerical variable that no longer requires multiple dummy variables.

4.8   PRINCIPAL COMPONENTS ANALYSIS

Principal components analysis (PCA) is a useful method for dimension reduction, especially when the number of variables is large. PCA is especially valuable when we have subsets of measurements that are measured on the same scale and are highly correlated. In that case it provides a few variables (often as few as three) that are weighted linear combinations of the original variables, and that retain the majority of the information of the full original set. PCA is intended for use with quantitative variables. For categorical variables, other methods, such as correspondence analysis, are more suitable.

Example 2: Breakfast Cereals

Data were collected on the nutritional information and consumer rating of 77 breakfast cereals.[1] The consumer rating is a rating of cereal "healthiness" for consumer information (not a rating by consumers). For each cereal the data include 13 numerical variables, and we are interested in reducing this dimension. For each cereal the information is based on a bowl of cereal rather than a serving size, since most people simply fill a cereal bowl (resulting in constant volume, but not weight). A snapshot of these data is given in Figure 4.6, and the description of the different variables is given in Table 4.3.

We focus first on two variables: *calories* and *consumer rating*. These are given in Table 4.4. The average calories across the 77 cereals is 106.88 and the average consumer rating is 42.67. The estimated covariance matrix between the two variables is

$$S = \begin{bmatrix} 379.63 & -188.68 \\ -188.68 & 197.32 \end{bmatrix}.$$

Cereal Name	mfr	type	calories	protein	fat	sodium	fiber	carbo	sugars	potass	vitamins
100% Bran	N	C	70	4	1	130	10	5	6	280	25
100% Natural Bran	Q	C	120	3	5	15	2	8	8	135	0
All-Bran	K	C	70	4	1	260	9	7	5	320	25
All-Bran with Extra Fiber	K	C	50	4	0	140	14	8	0	330	25
Almond Delight	R	C	110	2	2	200	1	14	8		25
Apple Cinnamon Cheerios	G	C	110	2	2	180	1.5	10.5	10	70	25
Apple Jacks	K	C	110	2	0	125	1	11	14	30	25
Basic 4	G	C	130	3	2	210	2	18	8	100	25
Bran Chex	R	C	90	2	1	200	4	15	6	125	25
Bran Flakes	P	C	90	3	0	210	5	13	5	190	25
Cap'n'Crunch	Q	C	120	1	2	220	0	12	12	35	25
Cheerios	G	C	110	6	2	290	2	17	1	105	25
Cinnamon Toast Crunch	G	C	120	1	3	210	0	13	9	45	25
Clusters	G	C	110	3	2	140	2	13	7	105	25
Cocoa Puffs	G	C	110	1	1	180	0	12	13	55	25
Corn Chex	R	C	110	2	0	280	0	22	3	25	25
Corn Flakes	K	C	100	2	0	290	1	21	2	35	25
Corn Pops	K	C	110	1	0	90	1	13	12	20	25
Count Chocula	G	C	110	1	1	180	0	12	13	65	25
Cracklin' Oat Bran	K	C	110	3	3	140	4	10	7	160	25

FIGURE 4.6   SAMPLE FROM THE 77 BREAKFAST CEREALS DATASET

---

[1] The data are available at http://lib.stat.cmu.edu/DASL/Stories/HealthyBreakfast.html.

<table>
<tr><th colspan="2">TABLE 4.3   DESCRIPTION OF THE VARIABLES IN THE BREAKFAST CEREAL DATASET</th></tr>
</table>

Variable	Description
mfr	Manufacturer of cereal (American Home Food Products, General Mills, Kellogg, etc.)
type	Cold or hot
calories	Calories per serving
protein	Grams of protein
fat	Grams of fat
sodium	Milligrams of sodium
fiber	Grams of dietary fiber
carbo	Grams of complex carbohydrates
sugars	Grams of sugars
potass	Milligrams of potassium
vitamins	Vitamins and minerals: 0, 25, or 100, indicating the typical percentage of FDA recommended
shelf	Display shelf (1, 2, or 3, counting from the floor)
weight	Weight in ounces of one serving
cups	Number of cups in one serving
rating	Rating of the cereal calculated by Consumer Reports

It can be seen that the two variables are strongly correlated with a negative correlation of

$$-0.69 = \frac{-188.68}{\sqrt{(379.63)(197.32)}}.$$

Roughly speaking, 69% of the total variation in both variables is actually "co-variation," or variation in one variable that is duplicated by similar variation in the other variable. Can we use this fact to reduce the number of variables, while making maximum use of their unique contributions to the overall variation? Since there is redundancy in the information that the two variables contain, it might be possible to reduce the two variables to a single variable without losing too much information. The idea in PCA is to find a linear combination of the two variables that contains most, even if not all, of the information, so that this new variable can replace the two original variables. Information here is in the sense of variability: What can explain the most variability *among* the 77 cereals? The total variability here is the sum of the variances of the two variables, which in this case is 379.63 + 197.32 = 577. This means that *calories* accounts for 66% = 379.63/577 of the total variability, and *rating* for the remaining 34%. If we drop one of the variables for the sake of dimension reduction, we lose at least 34% of the total variability. Can we redistribute the total variability between two new variables in a more polarized way? If so, it might be possible to keep only the one new variable that (hopefully) accounts for a large portion of the total variation.

Figure 4.7 shows a scatter plot of *rating* vs. *calories*. The line z_1 is the direction in which the variability of the points is largest. It is the line that captures the most

| TABLE 4.4 | CEREAL CALORIES AND RATINGS |

Cereal	Calories	Rating	Cereal	Calories	Rating
100% Bran	70	68.40297	Just Right Fruit & Nut	140	36.471512
100% Natural Bran	120	33.98368	Kix	110	39.241114
All-Bran	70	59.42551	Life	100	45.328074
All-Bran with Extra Fiber	50	93.70491	Lucky Charms	110	26.734515
Almond Delight	110	34.38484	Maypo	100	54.850917
Apple Cinnamon Cheerios	110	29.50954	Muesli Raisins, Dates & Almonds	150	37.136863
Apple Jacks	110	33.17409	Muesli Raisins, Peaches & Pecans	150	34.139765
Basic 4	130	37.03856	Mueslix Crispy Blend	160	30.313351
Bran Chex	90	49.12025	Multi-Grain Cheerios	100	40.105965
Bran Flakes	90	53.31381	Nut&Honey Crunch	120	29.924285
Cap'n'Crunch	120	18.04285	Nutri-Grain Almond-Raisin	140	40.69232
Cheerios	110	50.765	Nutri-grain Wheat	90	59.642837
Cinnamon Toast Crunch	120	19.82357	Oatmeal Raisin Crisp	130	30.450843
Clusters	110	40.40021	Post Nat. Raisin Bran	120	37.840594
Cocoa Puffs	110	22.73645	Product 19	100	41.50354
Corn Chex	110	41.44502	Puffed Rice	50	60.756112
Corn Flakes	100	45.86332	Puffed Wheat	50	63.005645
Corn Pops	110	35.78279	Quaker Oat Squares	100	49.511874
Count Chocula	110	22.39651	Quaker Oatmeal	100	50.828392
Cracklin' Oat Bran	110	40.44877	Raisin Bran	120	39.259197
Cream of Wheat (Quick)	100	64.53382	Raisin Nut Bran	100	39.7034
Crispix	110	46.89564	Raisin Squares	90	55.333142
Crispy Wheat & Raisins	100	36.1762	Rice Chex	110	41.998933
Double Chex	100	44.33086	Rice Krispies	110	40.560159
Froot Loops	110	32.20758	Shredded Wheat	80	68.235885
Frosted Flakes	110	31.43597	Shredded Wheat 'n'Bran	90	74.472949
Frosted Mini-Wheats	100	58.34514	Shredded Wheat spoon size	90	72.801787
Fruit & Fibre Dates, Walnuts & Oats	120	40.91705	Smacks	110	31.230054
Fruitful Bran	120	41.01549	Special K	110	53.131324
Fruity Pebbles	110	28.02577	Strawberry Fruit Wheats	90	59.363993
Golden Crisp	100	35.25244	Total Corn Flakes	110	38.839746
Golden Grahams	110	23.80404	Total Raisin Bran	140	28.592785
Grape Nuts Flakes	100	52.0769	Total Whole Grain	100	46.658844
Grape-Nuts	110	53.37101	Triples	110	39.106174
Great Grains Pecan	120	45.81172	Trix	110	27.753301
Honey Graham Ohs	120	21.87129	Wheat Chex	100	49.787445
Honey Nut Cheerios	110	31.07222	Wheaties	100	51.592193
Honey-comb	110	28.74241	Wheaties Honey Gold	110	36.187559
Just Right Crunchy Nuggets	110	36.52368			

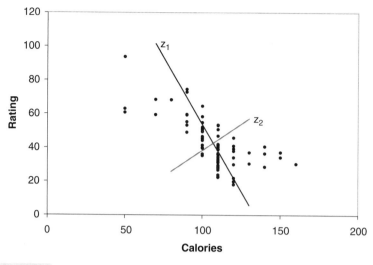

FIGURE 4.7 SCATTER PLOT OF *RATING* VS. *CALORIES* FOR 77 BREAKFAST CEREALS, WITH THE TWO PRINCIPAL COMPONENT DIRECTIONS

variation in the data if we decide to reduce the dimensionality of the data from two to one. Among all possible lines, it is the line for which, if we project the points in the dataset orthogonally to get a set of 77 (one-dimensional) values, the variance of the z_1 values will be maximal. This is called the *first principal component*. It is also the line that minimizes the sum-of-squared perpendicular distances from the line. The z_2-axis is chosen to be perpendicular to the z_1-axis. In the case of two variables, there is only one line that is perpendicular to z_1, and it has the second largest variability, but its information is uncorrelated with z_1. This is called the *second principal component*. In general, when we have more than two variables, once we find the direction z_1 with the largest variability, we search among all the orthogonal directions to z_1 for the one with the next-highest variability. That is z_2. The idea is then to find the coordinates of these lines and to see how they redistribute the variability.

Figure 4.8 shows the XLMiner output from running PCA on these two variables. The principal components table gives the weights that are used to project the original points onto the two new directions. The weights for z_1 are given by $(-0.847, 0.532)$, and for z_2 they are given by $(-0.532, -0.847)$. Figure 4.8 gives the reallocated variance: z_1 accounts for 86% of the total variability and z_2 for the remaining 14%. Therefore, if we drop z_2, we still maintain 86% of the total variability.

The weights are used to compute principal component scores, which are the projected values of *calories* and *rating* onto the new axes (after subtracting the means). Figure 4.9 shows the scores for the two dimensions. The first column is the projection onto z_1 using the weights $(-0.847, 0.532)$. The second column

Principal Components		
Feature\Component	1	2
calories	−0.847053	−0.531508
rating	0.5315077	−0.847053

Variances		
	1	2
Variance	498.02448	78.932739
Variance %	86.319135	13.680865
Cumulative Variance %	86.319135	100

FIGURE 4.8   OUTPUT FROM PRINCIPAL COMPONENTS ANALYSIS OF *CALORIES* AND *RATING*

is the projection onto z_2 using the weights $(-0.532, -0.847)$. For example, the first score for the 100% Bran cereal (with 70 calories and a rating of 68.4) is $(-0.847)(70-106.88)+(0.532)(68.4-42.67) = 44.92$. Note that the means of the new variables z_1 and z_2 are zero, since we've subtracted the mean of each variable. The sum of the variances $\text{var}(z_1) + \text{var}(z_2)$ is equal to the sum of the variances of the original variables, $\text{var}(calories) + \text{var}(rating)$. Furthermore the variances of z_1 and z_2 are 498 and 79, respectively, so the first principal component, z_1, accounts for 86% of the total variance. Since it captures most of the variability in the data, it seems reasonable to use one variable, the first principal score, to represent the two variables in the original data. Next, we generalize these ideas to more than two variables.

Principal Components

Let us formalize the procedure described above so that it can easily be generalized to $p > 2$ variables. Denote by $X_1, X_2, \ldots, X_p$ the original p variables. In PCA we are looking for a set of new variables $Z_1, Z_2, \ldots, Z_p$ that are weighted averages of the original variables (after subtracting their mean):

$$Z_i = a_{i,1}(X_1 - \bar{X}_1) + a_{i,2}(X_2 - \bar{X}_2) + \cdots + a_{i,p}(X_p - \bar{X}_p), \quad i = 1, \ldots, p, \text{ (4.1)}$$

where each pair of Z's has correlation $= 0$. We then order the resulting Z's by their variance, with Z_1 having the largest variance and Z_p having the smallest variance. The software computes the weights $a_{i,j}$, which are then used in computing the principal component scores.

A further advantage of the principal components compared to the original data is that they are uncorrelated (correlation coefficient $= 0$). If we construct regression models using these principal components as independent variables, we will not encounter problems of multicollinearity.

Scores		
Pattern\Component	1	2
100% Bran	44.921528	−2.197183
100% Natural Bran	−15.72526	0.3824165
All-Bran	40.149935	5.4072123
All-Bran with Extra Fiber	75.310772	−12.99913
Almond Delight	−7.041508	5.3576857
Apple Cinnamon Cheerios	−9.632769	9.4873273
Apple Jacks	−7.685031	6.3832549
Basic 4	−22.57211	−7.520309
Bran Chex	17.731545	3.5061586
Bran Flakes	19.960454	−0.046011
Cap'n'Crunch	−24.19794	13.88515
Cheerios	1.6646701	−8.517182
Cinnamon Toast Crun	−23.25147	12.376784
Clusters	−3.844296	0.2623499
Cocoa Puffs	−13.23272	15.224501
Corn Chex	−3.288971	−0.622661
Corn Flakes	7.5299271	0.949875

FIGURE 4.9   PRINCIPAL SCORES FROM PRINCIPAL COMPONENTS ANALYSIS OF *CALORIES* AND *RATING* FOR THE FIRST 17 CEREALS.

Let us return to the breakfast cereal dataset with all 15 variables, and apply PCA to the 13 numerical variables. The resulting output is shown in Figure 4.10. For simplicity, we removed three cereals that contained missing values.

Principal Components													
Feature\Component	1	2	3	4	5	6	7	8	9	10	11	12	13
calories	−0.078	0.009	0.629	0.601	0.455	0.119	−0.094	0.026	−0.009	0.065	−0.009	−0.004	−0.042
protein	0.001	−0.009	0.001	−0.003	0.056	0.113	−0.258	−0.655	0.202	−0.256	0.045	0.005	0.616
fat	0.000	−0.003	0.016	0.025	−0.016	−0.132	−0.373	0.118	−0.124	−0.841	0.062	0.009	−0.318
sodium	−0.980	−0.141	−0.136	0.001	0.014	0.023	−0.005	−0.001	0.004	−0.001	0.000	0.000	−0.010
fiber	0.005	−0.031	−0.018	−0.020	0.014	0.263	−0.043	0.659	−0.227	−0.144	−0.021	−0.001	0.648
carbo	−0.017	0.017	0.017	−0.026	0.349	−0.538	0.672	−0.006	0.025	−0.300	0.042	−0.014	0.206
sugars	−0.003	0.000	0.098	0.115	−0.299	0.648	0.567	−0.103	0.117	−0.320	0.028	−0.019	−0.136
potass	0.135	−0.987	0.037	0.042	−0.047	−0.050	0.018	−0.015	−0.001	0.006	0.000	−0.001	−0.006
vitamins	−0.094	−0.017	0.692	−0.714	−0.037	0.016	−0.012	−0.004	−0.012	0.001	0.002	−0.001	−0.010
shelf	0.002	−0.004	0.012	−0.006	−0.008	−0.060	−0.092	0.328	0.935	−0.046	−0.068	0.010	0.000
weight	−0.001	−0.001	0.004	0.003	0.003	0.009	0.024	0.003	−0.002	0.006	0.093	0.995	0.000
cups	−0.001	0.002	0.001	−0.001	0.002	−0.010	0.020	−0.062	−0.054	−0.080	−0.989	0.092	0.000
rating	0.075	−0.072	−0.308	−0.335	0.758	0.413	−0.018	−0.012	0.036	−0.023	−0.003	−0.003	−0.188

Variances													
	1	2	3	4	5	6	7	8	9	10	11	12	13
Variance	7016.42	5028.83	512.739	367.93	70.951	4.3751	2.888	0.6078	0.4327	0.13722	0.0348	0.004	0.0000
Variance %	53.950	38.667	3.943	2.829	0.546	0.034	0.022	0.005	0.003	0.001	0.000	0.000	0.000
Cumulative Variance %	53.95	92.62	96.56	99.39	99.93	99.97	99.99	100.00	100.00	100.00	100.00	100.00	100.00

FIGURE 4.10   PCA OUTPUT USING ALL 13 NUMERICAL VARIABLES IN THE BREAKFAST CEREALS DATASET.

Note that the first three components account for more than 96% of the total variation associated with all 13 of the original variables. This suggests that we can capture most of the variability in the data with less than 25% of the number of original dimensions in the data. In fact, the first two principal components alone capture 92.6% of the total variation. However, these results are influenced by the scales of the variables, as we describe next.

Normalizing the Data

A further use of PCA is to understand the structure of the data. This is done by examining the weights to see how the original variables contribute to the different principal components. In our example, it is clear that the first principal component is dominated by the sodium content of the cereal: it has the highest (in this case, positive) weight. This means that the first principal component is in fact measuring how much sodium is in the cereal. Similarly, the second principal component seems to be measuring the amount of potassium. Since both variables are measured in milligrams, whereas the other nutrients are measured in grams, the scale is obviously leading to this result. The variances of potassium and sodium are much larger than the variances of the other variables, and thus the total variance is dominated by these two variances. A solution is to normalize the data before performing the PCA. Normalization (or standardization) means replacing each original variable by a standardized version of the variable that has unit variance. This is easily accomplished by dividing each variable by its standard deviation. The effect of this normalization (standardization) is to give all variables equal importance in terms of the variability.

When should we normalize the data like this? It depends on the nature of the data. When the units of measurement are common for the variables (e.g., dollars), and when their scale reflects their importance (sales of jet fuel, sales of heating oil, etc.), it is probably best not to normalize (i.e., not to rescale the data so that they have unit variance). If the variables are measured in different units so that it is unclear how to compare the variability of different variables (e.g., dollars for some, parts per million for others) or if for variables measured in the same units, scale does not reflect importance (earnings per share, gross revenues), it is generally advisable to normalize. In this way the changes in units of measurement do not change the principal components' weights. In the rare situations where we can give relative weights to variables, we multiply the normalized variables by these weights before doing the principal components analysis.

Thus far we have calculated principal components using the covariance matrix. An alternative to normalizing and then performing PCA is to perform PCA on the correlation matrix instead of the covariance matrix. Most software programs allow the user to choose between the two. Remember, using the correlation matrix means that you are operating on the normalized data.

Returning to the breakfast cereal data, we normalize the 13 variables due to the different scales of the variables and then perform PCA (or equivalently, we use PCA applied to the correlation matrix). The output is shown in Figure 4.11. Now we find that we need 7 principal components to account for

Principal Components

Feature\Component	1	2	3	4	5	6	7	8	9	10	11	12	13
calories	−0.300	0.393	0.115	−0.204	0.204	−0.256	0.026	0.002	−0.030	0.500	−0.214	0.492	−0.234
protein	0.307	0.165	0.277	−0.301	0.320	0.121	−0.283	0.427	−0.535	−0.022	0.032	−0.100	0.186
fat	−0.040	0.346	−0.205	−0.187	0.587	0.348	0.051	−0.063	0.460	−0.145	−0.067	−0.291	−0.090
sodium	−0.183	0.137	0.389	−0.120	−0.338	0.664	0.284	−0.177	−0.215	−0.001	−0.087	−0.054	−0.239
fiber	0.453	0.180	0.070	−0.039	−0.255	0.064	−0.112	−0.216	0.244	0.295	−0.531	−0.059	0.442
carbo	−0.192	−0.149	0.562	−0.088	0.183	−0.326	0.260	−0.167	0.117	0.241	0.179	−0.476	0.225
sugars	−0.228	0.351	−0.355	0.023	−0.315	−0.152	−0.228	0.063	−0.225	0.252	−0.003	−0.614	−0.167
potass	0.402	0.301	0.068	−0.091	−0.148	0.025	−0.149	−0.262	0.167	0.177	0.729	0.121	−0.128
vitamins	−0.116	0.173	0.388	0.604	−0.049	0.129	−0.294	0.457	0.346	0.052	0.019	0.000	−0.060
shelf	0.171	0.265	−0.002	0.639	0.329	−0.052	0.175	−0.414	−0.416	−0.046	−0.059	−0.017	0.000
weight	−0.050	0.450	0.247	−0.153	−0.221	−0.399	−0.014	−0.075	0.065	−0.692	−0.113	0.031	0.000
cups	−0.295	−0.212	0.140	−0.047	0.121	0.099	−0.749	−0.499	−0.050	−0.077	−0.055	0.032	0.000
rating	0.438	−0.252	0.182	−0.038	0.058	−0.186	−0.063	−0.015	0.063	0.012	−0.276	−0.189	−0.743

Variances

	1	2	3	4	5	6	7	8	9	10	11	12	13
Variance	3.634	3.148	1.909	1.019	0.989	0.722	0.672	0.416	0.316	0.092	0.063	0.019	0.000
Variance %	27.95	24.22	14.69	7.84	7.61	5.55	5.17	3.20	2.43	0.71	0.49	0.15	0.00
Cumulative Variance %	27.95	52.17	66.85	74.70	82.31	87.86	93.03	96.23	98.66	99.36	99.85	100.00	100.00

FIGURE 4.11   PCA OUTPUT USING ALL *NORMALIZED* 13 NUMERICAL VARIABLES IN THE BREAKFAST CEREALS DATASET.

more than 90% of the total variability. The first 2 principal components account for only 52% of the total variability, and thus reducing the number of variables to 2 would mean losing a lot of information. Examining the weights, we see that the first principal component measures the balance between 2 quantities: (1) calories and cups (large negative weights) vs. (2) protein, fiber, potassium, and consumer rating (large positive weights). High scores on principal component 1 mean that the cereal is high in calories and the amount per bowl, and low in protein and potassium. Unsurprisingly, this type of cereal is associated with a low consumer rating. The second principal component is most affected by the weight of a serving, and the third principal component by the carbohydrate content. We can continue labeling the next principal components in a similar fashion to learn about the structure of the data.

When the data can be reduced to two dimensions, a useful plot is a scatter plot of the first vs. second principal scores with labels for the observations (if the dataset is not too large). To illustrate this, Figure 4.12 displays the first two principal component scores for the breakfast cereals.

We can see that as we move from right (bran cereals) to left, the cereals are less "healthy" in the sense of high calories, low protein and fiber, and so on. Also, moving from bottom to top, we get heavier cereals (moving from puffed rice to raisin bran). These plots are especially useful if interesting clusterings of

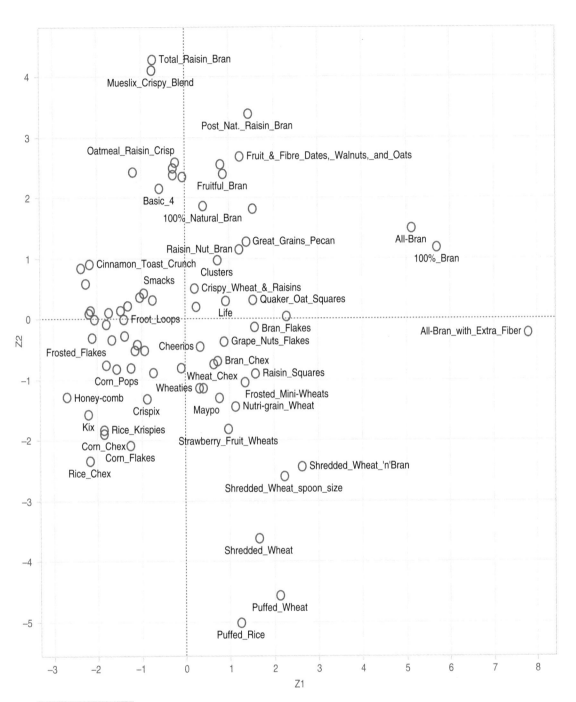

FIGURE 4.12   **SCATTER PLOT OF THE SECOND VS. FIRST PRINCIPAL COMPONENTS SCORES FOR THE NORMALIZED BREAKFAST CEREAL OUTPUT**

observations can be found. For instance, we see here that children's cereals are close together on the middle-right part of the plot.

Using Principal Components for Classification and Prediction

When the goal of the data reduction is to have a smaller set of variables that will serve as predictors, we can proceed as following: Apply PCA to the predictors using the training data. Use the output to determine the number of principal components to be retained. The predictors in the model now use the (reduced number of) principal scores columns. For the validation set we can use the weights computed from the training data to obtain a set of principal scores by applying the weights to the variables in the validation set. These new variables are then treated as the predictors.

One disadvantage of using a subset of principal components as predictors in a supervised task is that we might lose predictive information that is nonlinear (e.g., a quadratic effect of a predictor on the outcome or an interaction between predictors). This is because PCA produces linear transformations, thereby capturing linear relationships between the original variables.

4.9   DIMENSION REDUCTION USING REGRESSION MODELS

In this chapter we discussed methods for reducing the number of columns using summary statistics, plots, and principal components analysis. All these are considered exploratory methods. Some of them completely ignore the output variable (e.g., PCA), and in other methods we informally try to incorporate the relationship between the predictors and the output variable (e.g., combining similar categories, in terms of their behavior with y). Another approach to reducing the number of predictors, which directly considers the predictive or classification task, is by fitting a regression model. For prediction, a linear regression model is used (see Chapter 6) and for classification, a logistic regression model (see Chapter 10). In both cases we can employ subset selection procedures that algorithmically choose a subset of variables among the larger set (see details in the relevant chapters).

Fitted regression models can also be used to further combine similar categories: categories that have coefficients that are not statistically significant (i.e., have a high p-value) can be combined with the reference category because their distinction from the reference category appears to have no significant effect on the output variable. Moreover categories that have similar coefficient values (and the same sign) can often be combined because their effect on the output variable is similar. See the example in Chapter 10 on predicting delayed flights for an illustration of how regression models can be used for dimension reduction.

4.10 Dimension Reduction Using Classification and Regression Trees

Another method for reducing the number of columns and for combining categories of a categorical variable is by applying classification and regression trees (see Chapter 9). Classification trees are used for classification tasks and regression trees for prediction tasks. In both cases the algorithm creates binary splits on the predictors that best classify/predict the outcome (e.g., above/below age 30). Although we defer the detailed discussion to Chapter 9, we note here that the resulting tree diagram can be used for determining the important predictors. Predictors (numerical or categorical) that do not appear in the tree can be removed. Similarly, categories that do not appear in the tree can be combined.

PROBLEMS

4.1 **Breakfast Cereals.** Use the data for the breakfast cereals example in Section 4.8 to explore and summarize the data as follows. (Note that a few records contain missing values; since there are just a few, a simple solution is to remove them first. You can use the "Missing Data Handling" utility in XLMiner.)

   a. Which variables are quantitative/numerical? Which are ordinal? Which are nominal?

   b. Create a table with the average, median, min, max, and standard deviation for each of the quantitative variables. This can be done through Excel's functions or Excel's *Data → Data Analysis → Descriptive Statistics* menu.

   c. Use XLMiner to plot a histogram for each of the quantitative variables. Based on the histograms and summary statistics, answer the following questions:

   i. Which variables have the largest variability?

   ii. Which variables seem skewed?

   iii. Are there any values that seem extreme?

   d. Use XLMiner to plot a side-by-side boxplot comparing the calories in hot vs. cold cereals. What does this plot show us?

   e. Use XLMiner to plot a side-by-side boxplot of consumer rating as a function of the shelf height. If we were to predict consumer rating from shelf height, does it appear that we need to keep all three categories of shelf height?

   f. Compute the correlation table for the quantitative variable (use Excel's *Data → Data Analysis → Correlation* menu). In addition, use XLMiner to generate a matrix plot for these variables.

   i. Which pair of variables is most strongly correlated?

   ii. How can we reduce the number of variables based on these correlations?

   iii. How would the correlations change if we normalized the data first?

   g. Consider the first column on the left in Figure 4.10. Describe briefly what this column represents.

4.2 **Chemical Features of Wine.** Figure 4.13 shows the PCA output on data (nonnormalized) in which the variables represent chemical characteristics of wine, and each case is a different wine.

   a. The data are in the file `Wine.xls`. Consider the row near the bottom labeled "Variance." Explain why column 1's variance is so much greater than that of any other column.

   b. Comment on the use of normalization (standardization) in part (a).

4.3 **University Rankings.** The dataset on American college and university rankings (available from www.dataminingbook.com) contains information on 1302 American colleges and universities offering an undergraduate program. For each university there are 17 measurements that include continuous measurements (e.g., tuition and graduation rate) and categorical measurements (e.g., location by state and whether it is a private or a public school).

   a. Remove all categorical variables. Then remove all records with missing numerical measurements from the dataset (by creating a new worksheet).

	Components					
Variable	1	2	3	4	5	6
Alcohol	0.00165926	0.00120342	0.01687386	−0.14144674	0.02033708	0.19412018
Malic_Acid	−0.00068102	0.00215498	0.12200337	−0.16038956	−0.61288345	0.74247289
Ash	0.00019491	0.00459369	0.05198744	0.00977282	0.02017558	0.04175295
Ash_Alcalinity	−0.0046713	0.02645036	0.93859297	0.33096525	0.06435229	−0.02406531
Magnesium	0.01786801	0.99934423	−0.02978026	0.00539375	−0.00614938	−0.0019238
Total_Phenols	0.00098983	0.00087797	−0.04048461	0.07458466	0.31524512	0.2787168
Flavanoids	0.00156729	−0.00005184	−0.08544329	0.16908674	0.5247612	0.43359798
Nonflavanoid_Ph	−0.00012309	−0.00135448	0.01351078	−0.01080556	−0.02964753	−0.02195283
Proanthocyanins	0.00060061	0.0050044	−0.02465936	0.05012095	0.25118256	0.24188447
Color_Intensity	0.00232714	0.01510037	0.29139856	−0.87889373	0.33174714	0.00273963
Hue	0.00017138	−0.00076267	−0.02597765	0.06003497	0.05152407	−0.02377617
OD280_OD315	0.00070493	−0.00349536	−0.07032393	0.17820027	0.26063919	0.28891277
Proline	0.99982297	−0.01777381	0.00452868	0.00311292	−0.00229857	−0.00121226
Variance	99201.78906	172.5352631	9.43811321	4.99117851	1.22884524	0.84106386
Variance%	99.80912018	0.17359155	0.0094959	0.00502174	0.00123637	0.00084621
Cum%	99.80912018	99.98271179	99.99221039	99.99723053	99.99846649	99.99931335

FIGURE 4.13   PRINCIPAL COMPONENTS OF NONNORMALIZED WINE DATA

b. Conduct a principal components analysis on the cleaned data and comment on the results. Should the data be normalized? Discuss what characterizes the components you consider key.

4.4 Sales of Toyota Corolla Cars. The file ToyotaCorolla.xls contains data on used cars (Toyota Corollas) on sale during late summer of 2004 in The Netherlands. It has 1436 records containing details on 38 attributes, including *Price, Age, Kilometers, HP,* and other specifications. The goal will be to predict the price of a used Toyota Corolla based on its specifications.

a. Identify the categorical variables.

b. Explain the relationship between a categorical variable and the series of binary dummy variables derived from it.

c. How many dummy binary variables are required to capture the information in a categorical variable with *N* categories?

d. Using XLMiner's data utilities, convert the categorical variables in this dataset into dummy binaries, and explain in words, for one record, the values in the derived binary dummies.

e. Use Excel's correlation command (*Data → Data Analysis → Correlation* menu) to produce a correlation matrix and XLMiner's matrix plot to obtain a matrix of all scatter plots. Comment on the relationships among variables.

## PART III

Performance Evaluation

## CHAPTER 5

Evaluating Predictive Performance

In this chapter we discuss how the predictive performance of data mining methods can be assessed. We point out the danger of overfitting to the training data, and the need to test model performance on data that were not used in the training step. We discuss popular performance metrics. For prediction, metrics include Average Error, MAPE, and RMSE (based on the validation data). For classification tasks, metrics based on the classification matrix include overall accuracy, specificity and sensitivity and metrics that account for misclassification costs. We also show the relation between the choice of cutoff value and classification performance, and present the ROC curve, which is a popular chart for assessing method performance at different cutoff values. When the goal is to accurately classify the most interesting or important cases, called *ranking*, rather than accurately classify the entire sample (e.g., the 10% of customers most likely to respond to an offer, or the 5% of claims most likely to be fraudulent), lift charts are used to assess performance. We also discuss the need for oversampling rare classes and how to adjust performance metrics for the oversampling. Finally, we mention the usefulness of comparing metrics based on the validation data to those based on the training data for the purpose of detecting overfitting. While some differences are expected, extreme differences can be indicative of overfitting.

Data Mining for Business Analytics: Concepts, Techniques, and Applications with XLMiner®, Third Edition.
Galit Shmueli, Peter C. Bruce, and Nitin R. Patel.
© 2016 John Wiley & Sons Inc. Published 2016 by John Wiley & Sons, Inc.

5.1 Introduction

In supervised learning we are interested in predicting the outcome variable for new records. There are three main types of outcomes of interest:

Predicted numerical value—when the outcome variable is numerical (e.g., house price)

Predicted class membership—when the outcome variable is categorical (e.g., buyer/nonbuyer)

Propensity—the probability of class membership, when the outcome variable is categorical (e.g., the propensity to default)

Prediction methods are used for generating numerical predictions, while classification methods ("classifiers") are used for generating propensities, and using a cutoff value on the propensities, we can generate predicted class memberships.

A subtle distinction to keep in mind is the two distinct predictive uses of classifiers: one use, *classification*, is aimed at predicting class membership for new records. The other, *ranking*, is detecting among a set of new records the ones most likely to belong to a class of interest.

Let's now examine the approach for judging the usefulness of a prediction method used for generating numerical predictions (Section 5.2), a classifier used for classification (Section 5.3), and a classifier used for ranking (Section 5.4). In Section 5.5 we'll look at evaluating performance under the scenario of oversampling.

5.2 Evaluating Predictive Performance

First, let us emphasize that predictive accuracy is not the same as goodness-of-fit. Classical statistical measures of performance are aimed at finding a model that fits well to the data on which the model was trained. In data mining we are interested in models that have high predictive accuracy when applied to *new* records. Measures such as R^2 and standard error of estimate are common metrics in classical regression modeling, and residual analysis is used to gauge goodness-of-fit in that situation. However, these measures do not tell us much about the ability of the model to predict new cases.

For assessing prediction performance, several measures are used. In all cases the measures are based on the validation set, which serves as a more objective ground than the training set to assess predictive accuracy. This is because records in the validation set are more similar to the future records to be predicted, in the sense that they are not used to select predictors or to estimate the model coefficients. Models are trained on the training data, applied to the validation data, and measures of accuracy then use the prediction errors on that validation set.

Benchmark: The Average

The benchmark criterion in prediction is using the average outcome (thereby ignoring all predictor information). In other words, the prediction for a new record is simply the average outcome of the records in the training set ($\bar{y}$). A good predictive model should outperform the benchmark criterion in terms of predictive accuracy.

Prediction Accuracy Measures

The prediction error for record i is defined as the difference between its actual y value and its predicted y value: $e_i = y_i - \hat{y}_i$. A few popular numerical measures of predictive accuracy are:

- *MAE* or *MAD* (mean absolute error/deviation) $= \frac{1}{n} \sum_{i=1}^{n} |e_i|$. This gives the magnitude of the average absolute error.
- *Average error* $= \frac{1}{n} \sum_{i=1}^{n} e_i$. This measure is similar to MAD except that it retains the sign of the errors, so that negative errors cancel out positive errors of the same magnitude. It therefore gives an indication of whether the predictions are on average over- or underpredicting the response.
- *MAPE* (mean absolute percentage error) $= 100\% \times \frac{1}{n} \sum_{i=1}^{n} |e_i/y_i|$. This measure gives a percentage score of how predictions deviate (on average) from the actual values.
- *RMSE* (root-mean-squared error) $= \sqrt{\frac{1}{n} \sum_{i=1}^{n} e_i^2}$. This is similar to the standard error of estimate in linear regression, except that it is computed on the validation data rather than on the training data. It has the same units as the variable predicted.
- Total *SSE* (total sum of squared errors) $= \sum_{i=1}^{n} e_i^2$.

Such measures can be used to compare models and to assess their degree of prediction accuracy. Note that all these measures are influenced by outliers. To check outlier influence, we can compute median-based measures (and compare to the mean-based measures) or simply plot a histogram or boxplot of the errors. Plotting the prediction errors' distribution is in fact very useful and can highlight more information than the metrics alone.

To illustrate the use of predictive accuracy measures and charts of prediction error distribution, consider the error metrics and charts shown in Figures 5.1 and 5.2. These are the result of fitting a certain predictive model to prices of

Training Data Scoring—Summary Report

Total Sum of Squared Errors	RMS Error	Average Error
1093330981	1349.895663	0.000720405

Validation Data Scoring—Summary Report

Total Sum of Squared Errors	RMS Error	Average Error
795600925.2	1410.319933	110.9145714

FIGURE 5.1 PREDICTION ERROR METRICS FROM A MODEL FOR TOYOTA CAR PRICES. TRAINING (TOP) AND VALIDATION (BOTTOM).

used Toyota Corolla cars. The training set includes 600 cars and the validation set includes 400 cars. The prediction errors for the validation set are summarized in the bottom table of Figure 5.1. The histogram and boxplot corresponding to the validation set show a small number of outliers (all under-predictions), with most errors in the [−1000,1000] range.

Comparing Training and Validation Performance

Residuals that are based on the training set tell us about model fit, whereas those that are based on the validation set (called "prediction errors") measure the model's ability to predict new data (predictive performance). We expect training errors to be smaller than the validation errors (because the model was fitted using the training set), and the more complex the model, the greater is the likelihood that it will *overfit* the training data (indicated by a greater difference between the training and validation errors). In an extreme case of overfitting, the training errors would be zero (perfect fit of the model to the training data), and the validation errors would be nonzero and nonnegligible. For this reason, it is important to compare the error plots and metrics (RMSE, MAE, etc.) of the training and validation sets. Figure 5.1 illustrates this comparison: the RMSE and average error for the training set are slightly lower than those for the validation set, as expected, but are not drastically different from the validation metrics.

The charts reveal more than the metrics alone: looking at the charts in Figure 5.2 shows that the discrepancies are also due to some outliers, and especially the large negative training error. Last, the validation errors have slightly more high positive errors (underpredictions) than the training errors, as reflected by the medians and outliers.

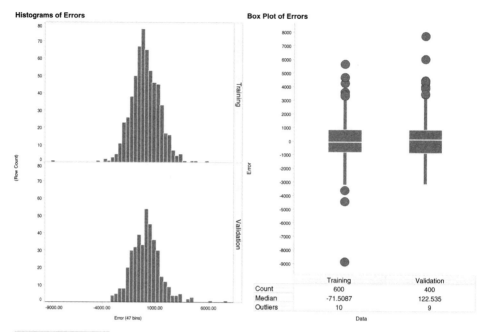

FIGURE 5.2 **HISTOGRAMS AND BOXPLOTS OF TOYOTA PRICE PREDICTION ERRORS FOR TRAINING AND VALIDATION SETS**

Lift Chart

In some applications the goal is to search, among a set of new records, for a subset of records that gives the highest cumulative predicted values. In such cases a graphical way to assess predictive performance is through a *lift chart*. This compares the model's predictive performance to a baseline model that has no predictors. A lift chart for a continuous response is relevant only when we are searching for a set of records that gives the highest cumulative predicted values. It is not relevant, if we are interested in predicting the outcome for each new record.

To illustrate this type of goal, consider a car rental firm that renews its fleet regularly so that customers drive late-model cars. This entails disposing of a large quantity of used vehicles on a continuing basis. Since the firm is not primarily in the used car sales business, it tries to dispose of as much of its fleet as possible through volume sales to used car dealers. However, it is profitable to sell a limited number of cars through its own channels. Its volume deals with the used car dealers allow it flexibility to pick and choose which cars to sell in this fashion, so it would like to have a model for selecting cars for resale through its own channels. Since all cars were purchased some time ago and the deals with the used car dealers are for fixed prices (specifying a given number of cars of a certain make and model class), the cars' costs are now irrelevant and the dealer is interested only in maximizing revenue. This is done by selecting for its own resale the cars likely to generate the most revenue. The lift chart in this case gives the predicted lift for revenue.

The lift chart is based on ordering the set of records of interest (typically, validation data) by their predicted value, from high to low. Then, we accumulate the actual values and plot their cumulative value on the y-axis as a function of the number of records accumulated (the x-axis value). This curve is compared to assigning a naive prediction ($\bar{y}$) to each record and accumulating these average values, which results in a diagonal line. The farther away the lift curve is from the diagonal benchmark line, the better the model is doing in separating records with high-value outcomes from those with low-value outcomes. The same information can be presented in a decile lift chart, where the ordered records are grouped into 10 deciles, and for each decile the chart presents the ratio of model lift to naive benchmark lift.

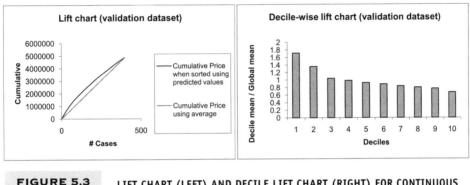

FIGURE 5.3   **LIFT CHART (LEFT) AND DECILE LIFT CHART (RIGHT) FOR CONTINUOUS RESPONSE (SALES OF TOYOTA CARS)**

Figure 5.3 shows a lift chart and decile lift chart based on fitting a linear regression model to the Toyota data. The charts are based on the validation data of 400 cars. It can be seen that the model's predictive performance in terms of lift is better than the baseline model, since its lift curve is higher than that of the baseline model. The lift and decile charts in Figure 5.3 would be useful in the following scenario: Choosing the top 10% of the cars that gave the highest predicted sales, for example, we would gain 1.7 times the amount of revenue, compared to choosing 10% of the cars at random. This can be seen from the decile chart (Figure 5.3). This number can also be computed from the lift chart by comparing the sales for 40 random cars (the value of the baseline curve at $x = 40$), which is \$486,871 (= the sum of the actual sales for the 400 validation set cars divided by 10) with the actual sales of the 40 cars that have the highest predicted values (the value of the lift curve at $x = 40$), \$835,883. The ratio between these numbers is 1.7.

5.3 JUDGING CLASSIFIER PERFORMANCE

The need for performance measures arises from the wide choice of classifiers and predictive methods. Not only do we have several different methods, but even within a single method there are usually many options that can lead to completely different results. A simple example is the choice of predictors used within a particular predictive algorithm. Before we study these various algorithms in detail and face decisions on how to set these options, we need to know how we will measure success.

A natural criterion for judging the performance of a classifier is the probability of making a *misclassification error*. Misclassification means that the observation belongs to one class but the model classifies it as a member of a different class. A classifier that makes no errors would be perfect, but we do not expect to be able to construct such classifiers in the real world due to "noise" and not having all the information needed to classify cases precisely. Is there a minimal probability of misclassification that we should require of a classifier?

Benchmark: The Naive Rule

A very simple rule for classifying a record into one of m classes, ignoring all predictor information $(x_1, x_2, \ldots, x_p)$ that we may have, is to classify the record as a member of the majority class. In other words, "classify as belonging to the most prevalent class." The *naive rule* is used mainly as a baseline or benchmark for evaluating the performance of more complicated classifiers. Clearly, a classifier that uses external predictor information (on top of the class membership allocation) should outperform the naive rule. There are various performance measures based on the naive rule that measure how much better than the naive rule a certain classifier performs. One example is the multiple R^2 reported by XLMiner, which measures the distance between the fit of the classifier to the data and the fit of the naive rule to the data (for further details, see Section 10.5).

Similar to using the sample mean $(\bar{y})$ as the naive benchmark in the numerical outcome case, the naive rule for classification relies solely on the y information and excludes any additional predictor information.

Class Separation

If the classes are well separated by the predictor information, even a small dataset will suffice in finding a good classifier, whereas if the classes are not separated at all by the predictors, even a very large dataset will not help. Figure 5.4 illustrates this for a two-class case. The top panel includes a small dataset ($n = 24$ observations) where two predictors (income and lot size) are used for separating owners from nonowners (we thank Dean Wichern for this example, described in Johnson and Wichern (2002)). Here the predictor information seems useful in

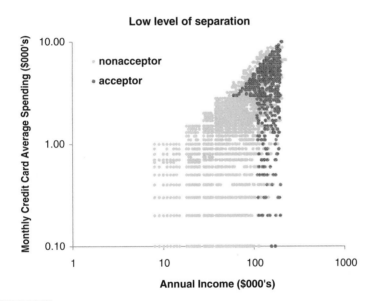

FIGURE 5.4   HIGH (TOP) AND LOW (BOTTOM) LEVELS OF SEPARATION BETWEEN TWO CLASSES, USING TWO PREDICTORS

that it separates the two classes (owners/nonowners). The bottom panel shows a much larger dataset ($n = 5000$ observations) where the two predictors (income and average credit card spending) do not separate the two classes well in most of the higher ranges (loan acceptors/nonacceptors).

The Classification Matrix

In practice, most accuracy measures are derived from the *classification matrix*, also called the *confusion matrix*. This matrix summarizes the correct and incorrect classifications that a classifier produced for a certain dataset. Rows and columns of

Classification Confusion Matrix		
	Predicted Class	
Actual Class	1	0
1	201	85
0	25	2689

FIGURE 5.5   CLASSIFICATION MATRIX BASED ON 3000 OBSERVATIONS AND TWO CLASSES

the classification matrix correspond to the true and predicted classes, respectively. Figure 5.5 shows an example of a classification (confusion) matrix for a two-class (0/1) problem resulting from applying a certain classifier to 3000 observations. The two diagonal cells (upper left, lower right) give the number of correct classifications, where the predicted class coincides with the actual class of the observation. The off-diagonal cells give counts of misclassification. The top right cell gives the number of class 1 members that were misclassified as 0's (in this example, there were 85 such misclassifications). Similarly, the lower left cell gives the number of class 0 members that were misclassified as 1's (25 such observations).

The classification matrix gives estimates of the true classification and misclassification rates. Of course, these are estimates and they may be incorrect, but if we have a large enough dataset and neither class is very rare, our estimates will be reliable. Sometimes we may be able to use public data such as US Census data to estimate these proportions. However, in most business settings, we will not know them.

Using the Validation Data

To obtain an honest estimate of future classification error, we use the classification matrix that is computed from the *validation data*. In other words, we first partition the data into training and validation sets by random selection of cases. We then construct a classifier using the training data, and apply it to the validation data. This will yield the predicted classifications for observations in the validation set (see Figure 2.2 in Chapter 2). We next summarize these classifications in a classification matrix. Although we can summarize our results in a classification matrix for training data as well, the resulting classification matrix is not useful for getting an honest estimate of the misclassification rate for new data due to the danger of overfitting.

In addition to examining the validation data classification matrix to assess the classification performance on new data, we compare the training data classification matrix to the validation data classification matrix, in order to detect

TABLE 5.1	CLASSIFICATION MATRIX: MEANING OF EACH CELL

	Predicted Class	
Actual Class	C_1	C_2
C_1	$n_{1,1}$ = number of C_1 cases classified correctly	$n_{1,2}$ = number of C_1 cases classified incorrectly as C_2
C_2	$n_{2,1}$ = number of C_2 cases classified incorrectly as C_1	$n_{2,2}$ = number of C_2 cases classified correctly

overfitting: although we expect somewhat inferior results on the validation data, a large discrepancy in training and validation performance might be indicative of overfitting.

Accuracy Measures

Different accuracy measures can be derived from the classification matrix. Consider a two-class case with classes C_1 and C_2 (e.g., buyer/nonbuyer). The schematic classification matrix in Table 5.1 uses the notation $n_{i,j}$ to denote the number of cases that are class C_i members and were classified as C_j members. Of course, if $i \neq j$, these are counts of misclassifications. The total number of observations is $n = n_{1,1} + n_{1,2} + n_{2,1} + n_{2,2}$.

A main accuracy measure is the *estimated misclassification rate*, also called the *overall error rate*. It is given by

$$\text{err} = \frac{n_{1,2} + n_{2,1}}{n},$$

where n is the total number of cases in the validation dataset. In the example in Figure 5.5, we get err $= (25+85)/3000 = 3.67\%$.

We can measure accuracy by looking at the correct classifications—the full half of the cup—instead of the misclassifications. The *overall accuracy* of a classifier is estimated by

$$\text{accuracy} = 1 - \text{err} = \frac{n_{1,1} + n_{2,2}}{n}.$$

In the example we have $(201+2689)/3000 = 96.33\%$.

Propensities and Cutoff for Classification

The first step in most classification algorithms is to estimate the probability that a case belongs to each of the classes. These probabilities are also called *propensities*. Propensities are typically used either as an interim step for generating predicted class membership (classification), or for rank-ordering the records by their probability of belonging to a class of interest. Let us consider their first use in this section. The second use is discussed in Section 5.4.

If overall classification accuracy (involving all the classes) is of interest, the case can be assigned to the class with the highest probability. In many cases, a single class is of special interest, so we will focus on that particular class and compare the propensity of belonging to that class to a *cutoff value* set by the analyst. This approach can be used with two classes or more than two classes, though it may make sense in such cases to consolidate classes so that you end up with two: the class of interest and all other classes. If the probability of belonging to the class of interest is above the cutoff, the case is assigned to that class.

CUTOFF VALUES FOR TRIAGE

In some cases it is useful to have two cutoffs, and allow a "cannot say" option for the classifier. In a two-class situation, this means that for a case, we can make one of three predictions: The case belongs to C_1, or the case belongs to C_2, or we cannot make a prediction because there is not enough information to pick C_1 or C_2 confidently. Cases that the classifier cannot classify are subjected to closer scrutiny either by using expert judgment or by enriching the set of predictor variables by gathering additional information that is perhaps more difficult or expensive to obtain. An example is classification of documents found during legal discovery (reciprocal forced document disclosure in a legal proceeding). Under traditional human-review systems, qualified legal personnel are needed to review what might be tens of thousands of documents to determine their relevance to a case. Using a classifier and a triage outcome, documents could be sorted into clearly relevant, clearly not relevant, and the gray area documents requiring human review. This substantially reduces the costs of discovery.

The default cutoff value in two-class classifiers is 0.5. Thus, if the probability of a record being a class C_1 member is greater than 0.5, that record is classified as a C_1. Any record with an estimated probability of less than 0.5 would be classified as a C_2. It is possible, however, to use a cutoff that is either higher or lower than 0.5. A cutoff greater than 0.5 will end up classifying fewer records as C_1's, whereas a cutoff less than 0.5 will end up classifying more records as C_1. Typically, the misclassification rate will rise in either case.

Consider the data in Table 5.2, showing the actual class for 24 records, sorted by the probability that the record is an "owner" (as estimated by a data mining

TABLE 5.2	24 RECORDS WITH THEIR ACTUAL CLASS AND THE PROBABILITY (PROPENSITY) OF THEM BEING CLASS "OWNER" MEMBERS, AS ESTIMATED BY A CLASSIFIER

Actual Class	Probability of Class "Owner"	Actual Class	Probability of Class "Owner"
owner	0.9959	owner	0.5055
owner	0.9875	nonowner	0.4713
owner	0.9844	nonowner	0.3371
owner	0.9804	owner	0.2179
owner	0.9481	nonowner	0.1992
owner	0.8892	nonowner	0.1494
owner	0.8476	nonowner	0.0479
nonowner	0.7628	nonowner	0.0383
owner	0.7069	nonowner	0.0248
owner	0.6807	nonowner	0.0218
owner	0.6563	nonowner	0.0161
nonowner	0.6224	nonowner	0.0031

algorithm). If we adopt the standard 0.5 as the cutoff, our misclassification rate is 3/24, whereas if we instead adopt a cutoff of 0.25, we classify more records as owners and the misclassification rate goes up (comprising more nonowners misclassified as owners) to 5/24. Conversely, if we adopt a cutoff of 0.75, we classify fewer records as owners. The misclassification rate goes up (comprising more owners misclassified as nonowners) to 6/24. All this can be seen in the classification tables in Figure 5.6.

To see the entire range of cutoff values and how the accuracy or misclassification rates change as a function of the cutoff, we can use one-variable tables in Excel (see the accompanying box), and then plot the performance measure of interest vs. the cutoff. The results for the data above are shown in Figure 5.7. We can see that the accuracy level is pretty stable around 0.8 for cutoff values between 0.2 and 0.8.

Why would we want to use cutoff values different from 0.5 if they increase the misclassification rate? The answer is that it might be more important to classify owners properly than nonowners, and we would tolerate a greater misclassification of the latter. Or the reverse might be true; in other words, the costs of misclassification might be asymmetric. We can adjust the cutoff value in such a case to classify more records as the high-value class, that is, accept more misclassifications where the misclassification cost is low. Keep in mind that we are doing so after the data mining model has already been selected—we are not changing that model. It is also possible to incorporate costs into the picture before deriving the model. These subjects are discussed in greater detail below.

Cutoff Prob.Val. for Success (Updatable)	0.5

Classification Confusion Matrix		
	Predicted Class	
Actual Class	owner	nonowner
owner	11	1
nonowner	2	10

Cutoff Prob.Val. for Success (Updatable)	0.25

Classification Confusion Matrix		
	Predicted Class	
Actual Class	owner	nonowner
owner	11	1
nonowner	4	8

Cutoff Prob.Val. for Success (Updatable)	0.75

Classification Confusion Matrix		
	Predicted Class	
Actual Class	owner	nonowner
owner	7	5
nonowner	1	11

FIGURE 5.6   CLASSIFICATION MATRICES BASED ON CUTOFFS OF 0.5, 0.25, AND 0.75

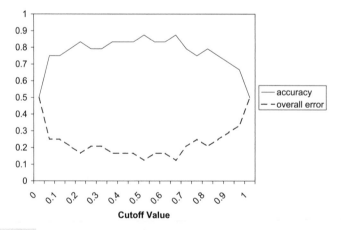

FIGURE 5.7   PLOTTING RESULTS FROM ONE-WAY TABLE: ACCURACY AND OVERALL ERROR AS A FUNCTION OF THE CUTOFF VALUE

ONE-VARIABLE TABLES IN EXCEL

Excel's one-variable data tables are very useful for studying how the cutoff affects different performance measures. They will change the cutoff values to values in a user-specified column and calculate different functions based on the corresponding classification matrix. To create a one-variable data table (see Figure 5.8):

1. In the top row, create column names for each of the measures you wish to compute. (We created "overall error" and "accuracy" in B11 and C11,) The leftmost column should be titled "cutoff" (A11).

2. In the row below, add formulas, using references to the relevant classification matrix cells. [The formula in B12 is $= (B6 + C7)/(B6 + C6 + B7 + C7)$.]

3. In the leftmost column, list the cutoff values that you want to evaluate. (We chose $0, 0.05, \ldots, 1$ in B13 to B33.)

4. Select the range excluding the first row (A12:C33). In Excel 2007 and up go to *Data > Whatif Analysis > Data Table*

5. In "column input cell," select the cell that changes (here, the cell with the cutoff value, *D*1). Click OK.

6. The table will now be automatically completed.

Performance in Unequal Importance of Classes

Suppose that it is more important to predict membership correctly in class C_1 than in class C_2. An example is predicting the financial status (bankrupt/solvent) of firms. It may be more important to predict correctly a firm that is going bankrupt than to predict correctly a firm that is going to remain solvent. The classifier is essentially used as a system for detecting or signaling bankruptcy. In such a case the overall accuracy is not a good measure for evaluating the classifier. Suppose that the important class is C_1. The following pair of accuracy measures are the most popular:

The **sensitivity** of a classifier is its ability to detect the important class members correctly. This is measured by $n_{1,1}/(n_{1,1} + n_{1,2})$, the percentage of C_1 members classified correctly.

The **specificity** of a classifier is its ability to rule out C_2 members correctly. This is measured by $n_{2,2}/(n_{2,1} + n_{2,2})$, the percentage of C_2 members classified correctly.

It can be useful to plot these measures vs. the cutoff value (using one-variable tables in Excel, as described above) in order to find a cutoff value that balances these measures.

FIGURE 5.8 CREATING ONE-VARIABLE TABLES IN EXCEL. ACCURACY AND OVERALL ERROR ARE COMPUTED FOR DIFFERENT VALUES OF THE CUTOFF

ROC Curve A more popular method for plotting the two measures is through *ROC* (Receiver Operating Characteristic) *curves*. The ROC curve plots the pairs {sensitivity, 1-specificity} as the cutoff value increases from 0 and 1. Better performance is reflected by curves that are closer to the top left corner. The comparison curve is the diagonal, which reflects the performance of the naive rule, using varying cutoff values (i.e., setting different thresholds on the level of majority used by the majority rule). A common metric is "area under the curve (AUC)," which ranges from 1 (perfect discrimination between classes) to 0.5 (no better than the naive rule). The ROC curve for our 24-case example above is shown in Figure 5.9.

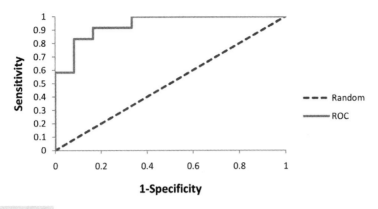

FIGURE 5.9   ROC CURVE FOR THE MOWER EXAMPLE

FALSE-POSITIVE AND FALSE-NEGATIVE RATES

Sensitivity and specificity measure the performance of a classifier from the point of view of the "classifying agency" (e.g., a company classifying customers or a hospital classifying patients). They answer the question "how well does the classifier segregate the important class members?" It is also possible to measure accuracy from the perspective of the entity that is being predicted (e.g., the customer or the patient), who asks "what is my chance of belonging to the important class?" This question is usually less relevant in a data mining application. The terms "false-positive rate" and "false-negative rate," which are sometimes used erroneously to describe 1-sensitivity and 1-specificity, are measures of performance from the perspective of the individual entity. If C_1 is the important class, then they are defined as:

The false-positive rate is the proportion of C_1 predictions that is wrong: $n_{2,1}/(n_{1,1} + n_{2,1})$. Notice that this is a ratio within the column of C_1 predictions (i.e., it uses only records that were classified as C_1).

The false-negative rate is the proportion of C_2 predictions that is wrong: $n_{1,2}/(n_{1,2} + n_{2,2})$. Notice that this is a ratio within the column of C_2 predictions (i.e., it uses only records that were classified as C_2).

Asymmetric Misclassification Costs

Implicit in our discussion of the lift curve, which measures how effective we are in identifying the members of one particular class, is the assumption that the error of misclassifying a case belonging to one class is more serious than for the other class. For example, misclassifying a household as unlikely to respond to a sales offer when it belongs to the class that would respond incurs a greater cost (the opportunity cost of the forgone sale) than the converse error. In the former case, you are missing out on a sale worth perhaps tens or hundreds of dollars. In the latter, you are incurring the costs of mailing a letter to someone who will not purchase. In such a scenario, using the misclassification rate as a criterion can be misleading.

Note that we are assuming that the cost (or benefit) of making correct classifications is zero. At first glance, this may seem incomplete. After all, the benefit (negative cost) of classifying a buyer correctly as a buyer would seem substantial. And in other circumstances (e.g., scoring our classification algorithm to fresh data to implement our decisions), it will be appropriate to consider the actual net dollar impact of each possible classification (or misclassification). Here, however, we are attempting to assess the value of a classifier in terms of classification error, so it greatly simplifies matters if we can capture all cost/benefit information in the misclassification cells. So, instead of recording the benefit of classifying a respondent household correctly, we record the cost of failing to classify it as a respondent household. It amounts to the same thing and our goal becomes the minimization of costs, whether the costs are actual costs or missed benefits (opportunity costs).

Consider the situation where the sales offer is mailed to a random sample of people for the purpose of constructing a good classifier. Suppose that the offer is accepted by 1% of those households. For these data, if a classifier simply classifies every household as a nonresponder, it will have an error rate of only 1%, but it will be useless in practice. A classifier that misclassifies 2% of buying households as nonbuyers and 20% of the nonbuyers as buyers would have a higher error rate but would be better if the profit from a sale is substantially higher than the cost of sending out an offer. In these situations, if we have estimates of the cost of both types of misclassification, we can use the classification matrix to compute the expected cost of misclassification for each case in the validation data. This enables us to compare different classifiers using overall expected costs (or profits) as the criterion.

Suppose that we are considering sending an offer to 1000 more people, 1% of whom respond (1), on average. Naively classifying everyone as a 0 has an error rate of only 1%. Using a data mining routine, suppose that we can produce these classifications:

	Predict Class 0	Predict Class 1
Actual 0	970	20
Actual 1	2	8

These classifications have an error rate of $100 \times (20 + 2)/1000 = 2.2\%$—higher than the naive rate.

Now suppose that the profit from a 1 is $10 and the cost of sending the offer is $1. Classifying everyone as a 0 still has a misclassification rate of only 1% but yields a profit of $0. Using the data mining routine, despite the higher misclassification rate, yields a profit of $60.

The matrix of profit is as follows (nothing is sent to the predicted 0's, so there are no costs or sales in that column):

Profit	Predict Class 0	Predict Class 1
Actual 0	0	– $20
Actual 1	0	$80

Looked at purely in terms of costs, when everyone is classified as a 0, there are no costs of sending the offer; the only costs are the opportunity costs of failing to make sales to the ten 1's = $100. The cost (actual costs of sending the offer plus the opportunity costs of missed sales) of using the data mining routine to select people to send the offer to is only $48, as follows:

Costs	Predict Class 0	Predict Class 1
Actual 0	0	$20
Actual 1	$20	$8

However, this does not improve the actual classifications. A better method is to change the classification rules (and hence the misclassification rates), as discussed in the preceding section, to reflect the asymmetric costs.

A popular performance measure that includes costs is the *average misclassification cost*, which measures the average cost of misclassification per classified observation. Denote by q_1 the cost of misclassifying a class C_1 observation (as belonging to class C_2) and by q_2 the cost of misclassifying a class C_2 observation (as belonging to class C_1). The average misclassification cost is

$$\frac{q_1 n_{1,2} + q_2 n_{2,1}}{n}$$

Thus we are looking for a classifier that minimizes this quantity. This can be computed, for instance, for different cutoff values.

It turns out that the optimal parameters are affected by the misclassification costs only through the ratio of these costs. This can be seen if we write the foregoing measure slightly differently:

$$\frac{q_1 n_{1,2} + q_2 n_{2,1}}{n} = \frac{n_{1,2}}{n_{1,1} + n_{1,2}} \frac{n_{1,1} + n_{1,2}}{n} q_1 + \frac{n_{2,1}}{n_{2,1} + n_{2,2}} \frac{n_{2,1} + n_{2,2}}{n} q_2.$$

Minimizing this expression is equivalent to minimizing the same expression divided by a constant. If we divide by q_1, it can be seen clearly that the minimization depends only on q_2/q_1 and not on their individual values. This is very practical, since in many cases it is difficult to assess the costs associated with misclassifying a C_1 member and a C_2 member, but estimating the ratio is easier.

This expression is a reasonable estimate of future misclassification cost if the proportions of classes C_1 and C_2 in the sample data are similar to the proportions of classes C_1 and C_2 that are expected in the future. If, instead of a random sample, we draw a sample such that one class is oversampled (as described in the next section), then the sample proportions of C_1's and C_2's will be distorted compared to the future or population. We can then correct the average misclassification cost measure for the distorted sample proportions by incorporating estimates of the true proportions (from external data or domain knowledge), denoted by $p(C_1)$ and $p(C_2)$, into the formula

$$\frac{n_{1,2}}{n_{1,1} + n_{1,2}} p(C_1) q_1 + \frac{n_{2,1}}{n_{2,1} + n_{2,2}} p(C_2) q_2.$$

Using the same logic as above, it can be shown that optimizing this quantity depends on the costs only through their ratio (q_2/q_1) and on the prior probabilities only through their ratio $[p(C_2)/p(C_1)]$. This is why software packages that incorporate costs and prior probabilities might prompt the user for ratios rather than actual costs and probabilities.

Generalization to More Than Two Classes

All the comments made above about two-class classifiers extend readily to classification into more than two classes. Let us suppose that we have m classes $C_1, C_2, \dots, C_m$. The classification matrix has m rows and m columns. The misclassification cost associated with the diagonal cells is, of course, always zero. Incorporating prior probabilities of the various classes (where now we have m such numbers) is still done in the same manner. However, evaluating misclassification costs becomes much more complicated: For an m-class case we have $m(m - 1)$ types of misclassifications. Constructing a matrix of misclassification costs thus becomes prohibitively complicated.

5.4 JUDGING RANKING PERFORMANCE

We now turn to the predictive goal of detecting, among a set of new records, the ones most likely to belong to a class of interest. Recall that this differs from the goal of predicting class membership for each new record.

Lift Charts for Binary Data

We already introduced lift charts in the context of a numerical outcome (Section 5.2). We now describe lift charts, also called *lift curves*, *gains curves*, or *gains charts*, for a binary outcome. This is a more common usage than for predicted continuous outcomes. The lift curve helps us determine how effectively we can "skim the cream" by selecting a relatively small number of cases and getting a relatively large portion of the responders. The input required to construct a lift curve is a validation dataset that has been "scored" by appending to each case the propensity that it will belong to a given class.

Let's continue with the case in which a particular class is relatively rare and of much more interest than the other class: tax cheats, debt defaulters, or responders to a mailing. We would like our classification model to sift through the records and sort them according to which ones are most likely to be tax cheats, responders to the mailing, and so on. We can then make more informed decisions. For example, we can decide how many and which tax returns to examine if looking for tax cheats. The model will give us an estimate of the extent to which we will encounter more and more noncheaters as we proceed through the sorted data starting with the records most likely to be tax cheats. Or we can use the sorted data to decide to which potential customers a limited-budget mailing should be targeted. In other words, we are describing the case where our goal is to obtain a rank ordering among the records according to their class membership propensities.

Sorting by Propensity   To construct a lift chart, we sort the set of records by propensity, in descending order. This is the propensity to belong to the important class, say C_1. Then in each row we compute the cumulative number of C_1 members (Actual Class = C_1). For example, the table in Figure 5.10 shows the 24 records ordered in descending class "1" propensity. The rightmost column accumulates the number of actual 1's. The lift chart then plots this cumulative column against the number of records, as shown in the lift chart at the bottom of Figure 5.10.

Interpreting the Lift Chart   What is considered good or bad performance? The ideal ranking performance would place all the 1's at the beginning (the actual 1's would have the highest propensities and be at the top of the table) and all the 0's at the end. A lift chart corresponding to this ideal case would be a diagonal line with slope 1 that turns into a horizontal line (once all the 1's were accumulated). In the example, the lift curve for the best possible classifier—a classifier that makes no errors—would overlap the existing curve at the start, continue with a slope of 1 until it reaches all the 12 1's, then continue horizontally to the right.

Serial Number	Predicted Probability of 1	Actual Class	Cumulative Actual class
1	0.995976726	1	1
2	0.987533139	1	2
3	0.984456382	1	3
4	0.980439587	1	4
5	0.948110638	1	5
6	0.889297203	1	6
7	0.847631864	1	7
8	0.762806287	0	7
9	0.706991915	1	8
10	0.680754087	1	9
11	0.656343749	1	10
12	0.622419543	0	10
13	0.505506928	1	11
14	0.47134045	0	11
15	0.337117362	0	11
16	0.21796781	1	12
17	0.199240432	0	12
18	0.149482655	0	12
19	0.047962588	0	12
20	0.038341401	0	12
21	0.024850999	0	12
22	0.021806029	0	12
23	0.016129906	0	12
24	0.003559986	0	12

FIGURE 5.10 **TABLE AND LIFT CHART FOR THE MOWER EXAMPLE ("PREDICTED VALUES" = PROPENSITIES)**

In contrast, a useless model would be one that randomly assigns propensities (shuffling the 1's and 0's randomly in the Actual Class column). Such behavior would increase the cumulative number of 1's, on average, by $\frac{\#1's}{n}$ in each row. And in fact, this is the diagonal line joining the points $(0, 0)$ to $(24, 12)$ seen in Figure 5.10. This serves as a reference line. For any given number of cases (the x-axis value), it represents the expected number of 1 classifications if we did not have a model but simply selected cases at random. It provides a benchmark against which we can evaluate the ranking performance of the model. In this example, although our model is not perfect, it seems to perform much better than the random benchmark.

How do we read a lift chart? For a given number of cases (x-axis), the lift curve value on the y-axis tells us how much better we are doing compared to random assignment. For example, looking at Figure 5.10, if we use our model to choose the top 10 records, the lift curve tells us that we would be right for about nine of them. If we simply select 10 cases at random, we expect to be right for $10 \times 12/24 = 5$ cases. The model gives us a "lift" in detecting class 1 members of $9/5 = 1.8$. The lift will vary with the number of cases we choose to act on. A good classifier will give us a high lift when we act on only a few cases. As we include more cases, the lift will decrease.

Decile Lift Charts

The information from the lift chart can be portrayed as a *decile chart*, as shown in Figure 5.11, which is widely used in direct marketing predictive modeling. The decile chart aggregates all the lift information into 10 buckets. The bars show, on the y-axis, the factor by which our model outperforms a random assignment of 0's and 1's, taking one decile at a time. Reading the first bar on the left, we see that taking the 10% of the records that are ranked by the model as "the most probable 1's" (having the highest propensities) yields twice as many 1's as would a random selection of 10% of the records. In this example the decile chart indicates that we can even use the model to select the top 30% records with the highest propensities and still perform twice as well as random.

> XLMiner automatically creates lift (and decile) charts from probabilities (propensities) predicted by classifiers for both training and validation data. Of course, the lift curve based on the validation data is a better estimator of performance for new cases.

Beyond Two Classes

A lift chart cannot be used with a multiclass classifier unless a single "important class" is defined and the classifications are reduced to "important" and "unimportant" classes.

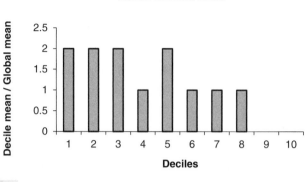

FIGURE 5.11 DECILE LIFT CHART

Lift Charts Incorporating Costs and Benefits

When the benefits and costs of correct and incorrect classification are known or can be estimated, the lift chart is still a useful presentation and decision tool. As before, we need a classifier that assigns to each record a propensity that it belongs to a particular class. The procedure is then as follows:

1. Sort the records in descending order of predicted probability of success (where *success* = belonging to the class of interest).

2. For each record, record the cost (benefit) associated with the actual outcome.

3. For the highest propensity (i.e., first) record, its *x*-axis value is 1 and its *y*-axis value is its cost or benefit (computed in step 2) on the lift curve.

4. For the next record, again calculate the cost (benefit) associated with the actual outcome. Add this to the cost (benefit) for the previous record. This sum is the *y*-axis coordinate of the second point on the lift curve. Its *x*-axis value is 2.

5. Repeat step 4 until all records have been examined. Connect all the points, and this is the lift curve.

6. The reference line is a straight line from the origin to the point *y* = total net benefit and $x = n(n =$ number of records).

Note: It is entirely possible for a reference line that incorporates costs and benefits to have a negative slope if the net value for the entire dataset is negative. For example, if the cost of mailing to a person is $0.65, the value of a responder is $25, and the overall response rate is 2%, the expected net value of mailing to a list of 10,000 is $(0.02 \times \$25 \times 10,000) - (\$0.65 \times 10,000) = \$5000 - \$6500 = -\$1500$. Hence the *y*-value at the far right of the lift curve (x = 10,000) is

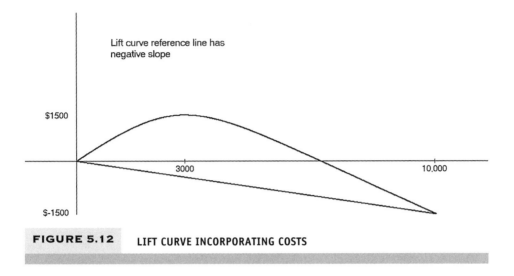

Lift curve reference line has
negative slope

FIGURE 5.12   **LIFT CURVE INCORPORATING COSTS**

−1500, and the slope of the reference line from the origin will be negative. The optimal point will be where the lift curve is at a maximum (i.e., mailing to about 3000 people) in Figure 5.12.

Lift as Function of Cutoff

We could also plot the lift as a function of the cutoff value. The only difference is the scale on the x-axis. When the goal is to select the top records based on a certain budget, the lift vs. number of records is preferable. In contrast, when the goal is to find a cutoff that distinguishes well between the two classes, the lift vs. cutoff value is more useful.

5.5 OVERSAMPLING

As we saw briefly in Chapter 2, when classes are present in very unequal proportions, simple random sampling may produce too few of the rare class to yield useful information about what distinguishes them from the dominant class. In such cases stratified sampling is often used to oversample the cases from the rarer class and improve the performance of classifiers. It is often the case that the rarer events are the more interesting or important ones: responders to a mailing, those who commit fraud, defaulters on debt, and the like. This same stratified sampling procedure is sometimes called *weighted sampling* or *undersampling*, the latter referring to the fact that the more plentiful class is undersampled, relative to the rare class. We will stick to the term *oversampling*.

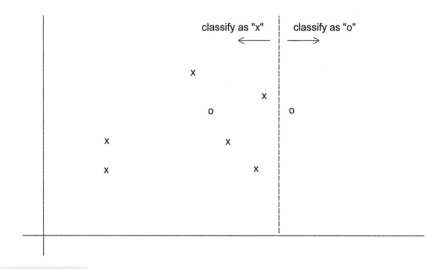

FIGURE 5.13 CLASSIFICATION ASSUMING EQUAL COSTS OF MISCLASSIFICATION

In all discussions of *oversampling*, we assume the common situation in which there are two classes, one of much greater interest than the other. Data with more than two classes do not lend themselves to this procedure.

Consider the data in Figure 5.13, where × represents nonresponders, and o, responders. The two axes correspond to two predictors. The dashed vertical line does the best job of classification under the assumption of equal costs: It results in just one misclassification (one o is misclassified as an ×). If we incorporate more realistic misclassification costs—let's say that failing to catch a o is five times as costly as failing to catch an ×— the costs of misclassification jump to 5. In such a case a horizontal line as shown in Figure 5.14 does a better job: It results in misclassification costs of just 2.

Oversampling is one way of incorporating these costs into the training process. In Figure 5.15 we can see that classification algorithms would automatically determine the appropriate classification line if four additional o's were present at each existing o. We can achieve appropriate results either by taking five times as many o's as we would get from simple random sampling (by sampling with replacement if necessary) or by replicating the existing o's fourfold.

Oversampling without replacement in accord with the ratio of costs (the first option above) is the optimal solution, but that may not always be practical. There may not be an adequate number of responders to assure that there will be enough of them to fit a model if they constitute only a small proportion of the total. Also it is often the case that our interest in discovering responders is known

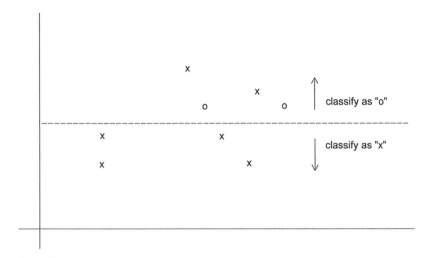

FIGURE 5.14   CLASSIFICATION ASSUMING UNEQUAL COSTS OF MISCLASSIFICATION

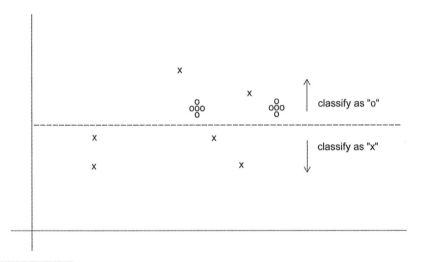

FIGURE 5.15   CLASSIFICATION USING OVERSAMPLING TO ACCOUNT FOR UNEQUAL COSTS

to be much greater than our interest in discovering nonresponders, but the exact ratio of costs is difficult to determine. When faced with very low response rates in a classification problem, practitioners often sample equal numbers of responders and nonresponders as a relatively effective and convenient approach. Whatever approach is used, when it comes time to assess and predict model performance, we will need to adjust for the oversampling in one of two ways:

1. Score the model to a validation set that has been selected without over-sampling (i.e., via simple random sampling).

2. Score the model to an oversampled validation set, and reweight the results to remove the effects of oversampling.

The first method is more straightforward and easier to implement. We describe how to oversample and how to evaluate performance for each of the two methods.

When classifying data with very low response rates, practitioners typically:

- Train models on data that are 50% responder, 50% nonresponder.
- Validate the models with an unweighted (simple random) sample from the original data.

Oversampling the Training Set

How is weighted sampling done? When responders are sufficiently scarce that you will want to use all of them, one common procedure is as follows:

1. First, the response and nonresponse data are separated into two distinct sets, or *strata*.
2. Records are then randomly selected for the training set from each stratum. Typically, one might select half the (scarce) responders for the training set, then an equal number of nonresponders.
3. The remaining responders are put in the validation set.
4. Nonresponders are randomly selected for the validation set in sufficient numbers to maintain the original ratio of responders to nonresponders.
5. If a test set is required, it can be taken randomly from the validation set.

Note: XLMiner has a utility for this purpose.

Evaluating Model Performance Using a Non-oversampled Validation Set

Although the oversampled data can be used to train models, they are often not suitable for evaluating model performance because the number of responders will (of course) be exaggerated. The most straightforward way of gaining an unbiased estimate of model performance is to apply the model to regular data (i.e., data not oversampled). In short: Train the model on oversampled data, but validate it with regular data.

Evaluating Model Performance If Only Oversampled Validation Set Exists

In some cases very low response rates may make it more practical to use oversampled data not only for the training data but also for the validation data.

This might happen, for example, if an analyst is given a dataset for exploration and prototyping that is already oversampled to boost the proportion with the rare response of interest (perhaps because it is more convenient to transfer and work with a smaller dataset). In such cases it is still possible to assess how well the model will do with real data, but this requires the oversampled validation set to be reweighted, in order to restore the class of observations that were underrepresented in the sampling process. This adjustment should be made to the classification matrix and to the lift chart in order to derive good accuracy measures. These adjustments are described next.

I. Adjusting the Confusion Matrix for Oversampling   Suppose that the response rate in the data as a whole is 2%, and that the data were oversampled, yielding a sample in which the response rate is 25 times higher (50% responders). The relationship is as follows:

Responders: 2% of the whole data; 50% of the sample

Nonresponders: 98% of the whole data, 50% of the sample

Each responder in the whole data is worth 25 responders in the sample (50/2). Each nonresponder in the whole data is worth 0.5102 nonresponders in the sample (50/98). We call these values *oversampling weights*.

Assume that the validation classification matrix looks like this:

CLASSIFICATION MATRIX, OVERSAMPLED DATA (VALIDATION)

	Predicted 0	Predicted 1	Total
Actual 0	390	110	500
Actual 1	80	420	500
Total	470	530	1000

CLASSIFICATION MATRIX, REWEIGHTED

	Predicted 0	Predicted 1	Total
Actual 0	$390/0.5102 = 764.4$	$110/0.5102 = 215.6$	980
Actual 1	$19,180/25 = 3.2$	$5,420/25 = 16.8$	20
Total			1000

At this point the misclassification rate appears to be $(80 + 110)/1000 = 19\%$, and the model ends up classifying 53% of the records as 1's. However, this reflects the performance on a sample where 50% are responders.

To estimate predictive performance when this model is used to score the original population (with 2% responders), we need to undo the effects of the oversampling. The actual number of responders must be divided by 25, and the actual number of nonresponders divided by 0.5102.

The revised classification matrix is as follows:

The adjusted misclassification rate is $(3.2 + 215.6)/1000 = 21.9\%$. The model ends up classifying $(215.6 + 16.8)/1000 = 23.24\%$ of the records as 1's, when we assume 2% responders.

II. Adjusting the Lift Curve for Oversampling   The lift curve is likely to be a more useful measure in low-response situations, where our interest lies not so much in classifying all the records correctly as in finding a model that guides us toward those records most likely to contain the response of interest (under the assumption that scarce resources preclude examining or contacting all the records). Typically, our interest in such a case is in maximizing value or minimizing cost, so we will show the adjustment process incorporating the benefit/cost element. The following procedure can be used (and easily implemented in Excel):

1. Sort the validation records in order of the predicted probability of success (where success = belonging to the class of interest).

2. For each record, record the cost (benefit) associated with the actual outcome.

3. Divide that value by the oversampling rate. For example, if responders are overweighted by a factor of 25, divide by 25.

4. For the highest probability (i.e., first) record, the value above is the y-coordinate of the first point on the lift chart. The x-coordinate is index number 1.

5. For the next record, again calculate the adjusted value associated with the actual outcome. Add this to the adjusted cost (benefit) for the previous record. This sum is the y-coordinate of the second point on the lift curve. The x-coordinate is index number 2.

6. Repeat step 5 until all records have been examined. Connect all the points, and this is the lift curve.

7. The reference line is a straight line from the origin to the point y = total net benefit and $x = n$ (n = number of records).

PROBLEMS

5.1 A data mining routine has been applied to a transaction dataset and has classified 88 records as fraudulent (30 correctly so) and 952 as nonfraudulent (920 correctly so). Construct the classification matrix and calculate the error rate.

5.2 Suppose that this routine has an adjustable cutoff (threshold) mechanism by which you can alter the proportion of records classified as fraudulent. Describe how moving the cutoff up or down would affect

   a. The classification error rate for records that are truly fraudulent.

   b. The classification error rate for records that are truly nonfraudulent.

5.3 FiscalNote is a startup founded by a Washington, DC, entrepreneur and funded by a Singapore sovereign wealth fund, the Winklevoss twins of Facebook fame, and others. It uses machine learning and data mining techniques to predict for its clients whether legislation in the US Congress and in US state legislatures will pass. The company reports 94% accuracy. (*Washington Post*, Nov. 21, 2014, "Capital Business")

   Considering just bills introduced in the US Congress, do a bit of Internet research to learn about numbers of bills introduced and passage rates. Identify the possible types of misclassifications, and comment on the use of overall accuracy as a metric. Include a discussion of other possible metrics and the potential role of propensities.

5.4 Consider Figure 5.16, the decile-wise lift chart for the transaction data model, applied to new data.

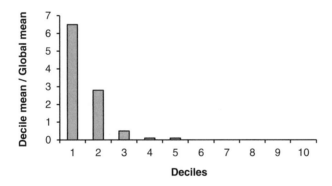

FIGURE 5.16   DECILE-WISE LIFT CHART FOR TRANSACTION DATA

   a. Interpret the meaning of the first and second bars from the left.

   b. Explain how you might use this information in practice.

   c. Another analyst comments that you could improve the accuracy of the model by classifying everything as nonfraudulent. If you do that, what is the error rate?

   d. Comment on the usefulness, in this situation, of these two metrics of model performance (error rate and lift).

5.5 A large number of insurance records are to be examined to develop a model for predicting fraudulent claims. Of the claims in the historical database, 1% were judged to be fraudulent. A sample is taken to develop a model, and oversampling is used to provide a balanced sample in light of the very low response rate. When applied to this sample ($n = 800$), the model ends up correctly classifying 310 frauds and 270 nonfrauds. It missed 90 frauds, and classified 130 records incorrectly as frauds when they were not.

a. Produce the classification matrix for the sample as it stands.

b. Find the adjusted misclassification rate (adjusting for the oversampling).

c. What percentage of new records would you expect to be classified as fraudulent?

5.6 A firm that sells software services has been piloting a new product and has records of 500 customers who have either bought the services or decided not to. The target value is the estimated profit from each sale (excluding sales costs). The global mean is $2128. However, the cost of the sales effort is not cheap—the company figures it comes to $2500 for each of the 500 customers (whether or not they buy). The firm developed a predictive model to identify the top spenders in the future. The lift and decile charts are shown in Figure 5.17.

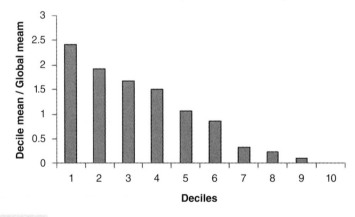

FIGURE 5.17 LIFT AND DECILE-WISE LIFT CHARTS FOR SOFTWARE SERVICES PRODUCT PROFIT

a. If the company begins working with a new set of 1000 leads to sell the same services, similar to the 500 in the pilot study, without any use of predictive modeling to target sales efforts, what is the estimated profit?

b. If the firm wants the average profit on each sale to at least double the sales effort cost, and applies an appropriate cutoff with this predictive model to a new set of 1000 leads, how far down the new list of 1000 should it proceed (how many deciles)?

c. Still considering the new list of 1000 leads, if the company applies this predictive model with a lower cutoff of $2500, how far should it proceed down the ranked leads, in terms of deciles?

d. Why use this two-stage process for predicting profit – why not simply develop a model for predicting profit for the 1000 new leads?

5.7 Table 5.5 shows a small set of predictive model validation results for a classification model, with both actual values and propensities.

a. Calculate error rates, sensitivity and specificity using cutoffs of 0.25, 0.5, and 0.75.

b. Using a cutoff of 0.5, create a decile lift chart. First, create a table where column 1 is the decile and column 2 is the lift ratio for each decile. Then, use a bar chart to plot the decile lift chart using the two columns of the table.

TABLE 5.5	PROPENSITIES AND ACTUAL CLASS MEMBERSHIP FOR VALIDATION DATA
Propensity	**Actual**
0.03	0
0.52	0
0.38	0
0.82	1
0.33	0
0.42	0
0.55	1
0.59	0
0.09	0
0.21	0
0.43	0
0.04	0
0.08	0
0.13	0
0.01	0
0.79	1
0.42	0
0.29	0
0.08	0
0.02	0

Prediction and Classification Methods

## CHAPTER 6

Multiple Linear Regression

In this chapter we introduce linear regression models for the purpose of prediction. We discuss the differences between fitting and using regression models for the purpose of inference (as in classical statistics) and for prediction. A predictive goal calls for evaluating model performance on a validation set, and for using predictive metrics. We then raise the challenges of using many predictors and describe variable selection algorithms that are often implemented in linear regression procedures.

6.1 INTRODUCTION

The most popular model for making predictions is the *multiple linear regression model* encountered in most introductory statistics courses and textbooks. This model is used to fit a relationship between a quantitative *dependent variable Y* (also called the *outcome, target,* or *response variable*) and a set of *predictors* $X_1, X_2, \ldots, X_p$ (also referred to as *independent variables, input variables, regressors,* or *covariates*). The assumption is that the following function approximates the relationship between the input and outcome variables:

$$Y = \beta_0 + \beta_1 x_1 + \beta_2 x_2 + \cdots + \beta_p x_p + \varepsilon, \tag{6.1}$$

where $\beta_0, \ldots, \beta_p$ are *coefficients* and ε is the *noise* or *unexplained* part. Data are then used to estimate the coefficients and to quantify the noise. In predictive modeling, the data are also used to evaluate model performance.

Data Mining for Business Analytics: Concepts, Techniques, and Applications with XLMiner®, Third Edition.
Galit Shmueli, Peter C. Bruce, and Nitin R. Patel.
© 2016 John Wiley & Sons Inc. Published 2016 by John Wiley & Sons, Inc.

Regression modeling means not only estimating the coefficients but also choosing which input variables to include and in what form. For example, a numerical input can be included as-is, or in logarithmic form (log X), or in a binned form (e.g., age group). Choosing the right form depends on domain knowledge, data availability and needed predictive power.

Multiple linear regression is applicable to numerous predictive modeling situations. Examples are predicting customer activity on credit cards from their demographics and historical activity patterns, predicting the time to failure of equipment based on utilization and environment conditions, predicting expenditures on vacation travel based on historical frequent flyer data, predicting staffing requirements at help desks based on historical data and product and sales information, predicting sales from cross selling of products from historical information, and predicting the impact of discounts on sales in retail outlets.

6.2 EXPLANATORY VS. PREDICTIVE MODELING

Before introducing the use of linear regression for prediction, we must clarify an important distinction that often escapes those with earlier familiarity with linear regression from courses in statistics. In particular, the two popular but different objectives behind fitting a regression model are:

1. Explaining or quantifying the average effect of inputs on an output (explanatory or descriptive task, respectively)
2. Predicting the outcome value for new records, given their input values (predictive task)

The classical statistical approach is focused on the first objective. In that scenario, the data are treated as a random sample from a larger population of interest. The regression model estimated from this sample is an attempt to capture the *average* relationship in the larger population. This model is then used in decision making to generate statements such as "a unit increase in service speed (X_1) is associated with an average increase of 5 points in customer satisfaction (Y), all other factors $(X_2, X_3, \ldots, X_p)$ being equal." If X_1 is known to *cause* Y, then such a statement indicates actionable policy changes—this is called explanatory modeling. When the causal structure is unknown, then this model quantifies the degree of *association* between the inputs and output, and the approach is called descriptive modeling.

In predictive analytics, however, the focus is typically on the second goal: predicting new individual observations. Here we are not interested in the coefficients themselves, nor in the "average record," but rather in the predictions that this model can generate for new records. In this scenario, the model is used for micro-decision-making at the record level. In our previous example, we would use the regression model to predict customer satisfaction for each new customer of interest.

Both explanatory and predictive modeling involve using a dataset to fit a model (i.e., to estimate coefficients), checking model validity, assessing its performance, and comparing to other models. However, the modeling steps and performance assessment differ in the two cases, usually leading to different final models. Therefore the choice of model is closely tied to whether the goal is explanatory or predictive.

In explanatory and descriptive modeling, where the focus is on modeling the average record, we try to fit the best model to the data in an attempt to learn about the underlying relationship in the population. In contrast, in predictive modeling (data mining), the goal is to find a regression model that best predicts new individual records. A regression model that fits the existing data too well is not likely to perform well with new data. Hence we look for a model that has the highest predictive power by evaluating it on a holdout set and using predictive metrics (see Chapter 5).

Let us summarize the main differences in using a linear regression in the two scenarios:

1. A good explanatory model is one that fits the data closely, whereas a good predictive model is one that predicts new cases accurately. Choices of input variables and their form can therefore differ.

2. In explanatory models, the entire dataset is used for estimating the best-fit model, to maximize the amount of information that we have about the hypothesized relationship in the population. When the goal is to predict outcomes of new individual cases, the data are typically split into a training set and a validation set. The training set is used to estimate the model, and the validation or *holdout set* is used to assess this model's predictive performance on new, unobserved data.

3. Performance measures for explanatory models measure how close the data fit the model (how well the model approximates the data) and how strong the average relationship is, whereas in predictive models performance is measured by predictive accuracy (how well the model predicts new individual cases).

4. In explanatory models the focus is on the coefficients (β), whereas in predictive models the focus is on the predictions ($\hat{y}$).

For these reasons it is extremely important to know the goal of the analysis before beginning the modeling process. A good predictive model can have a looser fit to the data on which it is based, and a good explanatory model can have low prediction accuracy. In the remainder of this chapter we focus on predictive models because these are more popular in data mining and because most statistics textbooks focus on explanatory modeling.

6.3   ESTIMATING THE REGRESSION EQUATION AND PREDICTION

Once we determine the input variables to include and their form, we estimate the coefficients of the regression formula from the data using a method called *ordinary least squares* (OLS). This method finds values $\hat{\beta}_0, \hat{\beta}_1, \hat{\beta}_2, \ldots, \hat{\beta}_p$ that minimize the sum of squared deviations between the actual values (Y) and their predicted values based on that model $(\hat{Y})$.

To predict the value of the output variable for a record with input values $x_1, x_2, \ldots, x_p$, we use the equation

$$\hat{Y} = \hat{\beta}_0 + \hat{\beta}_1 x_1 + \hat{\beta}_2 x_2 + \cdots + \hat{\beta}_p x_p. \tag{6.2}$$

Predictions based on this equation are the best predictions possible in the sense that they will be unbiased (equal to the true values on average) and will have the smallest average squared error compared to any unbiased estimates *if* we make the following assumptions:

1. The noise ε (or equivalently, Y) follows a normal distribution.
2. The choice of variables and their form is correct (*linearity*).
3. The cases are independent of each other.
4. The variability in Y values for a given set of predictors is the same regardless of the values of the predictors (*homoskedasticity*).

An important and interesting fact for the predictive goal is that *even if we drop the first assumption and allow the noise to follow an arbitrary distribution, these estimates are very good for prediction,* in the sense that among all linear models, as defined by equation (6.1), the model using the least squares estimates, $\hat{\beta}_0, \hat{\beta}_1, \hat{\beta}_2, \ldots, \hat{\beta}_p$, will have the smallest average squared errors. The assumption of a normal distribution is required in explanatory modeling, where it is used for constructing confidence intervals and statistical tests for the model parameters.

Even if the other assumptions are violated, it is still possible that the resulting predictions are sufficiently accurate and precise for the purpose they are intended for. The key is to evaluate predictive performance of the model, which is the main priority. Satisfying assumptions is of secondary interest and residual analysis can give clues to potential improved models to examine.

Example: Predicting the Price of Used Toyota Corolla Cars

A large Toyota car dealership offers purchasers of new Toyota cars the option to buy their used car as part of a trade-in. In particular, a new promotion promises to pay high prices for used Toyota Corolla cars for purchasers of a new car. The dealer then sells the used cars for a small profit. To ensure a reasonable profit, the dealer needs to be able to predict the price that the dealership will get for the used cars. For that reason data were collected on all previous sales of used Toyota Corollas at the dealership. The data include the sales price and other information on the car, such as its age, mileage, fuel type, and engine size. A description of each of these variables is given in Table 6.1. A sample of this dataset is shown in Table 6.2.

TABLE 6.1   VARIABLES IN THE TOYOTA COROLLA EXAMPLE

Variable	Description
Price	Offer price in euros
Age_08_04	Age in months as of August 2004
KM	Accumulated kilometers on odometer
Fuel_Type	Fuel type (*Petrol, Diesel, CNG*)
HP	Horsepower
Metallic	Metallic color? (Yes = 1, No = 0)
Automatic	Automatic (Yes = 1, No = 0)
CC	Cylinder volume in cubic centimeters
Doors	Number of doors
Quarterly_Tax	Quarterly road tax in euros
Weight	Weight in kilograms

The total number of records in the dataset is 1000 cars (we use the first 1000 cars from the dataset `ToyotoCorolla.xls`). After partitioning the data into training (60%) and validation (40%) sets, we fit a multiple linear regression model between price (the output variable) and the other variables (as predictors) using only the training set. Figure 6.1 shows the estimated coefficients, as computed by XLMiner.[1] Notice that the Fuel Type predictor has three categories (*Petrol, Diesel,* and *CNG*). We therefore have two dummy variables in the model: Petrol (0/1) and Diesel (0/1); the third, CNG (0/1), is redundant given the information on the first two dummies. Inclusion of this redundant variable will cause typical regression software to fail due to a *multicollinearity* error since the redundant variable will be a perfect linear combination of the other two (see Section 4.5).

The regression coefficients are then used to predict prices of individual used Toyota Corolla cars based on their age, mileage, and so on. Figure 6.2 shows a sample of predicted prices for 20 cars in the validation set, using the estimated

---

[1] In some versions of XLMiner, the intercept in the coefficients table is called "constant term."

TABLE 6.2   PRICES AND ATTRIBUTES FOR USED TOYOTA COROLLA CARS (SELECTED ROWS AND COLUMNS ONLY)

Price	Age	Kilometers	Fuel Type	HP	Metallic	Auto-matic	CC	Doors	Quart Tax	Weight
13500	23	46986	Diesel	90	1	0	2000	3	210	1165
13750	23	72937	Diesel	90	1	0	2000	3	210	1165
13950	24	41711	Diesel	90	1	0	2000	3	210	1165
14950	26	48000	Diesel	90	0	0	2000	3	210	1165
13750	30	38500	Diesel	90	0	0	2000	3	210	1170
12950	32	61000	Diesel	90	0	0	2000	3	210	1170
16900	27	94612	Diesel	90	1	0	2000	3	210	1245
18600	30	75889	Diesel	90	1	0	2000	3	210	1245
21500	27	19700	Petrol	192	0	0	1800	3	100	1185
12950	23	71138	Diesel	69	0	0	1900	3	185	1105
20950	25	31461	Petrol	192	0	0	1800	3	100	1185
19950	22	43610	Petrol	192	0	0	1800	3	100	1185
19600	25	32189	Petrol	192	0	0	1800	3	100	1185
21500	31	23000	Petrol	192	1	0	1800	3	100	1185
22500	32	34131	Petrol	192	1	0	1800	3	100	1185
22000	28	18739	Petrol	192	0	0	1800	3	100	1185
22750	30	34000	Petrol	192	1	0	1800	3	100	1185
17950	24	21716	Petrol	110	1	0	1600	3	85	1105
16750	24	25563	Petrol	110	0	0	1600	3	19	1065
16950	30	64359	Petrol	110	1	0	1600	3	85	1105
15950	30	67660	Petrol	110	1	0	1600	3	85	1105
16950	29	43905	Petrol	110	0	1	1600	3	100	1170
15950	28	56349	Petrol	110	1	0	1600	3	85	1120
16950	28	32220	Petrol	110	1	0	1600	3	85	1120
16250	29	25813	Petrol	110	1	0	1600	3	85	1120
15950	25	28450	Petrol	110	1	0	1600	3	85	1120
17495	27	34545	Petrol	110	1	0	1600	3	85	1120
15750	29	41415	Petrol	110	1	0	1600	3	85	1120
11950	39	98823	CNG	110	1	0	1600	5	197	1119

Regression Model

Input Variables	Coefficient	Std. Error	t-Statistic	P-Value	CI Lower	CI Upper	RSS Reduction
Intercept	−4200.5552	1438.422561	−2.920251213	0.00359	−7023.832	−1377.279	99626055314
Age_08_04	−121.60739	3.320627169	−36.62181408	5.76E-177	−128.125	−115.0898	8611425804
KM	−0.0192819	0.001699871	−11.34317078	7.202E-28	−0.022618	−0.015945	264977161.1
HP	29.756752	4.224060494	7.044584823	3.842E-12	21.465941	38.047564	207959275
Met_Color	101.16087	97.8079189	1.034280977	0.301299	−90.81248	293.13422	567466.6314
Automatic	456.65045	199.4455759	2.289599302	0.0222889	65.186892	848.11401	30074032.84
CC	−0.0223049	0.091510026	−0.243742598	0.807489	−0.201917	0.1573072	26418651.34
Doors	−118.3781	50.63479867	−2.337880406	0.0196246	−217.762	−18.99421	11322463.67
Quarterly_Tax	11.750228	2.201777534	5.336700883	1.215E-07	7.4286699	16.071786	281862947.6
Weight	16.067553	1.414469773	11.35941787	6.129E-28	13.29129	18.843816	256344071.7
Fuel_Type_Diesel	1289.8549	449.7143035	2.868165105	0.0042308	407.17416	2172.5356	1951679.235
Fuel_Type_Petrol	2367.2552	448.7481365	5.275242422	1.683E-07	1486.4709	3248.0396	48198815.61

Residual DF	850
R²	0.868708552
Adjusted R²	0.867009486
Std. Error Estimate	1316.060216
RSS	1472212319

FIGURE 6.1   ESTIMATED COEFFICIENTS FOR REGRESSION MODEL OF PRICE VS. CAR ATTRIBUTES

(a) Validation Data Scoring

Predicted value	Actual value	Residual
16199	13750	-2449
16686	13950	-2736
16266	16900	634
16236	18600	2364
20534	20950	416
20520	19600	-920
19860	21500	1640
19504	22500	2996
20385	22000	1615
16993	16950	-43
16106	16950	844
16099	16250	151
15789	15750	-39
15590	15950	360
15660	14950	-710
15668	14750	-918
15300	16750	1450
17919	19000	1081
17242	17950	708
19148	21950	2802

(b) Validation Data Scoring—Summary Report

Total sum of squared errors	RMS error	Average error
795600925.2	1410.319933	110.9145714

FIGURE 6.2   *(A)* PREDICTED PRICES (AND ERRORS) FOR 20 CARS IN VALIDATION SET, AND *(B)* SUMMARY PREDICTIVE MEASURES FOR ENTIRE VALIDATION SET

model. It gives the predictions and their errors (relative to the actual prices) for these 20 cars. On the right we get overall measures of predictive accuracy. Note that the average error is $111. A boxplot of the residuals (Figure 6.3) shows that 50% of the errors are approximately ±$850. This error magnitude might be small relative to the car price, but should be taken into account when considering the profit. Another observation of interest is the large positive residuals (underpredictions), which may or may not be a concern, depending on the application. Measures such as the average error, and error percentiles. are used to assess the predictive performance of a model and to compare models. We discuss such measures in the next section. This example also illustrates the point about the relaxation of the normality assumption. A histogram or probability plot of prices shows a right-skewed distribution. In a descriptive/explanatory modeling case where the goal is to obtain a good fit to the data, the output variable would be transformed (e.g., by taking a logarithm) to achieve a more "normal" variable. Although the fit of such a model to the training data is expected to be better, it will not necessarily improve predictive performance. In this example the average error in a model of log(price) is −$160, compared to $111 in the original model for price.

FIGURE 6.3   BOXPLOT OF MODEL RESIDUALS (BASED ON VALIDATION SET)

6.4   VARIABLE SELECTION IN LINEAR REGRESSION

Reducing the Number of Predictors

A frequent problem in data mining is that of using a regression equation to predict the value of a dependent variable when we have many variables available to choose as predictors in our model. Given the high speed of modern algorithms for multiple linear regression calculations, it is tempting in such a situation to take a kitchen-sink approach: Why bother to select a subset? Just use all the variables in the model.

Another consideration favoring the inclusions of numerous variables is the hope that a previously hidden relationship will emerge. For example, a company found that customers who had purchased anti-scuff protectors for chair and table legs had lower credit risks. However, there are several reasons for exercising caution before throwing all possible variables into a model.

- It may be expensive or not feasible to collect a full complement of predictors for future predictions.

- We may be able to measure fewer predictors more accurately (e.g., in surveys).
- The more predictors there are, the higher the chance of missing values in the data. If we delete or impute cases with missing values, multiple predictors will lead to a higher rate of case deletion or imputation.
- *Parsimony* is an important property of good models. We obtain more insight into the influence of predictors in models with few parameters.
- Estimates of regression coefficients are likely to be unstable, due to *multicollinearity* in models with many variables. (Multicollinearity is the presence of two or more predictors sharing the same linear relationship with the outcome variable.) Regression coefficients are more stable for parsimonious models. One very rough rule of thumb is to have a number of cases n larger than $5(p + 2)$, where p is the number of predictors.
- It can be shown that using predictors that are uncorrelated with the dependent variable increases the variance of predictions.
- It can be shown that dropping predictors that are actually correlated with the dependent variable can increase the average error (bias) of predictions.

The last two points mean that there is a trade-off between too few and too many predictors. In general, accepting some bias can reduce the variance in predictions. This *bias–variance trade-off* is particularly important for large numbers of predictors, since in that case it is very likely that there are variables in the model that have small coefficients relative to the standard deviation of the noise and also exhibit at least moderate correlation with other variables. Dropping such variables will improve the predictions, as it reduces the prediction variance. This type of bias–variance trade-off is a basic aspect of most data mining procedures for prediction and classification. In light of this, methods for reducing the number of predictors p to a smaller set are often used.

How to Reduce the Number of Predictors

The first step in trying to reduce the number of predictors should always be to use domain knowledge. It is important to understand what the various predictors are measuring and why they are relevant for predicting the response. With this knowledge the set of predictors should be reduced to a sensible set that reflects the problem at hand. Some practical reasons for predictor elimination are the expense of collecting this information in the future, inaccuracy, high correlation with another predictor, many missing values, or simply irrelevance. Also helpful in examining potential predictors are summary statistics and graphs, such as frequency and correlation tables, predictor-specific summary statistics and plots, and missing value counts.

The next step makes use of computational power and statistical significance. In general, there are two types of methods for reducing the number of predictors in a model. The first is an *exhaustive search* for the "best" subset of predictors by fitting regression models with all the possible combinations of predictors. The second is to search through a partial set of models. We describe these two approaches next.

Exhaustive Search (Best Subset)   The idea here is to evaluate all subsets. Since the number of subsets for even moderate values of p is very large, after the algorithm creates the subsets and runs all the models, we need some way to examine the most promising subsets and to select from them. Criteria for evaluating and comparing models are based on metrics computed from the training data. One popular criterion is the *adjusted R^2*, which is defined as

$$R^2_{\text{adj}} = 1 - \frac{n-1}{n-p-1}(1 - R^2),$$

where R^2 is the proportion of explained variability in the model (in a model with a single predictor, this is the squared correlation). Like R^2, higher values of adjusted R^2 indicate better fit. Unlike R^2, which does not account for the number of predictors used, adjusted R^2 uses a penalty on the number of predictors. This avoids the artificial increase in R^2 that can result from simply increasing the number of predictors but not the amount of information. It can be shown that using R^2_{adj} to choose a subset is equivalent to picking the subset that minimizes $\hat{\sigma}^2$.

Another criterion that is often used for subset selection is known as *Mallow's C_p* (see formula below). This criterion assumes that the full model (with all predictors) is unbiased, although it may have predictors that, if dropped, would reduce prediction variability. With this assumption we can show that if a subset model is unbiased, the average C_p value equals the number of parameters $p+1$ (= number of predictors + 1), the size of the subset. So a reasonable approach to identifying subset models with small bias is to examine those with values of C_p that are near $p+1$. C_p is also an estimate of the error[2] for predictions at the x-values observed in the training set. Thus good models are those that have values of C_p near $p+1$ and that have small p (i.e., are of small size). C_p is computed from the formula

$$C_p = \frac{\text{SSE}}{\hat{\sigma}^2_{\text{full}}} + 2(p+1) - n, \tag{6.3}$$

---

[2] In particular, it is the sum of the MSE standardized by dividing by σ^2.

where $\hat{\sigma}^2_{\text{full}}$ is the estimated value of σ^2 in the full model that includes all predictors. It is important to remember that the usefulness of this approach depends heavily on the reliability of the estimate of σ^2 for the full model. This requires that the training set contain a large number of observations relative to the number of predictors. Finally, a useful point to note is that for a fixed size of subset, R^2, R^2_{adj}, and C_p all select the same subset. There is in fact no difference among them in the order of merit that they ascribe to subsets of a fixed size. This is good to know if comparing models with the same number of predictors, but often we want to compare models with different numbers of predictors.

Figure 6.4 gives the results of applying an exhaustive search ("best subsets" in XLMiner) on the Toyota Corolla price data (with the 11 predictors). It reports the best model with a single predictor, two predictors, and so on. It can be seen that the R^2_{adj} increases until six predictors are used (number of coefficients = 7) and then stabilizes. The C_p indicates that a model with 9 to 11 predictors is good. The dominant predictor in all models is the age of the car, with horsepower and mileage playing important roles as well.

#Coef	RSS	Cp	R²	Adj R²	Prob	1	2	3	4	5	6	7	8	9	10	11	12
1	11213314688	5614.1	0	0	0	Intercept											
2	2601888884	644.23	0.768	0.7677	0	Intercept	Age_08_04										
3	2216733640	423.86	0.802	0.8019	0	Intercept	Age_08_04								Weight		
4	1701016304	128.1	0.848	0.8478	0	Intercept	Age_08_04	KM							Weight		
5	1558132753	47.607	0.861	0.8604	0	Intercept	Age_08_04	KM	HP						Weight		
6	1541917997	40.245	0.863	0.8617	0	Intercept	Age_08_04	KM	HP					Quarterly_Tax	Weight		
7	1506759674	21.946	0.866	0.8647	0.0014	Intercept	Age_08_04	KM	HP					Quarterly_Tax	Weight		Fuel_Type_Petrol
8	1493326424	16.191	0.867	0.8657	0.0165	Intercept	Age_08_04	KM	HP					Quarterly_Tax	Weight	Fuel_Type_Diesel	Fuel_Type_Petrol
9	1482738091	12.077	0.868	0.8665	0.1088	Intercept	Age_08_04	KM	HP				Doors	Quarterly_Tax	Weight	Fuel_Type_Diesel	Fuel_Type_Petrol
10	1474146471	9.1167	0.869	0.8671	0.5724	Intercept	Age_08_04	KM	HP		Automatic		Doors	Quarterly_Tax	Weight	Fuel_Type_Diesel	Fuel_Type_Petrol
11	1472315219	10.059	0.869	0.8672	0.8075	Intercept	Age_08_04	KM	HP	Met_Color	Automatic		Doors	Quarterly_Tax	Weight	Fuel_Type_Diesel	Fuel_Type_Petrol
12	1472212319	12	0.869	0.867	1	Intercept	Age_08_04	KM	HP	Met_Color	Automatic	CC	Doors	Quarterly_Tax	Weight	Fuel_Type_Diesel	Fuel_Type_Petrol

FIGURE 6.4 **EXHAUSTIVE SEARCH RESULT FOR REDUCING PREDICTORS IN TOYOTA COROLLA PRICE EXAMPLE**

Popular Subset Selection Algorithms The second method of finding the best subset of predictors relies on a partial, iterative search through the space of all possible regression models. The end product is one best subset of predictors (although there do exist variations of these methods that identify several close-to-best choices for different sizes of predictor subsets). This approach is computationally cheaper, but it has the potential of missing "good" combinations of predictors. None of the methods guarantee that they yield the best subset for any criterion, such as adjusted R^2. They are reasonable methods for situations with large numbers of predictors, but for moderate numbers of predictors, the exhaustive search is preferable.

Three popular iterative search algorithms are *forward selection*, *backward elimination*, and *stepwise regression*. In *forward selection* we start with no predictors and then add predictors one by one. Each predictor added is the one (among all predictors) that has the largest contribution to R^2 on top of the predictors that are already in it. The algorithm stops when the contribution of additional predictors is not statistically significant. The main disadvantage of this method is that the algorithm will miss pairs or groups of predictors that perform very well together but perform poorly as single predictors. This is similar to interviewing job candidates for a team project one by one, thereby missing groups of candidates who perform superiorly together, but poorly on their own.

In *backward elimination* we start with all predictors and then at each step eliminate the least useful predictor (according to statistical significance). The algorithm stops when all the remaining predictors have significant contributions. The weakness of this algorithm is that computing the initial model with all predictors can be time consuming and unstable. *Stepwise regression* is like forward selection except that at each step we consider dropping predictors that are not statistically significant, as in backward elimination.

Note: In XLMiner, unlike other popular software packages (SAS, Minitab, etc.), these three algorithms, like "best subsets" (exhaustive search), yield a table of multiple models rather than a single model. This allows the user to decide on the subset size after reviewing all possible sizes based on criteria such as R^2_{adj} and C_p.

For the Toyota Corolla price example, forward selection yields exactly the same results as those found in an exhaustive search: For each number of predictors the same subset is chosen (it therefore gives a table identical to the one in Figure 6.4). Notice that this is not always the case. Backward elimination starts with the full model and then drops predictors one by one, first dropping CC then Met_Color (see Figure 6.5). The R^2_{adj} and C_p measures indicate exactly the same subsets as those suggested by the exhaustive search. In other words, it correctly identifies Doors, CC, Diesel, Metallic, and Automatic as the least useful predictors. Backward elimination would yield a different model than that of the exhaustive search only if we decided to use fewer than six predictors. For instance, if we were limited to two predictors, backward elimination would choose Age and Weight, whereas an exhaustive search shows that the best pair of predictors is actually Age and HP.

					Model							
#Coef	RSS	Cp	R²	Adj R²	Prob 1	2	3 4 5	6	7 8 9	10 11	12		
12	1472212319	12	0.8687	0.867	1 Intercept	Age_08_04 KM HP	Met_Color	Automatic CC	Doors Quarterly_Tax Weight	Fuel_Type_Diesel Fuel_Type_Petrol		
11	1472315219	10.1	0.8687	0.8672 0.8075 Intercept	Age_08_04 KM HP	Met_Color	Automatic	Doors Quarterly_Tax Weight	Fuel_Type_Diesel Fuel_Type_Petrol			
10	1474146471	9.12	0.8685	0.8671 0.5724 Intercept	Age_08_04 KM HP		Automatic	Doors Quarterly_Tax Weight	Fuel_Type_Diesel Fuel_Type_Petrol			

FIGURE 6.5   BACKWARD ELIMINATION RESULT FOR REDUCING PREDICTORS IN TOYOTA COROLLA PRICE EXAMPLE

The results for stepwise selection can be seen in Figure 6.6. It chooses the same subsets as forward selection for subset sizes of one to seven predictors. However, for eight to 10 predictors, it chooses a different subset than that chosen using the other methods: it decides to drop Doors, QuartTax, and Weight. This

#Coef	RSS	Cp	R²	Adj R²	Prob	Model 1	2	3	4	5	6	7	8	9	10	11	12
1	11213314688	5614.15	0	0	0	Intercept											
2	2601888884	644.233	0.77	0.7677	0	Intercept	Age_08_04										
3	2216733640	423.859	0.8	0.8019	0	Intercept	Age_08_04								Weight		
4	1701016304	128.103	0.85	0.8478	0	Intercept	Age_08_04	KM							Weight		
5	1558132753	47.6072	0.86	0.8604	0	Intercept	Age_08_04	KM	HP						Weight		
6	1541917997	40.2454	0.86	0.8617	0	Intercept	Age_08_04	KM	HP					Quarterly_Tax	Weight		
7	1506759674	21.9463	0.87	0.8647	0.0014	Intercept	Age_08_04	KM	HP					Quarterly_Tax	Weight		Fuel_Type_Petrol
8	1493326424	16.1905	0.87	0.8657	0.0165	Intercept	Age_08_04	KM	HP					Quarterly_Tax	Weight	Fuel_Type_Diesel	Fuel_Type_Petrol
9	1482738091	12.0772	0.87	0.8665	0.1088	Intercept	Age_08_04	KM	HP				Doors	Quarterly_Tax	Weight	Fuel_Type_Diesel	Fuel_Type_Petrol
10	1474146471	9.1167	0.87	0.8671	0.5724	Intercept	Age_08_04	KM	HP			Automatic	Doors	Quarterly_Tax	Weight	Fuel_Type_Diesel	Fuel_Type_Petrol

FIGURE 6.6   **STEPWISE SELECTION RESULT FOR REDUCING PREDICTORS IN TOYOTA COROLLA PRICE EXAMPLE**

means that it fails to detect the best subsets for eight to 10 predictors. R^2_{adj} is largest at six predictors (the same six as were selected by the other models), but C_p indicates that the full model with 11 predictors is the best fit.

This example shows that the search algorithms yield fairly good solutions, but we need to carefully determine the number of predictors to retain. It also shows the merits of running a few searches and using the combined results to determine the subset to choose. There is a popular (but false) notion that stepwise regression is superior to backward elimination and forward selection because of its ability to add and to drop predictors. This example shows clearly that it is not always so.

Additional ways to reduce the dimension of the data are by using principal components (Chapter 4) and regression trees (Chapter 9).

PROBLEMS

6.1 **Predicting Boston Housing Prices.** The file `BostonHousing.xls` contains information collected by the US Bureau of the Census concerning housing in the area of Boston, Massachusetts. The dataset includes information on 506 census housing tracts in the Boston area. The goal is to predict the median house price in new tracts based on information such as crime rate, pollution, and number of rooms. The dataset contains 12 predictors, and the response is the median house price (MEDV). Table 6.3 describes each of the predictors and the response.

TABLE 6.3	DESCRIPTION OF VARIABLES FOR BOSTON HOUSING EXAMPLE
CRIM	Per capita crime rate by town
ZN	Proportion of residential land zoned for lots over 25,000 ft^2
INDUS	Proportion of nonretail business acres per town
CHAS	Charles River dummy variable (= 1 if tract bounds river; = 0 otherwise)
NOX	Nitric oxide concentration (parts per 10 million)
RM	Average number of rooms per dwelling
AGE	Proportion of owner-occupied units built prior to 1940
DIS	Weighted distances to five Boston employment centers
RAD	Index of accessibility to radial highways
TAX	Full-value property-tax rate per $10,000
PTRATIO	Pupil/teacher ratio by town
LSTAT	% Lower status of the population
MEDV	Median value of owner-occupied homes in $1000s

a. Why should the data be partitioned into training and validation sets? What will the training set be used for? What will the validation set be used for?

b. Fit a multiple linear regression model to the median house price (MEDV) as a function of CRIM, CHAS, and RM. Write the equation for predicting the median house price from the predictors in the model.

c. Using the estimated regression model, what median house price is predicted for a tract in the Boston area that does not bound the Charles River, has a crime rate of 0.1, and where the average number of rooms per house is 6? What is the prediction error?

d. Reduce the number of predictors:

   i. Which predictors are likely to be measuring the same thing among the 13 predictors? Discuss the relationships among INDUS, NOX, and TAX.

   ii. Compute the correlation table for the 12 numerical predictors and search for highly correlated pairs. These have potential redundancy and can cause multicollinearity. Choose which ones to remove based on this table.

   iii. Use an exhaustive search to reduce the remaining predictors as follows: First, choose the top three models. Then run each of these models separately on the training set, and compare their predictive accuracy for the validation set. Compare RMSE and average error, as well as lift charts. Finally, describe the best model.

6.2 **Predicting Software Reselling Profits.** Tayko Software is a software catalog firm that sells games and educational software. It started out as a software manufacturer and then added third-party titles to its offerings. It recently revised its collection of items in a new catalog, which it mailed out to its customers. This mailing yielded 1000 purchases. Based on these data, Tayko wants to devise a model for predicting the spending amount that a purchasing customer will yield. The file `Tayko.xls` contains information on 1000 purchases. Table 6.4 describes the variables to be used in the problem (the Excel file contains additional variables).

TABLE 6.4	DESCRIPTION OF VARIABLES FOR TAYKO SOFTWARE EXAMPLE
FREQ	Number of transactions in the preceding year
LAST_UPDATE	Number of days since last update to customer record
WEB	Whether customer purchased by Web order at least once
GENDER	Male or Female
ADDRESS_RES	Whether it is a residential address
ADDRESS_US	Whether it is a US address
SPENDING (response)	Amount spent by customer in test mailing (in dollars)

a. Explore the spending amount by creating a pivot table for the categorical variables and computing the average and standard deviation of spending in each category.

b. Explore the relationship between spending and each of the two continuous predictors by creating two scatterplots (SPENDING vs. FREQ, and SPENDING vs. LAST_UPDATE). Does there seem to be a linear relationship?

c. To fit a predictive model for SPENDING:

i. Partition the 1000 records into training and validation sets.

ii. Run a multiple linear regression model for SPENDING vs. all six predictors. Give the estimated predictive equation.

iii. Based on this model, what type of purchaser is most likely to spend a large amount of money?

iv. If we used backward elimination to reduce the number of predictors, which predictor would be dropped first from the model?

v. Show how the prediction and the prediction error are computed for the first purchase in the validation set.

vi. Evaluate the predictive accuracy of the model by examining its performance on the validation set.

vii. Create a histogram of the model residuals. Do they appear to follow a normal distribution? How does this affect the predictive performance of the model?

6.3 **Predicting Airfare on New Routes**
The following problem takes place in the United States in the late 1990s, when many major US cities were facing issues with airport congestion, partly as a result of the 1978 deregulation of airlines. Both fares and routes were freed from regulation, and low-fare carriers such as Southwest began competing on existing routes and starting nonstop service on routes that previously lacked it. Building completely new airports is generally not feasible, but sometimes decommissioned military bases or smaller municipal airports can be reconfigured as regional or larger commercial airports. There are numerous

players and interests involved in the issue (airlines, city, state and federal authorities, civic groups, the military, airport operators), and an aviation consulting firm is seeking advisory contracts with these players. The firm needs predictive models to support its consulting service. One thing the firm might want to be able to predict is fares, in the event a new airport is brought into service. The firm starts with the file `Airfares.xls`, which contains real data that were collected between Q3-1996 and Q2-97. The variables in these data are listed in Table 6.5, and are believed to be important in predicting FARE. Some airport-to-airport data are available, but most data are at the city-to-city level. One question that will be of interest in the analysis is the effect that the presence or absence of Southwest (SW) has on FARE.

TABLE 6.5   DESCRIPTION OF VARIABLES FOR AIRFARE EXAMPLE

Variable	Description
S_CODE	Starting airport's code
S_CITY	Starting city
E_CODE	Ending airport's code
E_CITY	Ending city
COUPON	Average number of coupons (a one-coupon flight is a nonstop flight, a two-coupon flight is a one-stop flight, etc.) for that route
NEW	Number of new carriers entering that route between Q3-96 and Q2-97
VACATION	Whether (Yes) or not (No) a vacation route
SW	Whether (Yes) or not (No) Southwest Airlines serves that route
HI	Herfindahl index: measure of market concentration
S_INCOME	Starting city's average personal income
E_INCOME	Ending city's average personal income
S_POP	Starting city's population
E_POP	Ending city's population
SLOT	Whether or not either endpoint airport is slot controlled (this is a measure of airport congestion)
GATE	Whether or not either endpoint airport has gate constraints (this is another measure of airport congestion)
DISTANCE	Distance between two endpoint airports in miles
PAX	Number of passengers on that route during period of data collection
FARE	Average fare on that route

a. Explore the numerical predictors and response (FARE) by creating a correlation table and examining some scatterplots between FARE and those predictors. What seems to be the best single predictor of FARE?

b. Explore the categorical predictors (excluding the first four) by computing the percentage of flights in each category. Create a pivot table with the average fare in each category. Which categorical predictor seems best for predicting FARE?

c. Find a model for predicting the average fare on a new route:

   i. Convert categorical variables (e.g., SW) into dummy variables. Then partition the data into training and validation sets. The model will be fit to the training data and evaluated on the validation set.

   ii. Use stepwise regression to reduce the number of predictors. You can ignore the first four predictors (S_CODE, S_CITY, E_CODE, E_CITY). Report the estimated model selected.

   iii. Repeat (ii) using exhaustive search instead of stepwise regression. Compare the resulting best model to the one you obtained in (ii) in terms of the predictors that are in the model.

   iv. Compare the predictive accuracy of both models (ii) and (iii) using measures such as RMSE and average error and lift charts.

   v. Using model (iii), predict the average fare on a route with the following characteristics: COUPON = 1.202, NEW = 3, VACATION = No, SW = No, HI = 4442.141, S_INCOME = $28,760, E_INCOME = $27,664, S_POP = 4,557,004, E_POP = 3,195,503, SLOT = Free, GATE = Free, PAX = 12,782, DISTANCE = 1976 miles.

   vi. Using model (iii), predict the reduction in average fare on the route in (v) if Southwest decides to cover this route.

   vii. In reality, which of the factors will not be available for predicting the average fare from a new airport (i.e., before flights start operating on those routes)? Which ones can be estimated? How?

  viii. Select a model that includes only factors that are available before flights begin to operate on the new route. Use an exhaustive search to find such a model.

   ix. Use the model in (viii) to predict the average fare on a route with characteristics COUPON = 1.202, NEW = 3, VACATION = No, SW = No, HI = 4442.141, S_INCOME = $28,760, E_INCOME = $27,664, S_POP = 4,557,004, E_POP = 3,195,503, SLOT = Free, GATE = Free, PAX = 12,782, DISTANCE = 1976 miles.

   x. Compare the predictive accuracy of this model with model (iii). Is this model good enough, or is it worthwhile re-evaluating the model once flights begin on the new route?

d. In competitive industries, a new entrant with a novel business plan can have a disruptive effect on existing firms. If a new entrant's business model is sustainable, other players are forced to respond by changing their business practices. If the goal of the analysis was to evaluate the effect of Southwest Airlines' presence on the airline industry rather than predicting fares on new routes, how would the analysis be different? Describe technical and conceptual aspects.

6.4 **Predicting Prices of Used Cars.** The file `ToyotaCorolla.xls` contains data on used cars (Toyota Corolla) on sale during late summer of 2004 in the Netherlands. It has 1436 records containing details on 38 attributes, including Price, Age, Kilometers, HP, and other specifications. The goal is to predict the price of a used Toyota Corolla based on its specifications. (The example in Section 6.3 is a subset of this dataset.)

Create dummy variables for the categorical predictors Fuel_Type and Color. Split the data into training (50%), validation (30%), and test (20%) datasets.

Run a multiple linear regression with the output variable Price and input variables Age_08_04, KM, Fuel_Type, HP, Automatic, Doors, Quarterly_Tax, Mfr_Guarantee, Guarantee_Period, Airco, Automatic_Airco, CD_Player, Powered_Windows, Sport_Model, and Tow_Bar.

a. What appear to be the three or four most important car specifications for predicting the car's price?

b. Using metrics you consider useful, assess the performance of the model in predicting prices.

k-Nearest-Neighbors (*k*-NN)

In this chapter we describe the *k*-nearest-neighbors algorithm that can be used for classification (of a categorical outcome) or prediction (of a numerical outcome). To classify or predict a new record, the method relies on finding "similar" records in the training data. These "neighbors" are then used to derive a classification or prediction for the new record by voting (for classification) or averaging (for prediction). We explain how similarity is determined, how the number of neighbors is chosen, and how a classification or prediction is computed. *k*-NN is a highly automated data-driven method. We discuss the advantages and weaknesses of the *k*-NN method in terms of performance and practical considerations such as computational time.

7.1 THE *k*-NN CLASSIFIER (CATEGORICAL OUTCOME)

The idea in *k*-nearest-neighbors methods is to identify *k* records in the training dataset that are similar to a new record that we wish to classify. We then use these similar (neighboring) records to classify the new record into a class, assigning the new record to the predominant class among these neighbors. Denote by $(x_1, x_2, \ldots, x_p)$ the values of the predictors for this new record. We look for records in our training data that are similar or "near" the record to be classified in the predictor space (i.e., records that have values close to $x_1, x_2, \ldots, x_p$). Then, based on the classes to which those proximate records belong, we assign a class to the record that we want to classify.

Data Mining for Business Analytics: Concepts, Techniques, and Applications with XLMiner®, Third Edition.
Galit Shmueli, Peter C. Bruce, and Nitin R. Patel.
© 2016 John Wiley & Sons Inc. Published 2016 by John Wiley & Sons, Inc.

Determining Neighbors

The *k*-nearest-neighbors algorithm is a classification method that does not make assumptions about the form of the relationship between the class membership (*Y*) and the predictors $X_1, X_2, \ldots, X_p$. This is a nonparametric method because it does not involve estimation of parameters in an assumed function form, such as the linear form assumed in linear regression (Chapter 6). Instead, this method draws information from similarities between the predictor values of the records in the dataset.

A central question is how to measure the distance between records based on their predictor values. The most popular measure of distance is the Euclidean distance. The Euclidean distance between two records $(x_1, x_2, \ldots, x_p)$ and $(u_1, u_2, \ldots, u_p)$ is

$$\sqrt{(x_1 - u_1)^2 + (x_2 - u_2)^2 + \cdots + (x_p - u_p)^2}. \tag{7.1}$$

You will find a host of other distance metrics in Chapters 12 and 15 for both numerical and categorical variables. However, the *k*-NN algorithm relies on many distance computations (between each record to be predicted and every record in the training set), and therefore the Euclidean distance, which is computationally cheap, is the most popular in *k*-NN.

To equalize the scales that the various predictors may have, note that in most cases, predictors should first be standardized before computing a Euclidean distance.

Classification Rule

After computing the distances between the record to be classified and existing records, we need a rule to assign a class to the record to be classified, based on the classes of its neighbors. The simplest case is $k = 1$, where we look for the record that is closest (the nearest neighbor) and classify the new record as belonging to the same class as its closest neighbor. It is a remarkable fact that this simple, intuitive idea of using a single nearest neighbor to classify records can be very powerful when we have a large number of records in our training set. In turns out that the misclassification error of the 1-nearest-neighbor scheme has a misclassification rate that is no more than twice the error when we know exactly the probability density functions for each class.

The idea of the **1-nearest-neighbor** can be extended to $k > 1$ neighbors as follows:

1. Find the nearest *k* neighbors to the record to be classified.
2. Use a majority decision rule to classify the record, where the record is classified as a member of the majority class of the *k* neighbors.

Example: Riding Mowers

A riding-mower manufacturer would like to find a way of classifying families in a city into those likely to purchase a riding mower and those not likely to buy one. A pilot random sample is undertaken of 12 owners and 12 nonowners in the city. The data are shown in Table 7.1. We first partition the data into training data (18 households) and validation data (6 households). Obviously, this dataset is too small for partitioning, which can result in unstable results, but we will continue with this partitioning for illustration purposes. A scatter plot of the training data is shown in Figure 7.1.

TABLE 7.1	LOT SIZE, INCOME, AND OWNERSHIP OF A RIDING MOWER FOR 24 HOUSEHOLDS		
Household Number	Income ($000s)	Lot Size (000s ft^2)	Ownership of Riding Mower
1	60.0	18.4	Owner
2	85.5	16.8	Owner
3	64.8	21.6	Owner
4	61.5	20.8	Owner
5	87.0	23.6	Owner
6	110.1	19.2	Owner
7	108.0	17.6	Owner
8	82.8	22.4	Owner
9	69.0	20.0	Owner
10	93.0	20.8	Owner
11	51.0	22.0	Owner
12	81.0	20.0	Owner
13	75.0	19.6	Nonowner
14	52.8	20.8	Nonowner
15	64.8	17.2	Nonowner
16	43.2	20.4	Nonowner
17	84.0	17.6	Nonowner
18	49.2	17.6	Nonowner
19	59.4	16.0	Nonowner
20	66.0	18.4	Nonowner
21	47.4	16.4	Nonowner
22	33.0	18.8	Nonowner
23	51.0	14.0	Nonowner
24	63.0	14.8	Nonowner

Now consider a new household with $60,000 income and lot size 20,000 ft^2 (also shown in Figure 7.1). Among the households in the training set, the one closest to the new household (in Euclidean distance after normalizing income and lot size) is household 4, with $61,500 income and lot size 20,800 ft^2. If we use a 1-NN classifier, we would classify the new household as an owner, like household 4. If we use $k = 3$, the three nearest households are 4, 9, and 14. The

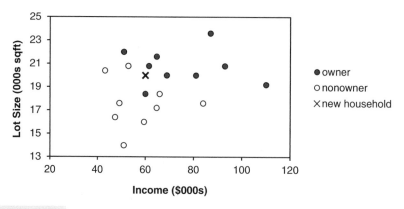

FIGURE 7.1   SCATTERPLOT OF *LOT SIZE* VS. *INCOME* FOR THE 18 HOUSEHOLDS IN THE TRAINING SET AND THE NEW HOUSEHOLD TO BE CLASSIFIED

first two are owners of riding mowers, and the last is a nonowner. The majority vote is therefore *owner*, and the new household would be classified as an owner.

Choosing *k*

The advantage of choosing $k > 1$ is that higher values of k provide smoothing that reduces the risk of overfitting due to noise in the training data. Generally speaking, if k is too low, we may be fitting to the noise in the data. However, if k is too high, we will miss out on the method's ability to capture the local structure in the data, one of its main advantages. In the extreme, $k = n = $ the number of records in the training dataset. In that case we simply assign all records to the majority class in the training data, regardless of the values of $(x_1, x_2, \ldots, x_p)$, which coincides with the naive rule! This is clearly a case of oversmoothing in the absence of useful information in the predictors about the class membership. In other words, we want to balance between overfitting to the predictor information and ignoring this information completely. A balanced choice depends greatly on the nature of the data. The more complex and irregular the structure of the data, the lower is the optimum value of k. Typically, values of k fall in the range 1 to 20. Often an odd number is chosen, to avoid ties.

So how is k chosen? Answer: We choose the k that has the best classification performance. We use the training data to classify the records in the validation data, then compute error rates for various choices of k. For our example, if, on the one hand, we choose $k = 1$, we will classify in a way that is very sensitive to the local characteristics of the training data. If, on the other hand, we choose a large value of k, such as $k = 18$, we will simply predict the most frequent class in the dataset in all cases. This is a very stable prediction, but it completely ignores the information in the predictors. To find a balance, we examine the misclassification rate (of the validation set) that results for different choices of k

Validation error log for different *k*

Value of *k*	% Error Training	% Error Validation	
1	0.00	33.33	
2	16.67	33.33	
3	11.11	33.33	
4	22.22	33.33	
5	11.11	33.33	
6	27.78	33.33	
7	22.22	33.33	
8	22.22	16.67	<--- Best *k*
9	22.22	16.67	
10	22.22	16.67	
11	16.67	33.33	
12	16.67	16.67	
13	11.11	33.33	
14	11.11	16.67	
15	5.56	33.33	
16	16.67	33.33	
17	11.11	33.33	
18	50.00	50.00	

FIGURE 7.2 MISCLASSIFICATION RATE OF VALIDATION SET FOR VARIOUS CHOICES OF *k*

between 1 and 18. This is shown in Figure 7.2. We would choose $k = 8$, which minimizes the misclassification rate in the validation set.[1] Note, however, that now the validation set is used as an addition to the training set and does not reflect a holdout set as before. Ideally, we would want a third test set to evaluate the performance of the method on data that it did not see.

Once k is chosen, the algorithm uses it to generate classifications of new records. An example is shown in Figure 7.3, where 8 neighbors are used to classify the new household.

Setting the Cutoff Value

k-NN uses a majority decision rule to classify a new record, where the record is classified as a member of the majority class of the k neighbors. The definition of "majority" is directly linked to the notion of a cutoff value applied to the class membership probabilities. Let us consider a binary outcome case. For a new

---

[1] Partitioning such a small dataset is unwise in practice, as results will heavily rely on the particular partition. For instance, if you use a different partitioning, you might obtain a different "optimal" k. The validation set in this example includes households 2,7,8,13,22 and 24.

XLMiner:*k*-Nearest-Neighbors—Classification of New Data (for *k* = 8)

Data range	['KNN Riding Mowers.xlsx']'Data_Partition1'!$G$20:$H$20			Back to Navigator

Cutoff Prob.Val. for Success (Updatable)	0.5	(Updating the value here will NOT update value in summary report)

Row ID	Predicted Class	Prob. for Owner	Actual # Nearest Neighbors	Income	Lot Size
1	Owner	0.625	8	60	20

FIGURE 7.3   CLASSIFYING A NEW HOUSEHOLD USING THE "BEST *k*" = 8

record, the proportion of class 1 members among its neighbors is an estimate of its propensity (probability) of belonging to class 1. In the riding mowers example with $k = 3$, we found that the three nearest neighbors to the new household (with income = \$60,000 and lot size = 20,000 ft^2) are households 4, 9, and 14. Since 4 and 9 are owners and 14 is a nonowner, we can estimate for the new household a probability of 2/3 of being an owner (and 1/3 for being a nonowner). Using a simple majority rule is equivalent to setting the cutoff value to 0.5. Another example can be seen in Figure 7.3, where $k = 8$ was used to classify the new household. The "Prob for Owner" of 0.625 was obtained because 5 of the 8 neighbors were owners. Using a cutoff of 0.5 leads to a classification of "owner" for the new household.

As mentioned in Chapter 5, changing the cutoff value affects the classification matrix (i.e., the error rates). Hence in some cases we might want to choose a cutoff other than the default 0.5 for the purpose of maximizing accuracy or for incorporating misclassification costs. In XLMiner this can be done by directly changing the cutoff value (as can be seen in Figure 7.3), which automatically changes the "Predicted Class" and the related classification matrices.

k-NN with More Than Two Classes

The k-NN classifier can easily be applied to an outcome with m classes, where $m > 2$. The "majority rule" means that a new record is classified as a member of the majority class of its k neighbors. An alternative, when there is a specific class that we are interested in identifying (and are willing to "overidentify" records as belonging to this class), is to calculate the proportion of the k neighbors that belong to this class of interest, use that as an estimate of the probability (propensity) that the new record belongs to that class, and then refer to a user-specified cutoff value to decide whether to assign the new record to that class. For more on the use of cutoff value in classification where there is a single class of interest, see Chapter 5.

Converting Categorical Variables to Binary Dummies

It usually does not make sense to calculate Euclidean distance between two non-numeric categories (e.g., cookbooks and maps in a bookstore). Therefore, before *k*-NN can be applied, categorical variables must be converted to binary dummies. In contrast to the situation with statistical models such as regression, all *m* binaries should be created and used with *k*-NN. While mathematically this is redundant, since $m - 1$ dummies contain the same information as *m* dummies, this redundant information does not create the multicollinearity problems that it does for linear models. Moreover, in *k*-NN, the use of $m - 1$ dummies can yield different classifications than the use of *m* dummies, and lead to an imbalance in the contribution of the different categories to the model.

7.2 *k*-NN FOR A NUMERICAL RESPONSE

The idea of *k*-NN can readily be extended to predicting a continuous value (as is our aim with multiple linear regression models). The first step of determining neighbors by computing distances remains unchanged. The second step, where a majority vote of the neighbors is used to determine class, is modified such that we take the average response value of the *k*-nearest-neighbors to determine the prediction. Often this average is a weighted average, with the weight decreasing with increasing distance from the point at which the prediction is required.

Another modification is in the error metric used for determining the "best *k*." Rather than the overall error rate used in classification, RMS error (or another prediction error metric) is used in prediction (see Chapter 5).

PANDORA

Pandora is an Internet music radio service that allows users to build customized "stations" that play music similar to a song or artist that they have specified. Pandora uses a *k*-NN style clustering/classification process called the Music Genome Project to locate new songs or artists that are close to the user-specified song or artist.

Pandora was the brainchild of Tim Westergren, who worked as a musician and a nanny when he graduated from Stanford in the 1980s. Together with Nolan Gasser, who was studying medieval music, he developed a "matching engine" by entering data about a song's characteristics into a spreadsheet. The first result was surprising—a Beatles song matched to a Bee Gees song, but they built a company around the concept. The early days were hard—Westergren racked up over $300,000 in personal debt, maxed out 11 credit cards, and ended up in the hospital once due to stress-induced heart palpitations. A venture capitalist finally invested funds in 2004 to rescue the firm, and as of 2013 it is listed on the NY Stock Exchange.

In simplified terms, the process works roughly as follows for songs:

1. Pandora has established hundreds of variables on which a song can be measured on a scale from 0–5. Four such variables from the beginning of the list are
   - Acid Rock Qualities
   - Accordion Playing
   - Acousti-Lectric Sonority
   - Acousti-Synthetic Sonority

2. Pandora pays musicians to analyze tens of thousands of songs, and rate each song on each of these attributes. Each song will then be represented by a row vector of values between 0 and 5. For example for Led Zeppelin's *Kashmir*:

   Kashmir 4 0 3 3 . . . (high on acid rock attributes, no accordion, etc.)

   This step represents a costly investment, and lies at the heart of Pandora's value because these variables have been tested and selected, because they accurately reflect the essence of a song, and because they provide a basis for defining highly individualized preferences.

3. The online user specifies a song that s/he likes (the song must be in Pandora's database).

4. Pandora then calculates the statistical distance[2] between the user's song, and the songs in its database. It selects a song that is close to the user-specified song and plays it.

5. The user then has the option of saying "I like this song," "I don't like this song," or saying nothing.

6. If "like" is chosen, the original song, plus the new song are merged into a 2-song cluster[3] that is represented by a single vector, comprised of means of the variables in the original two song vectors.

7. If "dislike" is chosen, the vector of the song that is not liked is stored for future reference. (If the user does not express an opinion about the song, in our simplified example here the new song is not used for further comparisons.)

8. Pandora looks in its database for a new song, one whose statistical distance is close to the "like" song cluster[4] and not too close to the "dislike" song. Depending on the user's reaction, this new song might be added to the "like" cluster or "dislike" cluster.

Over time, Pandora develops the ability to deliver songs that match a particular taste of a particular user. A single user might build up multiple stations around different song clusters. Clearly, this is a less limiting approach than selecting music in terms of which "genre" it belongs to.

While the process described above is a bit more complex than the basic "classification of new data" process described in this chapter, the fundamental process—classifying a record according to its proximity to other records—is the same at its core. Note the role of domain knowledge in this machine learning process—the variables have been tested and selected by the project leaders, and the measurements have been made by human experts.

Further reading: See www.pandora.com, Wikipedia's article on the Music Genome Project, and Joyce John's article "Pandora and the Music Genome Project," in the September 2006 *Scientific Computing* [23 (10): 14, pages 40–41].

---

[2]See Section 12.5 in Chapter 12 for an explanation of statistical distance.

[3]See Chapter 15 for more on clusters.

[4]See Case 21.6 "Segmenting Consumers of Bath Soap" for an exercise involving the identification of clusters, which are then used for classification purposes.

7.3 ADVANTAGES AND SHORTCOMINGS OF k-NN ALGORITHMS

The main advantage of k-NN methods is their simplicity and lack of parametric assumptions. In the presence of a large enough training set, these methods perform surprisingly well, especially when each class is characterized by multiple combinations of predictor values. For instance, in real-estate databases there are likely to be multiple combinations of {home type, number of rooms, neighborhood, asking price, etc.} that characterize homes that sell quickly vs. those that remain for a long period on the market.

There are three difficulties with the practical exploitation of the power of the k-NN approach. First, although no time is required to estimate parameters from the training data (as would be the case for parametric models such as regression), the time to find the nearest neighbors in a large training set can be prohibitive. A number of ideas have been implemented to overcome this difficulty. The main ideas are:

- Reduce the time taken to compute distances by working in a reduced dimension using dimension reduction techniques such as principal components analysis (Chapter 4).

- Use sophisticated data structures such as search trees to speed up identification of the nearest neighbor. This approach often settles for an "almost nearest" neighbor to improve speed. An example is using *bucketing*, where the records are grouped into buckets so that records within each bucket are close to each other. For a to-be-predicted record, buckets are ordered by their distance to the record. Starting from the nearest bucket, the distance to each of the records within the bucket is measured. The algorithm stops when the distance to a bucket is larger than the distance to the closest record thus far.

Second, the number of records required in the training set to qualify as large increases exponentially with the number of predictors *p*. This is because the expected distance to the nearest neighbor goes up dramatically with *p* unless the size of the training set increases exponentially with *p*. This phenomenon is known as the *curse of dimensionality*, a fundamental issue pertinent to all classification, prediction, and clustering techniques. This is why we often seek to reduce the number of predictors through methods such as selecting subsets of the predictors for our model or by combining them using methods such as principal components analysis, singular value decomposition, and factor analysis (see Chapter 4).

Third, *k*-NN is a "lazy learner": the time-consuming computation is deferred to the time of prediction. For every record to be predicted, we compute its distances from the entire set of training records only at the time of prediction. This behavior prohibits using this algorithm for real-time prediction of a large number of records simultaneously.

PROBLEMS

7.1 Calculating Distance with Categorical Predictors. This exercise with a tiny dataset illustrates the calculation of Euclidean distance and the creation of binary dummies. The online education company Statistics.com segments its customers and prospects into three main categories: IT professionals (IT), statisticians (Stat), and other (Other). It also tracks, for each customer, the number of years since first contact (years). Consider the following customers; information about whether they have taken a course or not (the outcome to be predicted) is included:

Customer 1: Stat, 1 year, did not take course
Customer 2: Other, 1.1 year, took course

a. Consider now the following new prospect:

Prospect 1: IT, 1 year

Using the information above on the two customers and one prospect, create one dataset for all three with the categorical predictor variable transformed into 2 binaries, and a similar dataset with the categorical predictor variable transformed into 3 binaries.

b. For each derived dataset, calculate the Euclidean distance between the prospect and each of the other two customers. (Note: While it is typical to normalize data for k-NN, this is not an iron-clad rule and you may proceed here without normalization.)

c. Using k-NN with $k = 1$, classify the prospect as taking or not taking a course using each of the two derived datasets. Does it make a difference whether you use 2 or 3 dummies?

7.2 Personal Loan Acceptance. Universal Bank is a relatively young bank growing rapidly in terms of overall customer acquisition. The majority of these customers are liability customers (depositors) with varying sizes of relationship with the bank. The customer base of asset customers (borrowers) is quite small, and the bank is interested in expanding this base rapidly to bring in more loan business. In particular, it wants to explore ways of converting its liability customers to personal loan customers (while retaining them as depositors).

A campaign that the bank ran last year for liability customers showed a healthy conversion rate of over 9% success. This has encouraged the retail marketing department to devise smarter campaigns with better target marketing. The goal is to use k-NN to predict whether a new customer will accept a loan offer.
This will serve as the basis for the design of a new campaign.

The file `UniversalBank.xls` contains data on 5000 customers. The data include customer demographic information (age, income, etc.), the customer's relationship with the bank (mortgage, securities account, etc.), and the customer response to the last personal loan campaign (*Personal Loan*). Among these 5000 customers, only 480 (= 9.6%) accepted the personal loan that was offered to them in the earlier campaign.

Transform categorial predictors with more than two categories into dummy variables, then, partition the data into training (60%) and validation (40%) sets.

a. Consider the following customer:

Age=40, Experience = 10, Income = 84, Family = 2, CCAvg = 2, Education_1 = 0, Education_2 = 1, Education_3 = 0, Mortgage = 0, Securities Account = 0, CD Account = 0, Online = 1, and Credit Card = 1. Perform a k-NN classification with all predictors except ID and ZIP code using $k = 1$. Specify the *success* class as 1 (loan acceptance), and use the default cutoff value of 0.5. How would this customer be classified?

b. What is a choice of *k* that balances between overfitting and ignoring the predictor information?

c. Show the classification matrix for the validation data that results from using the best *k*.

d. Consider the following customer: Age = 40, Experience = 10, Income = 84, Family = 2, CCAvg = 2, Education_1 = 0, Education_2 = 1, Education_3 = 0, Mortgage = 0, Securities Account = 0, CD Account = 0, Online = 1 and Credit Card = 1. Classify the customer using the best *k*.

e. Repartition the data, this time into training, validation, and test sets (50% : 30% : 20%). Apply the *k*-NN method with the *k* chosen above. Compare the classification matrix of the test set with that of the training and validation sets. Comment on the differences and their reason.

7.3 **Predicting Housing Median Prices.** The file `BostonHousing.xls` contains information on over 500 census tracts in Boston, where for each tract multiple variables are recorded. The last column (CAT.MEDV) was derived from MEDV, such that it obtains the value 1 if MEDV>30 and 0 otherwise. Consider the goal of predicting the median value (MEDV) of a tract, given the information in the first 12 columns.

Partition the data into training (60%) and validation (40%) sets.

a. Perform a *k*-NN prediction with all 12 predictors (ignore the CAT.MEDV column), trying values of *k* from 1 to 5. Make sure to normalize the data (click "normalize input data"). What is the best *k* chosen? What does it mean?

b. Predict the MEDV for a tract with the following information, using the best *k*:

CRIM	ZN	INDUS	CHAS	NOX	RM	AGE	DIS	RAD	TAX	PTRATIO	LSTAT
0.2	0	7	0	0.538	6	62	4.7	4	307	21	10

(In a new worksheet, create an identical table with these column names and values, and then in "Score new data" choose "from worksheet.")

c. Why is the error of the training data zero?

d. Why is the validation data error overly optimistic compared to the error rate when applying this *k*-NN predictor to new data?

e. If the purpose is to predict MEDV for several thousands of new tracts, what would be the disadvantage of using *k*-NN prediction? List the operations that the algorithm goes through in order to produce each prediction.

## CHAPTER 8

The Naive Bayes Classifier

In this chapter we introduce the naive Bayes classifier, which can be applied to data with categorical predictors. We review the concept of conditional probabilities, then present the complete, or exact, Bayesian classifier. We next see how it is impractical in most cases, and learn how to modify it and use instead the "naive Bayes" classifier, which is more generally applicable.

8.1 INTRODUCTION

The naive Bayes method (and, indeed, an entire branch of statistics) is named after the Reverend Thomas Bayes (1702–1761). To understand the naive Bayes classifier, we first look at the complete, or exact, Bayesian classifier. The basic principle is simple. For each record to be classified:

1. Find all the other records with the same predictor profile (i.e., where the predictor values are the same).
2. Determine what classes the records belong to and which class is most prevalent.
3. Assign that class to the new record.

Alternatively (or in addition), it may be desirable to tweak the method so that it answers the question: "What is the propensity of belonging to the class of interest?" instead of "Which class is the most probable?" Obtaining class probabilities allows using a sliding cutoff to classify a record as belonging to class

Data Mining for Business Analytics: Concepts, Techniques, and Applications with XLMiner®, Third Edition.
Galit Shmueli, Peter C. Bruce, and Nitin R. Patel.
© 2016 John Wiley & Sons Inc. Published 2016 by John Wiley & Sons, Inc.

i, even if i is not the most probable class for that record. This approach is useful when there is a specific class of interest that we are interested in identifying, and we are willing to "overidentify" records as belonging to this class. (See Chapter 5 for more details on the use of cutoffs for classification and on asymmetric misclassification costs.)

Cutoff Probability Method

1. Establish a cutoff probability for the class of interest above which we consider that a record belongs to that class.
2. Find all the training records with the same predictor profile as the new record (i.e., where the predictor values are the same).
3. Determine the probability that those records belong to the class of interest.
4. If that probability is above the cutoff probability, assign the new record to the class of interest.

Conditional Probability

Both procedures incorporate the concept of *conditional probability*, or the probability of event A given that event B has occurred [denoted $P(A|B)$]. In this case we will be looking at the probability of the record belonging to class i given that its predictor values are $x_1, x_2, \ldots, x_p$. In general, for a response with m classes $C_1, C_2, \ldots, C_m$, and the predictor values $x_1, x_2, \ldots, x_p$, we want to compute

$$P(C_i|x_1, \ldots, x_p). \tag{8.1}$$

To classify a record, we compute its probability of belonging to each of the classes in this way, then classify the record to the class that has the highest probability or use the cutoff probability to decide whether it should be assigned to the class of interest.

From this definition, we see that the Bayesian classifier works only with categorical predictors. If we use a set of numerical predictors, then it is highly unlikely that multiple records will have identical values on these numerical predictors. Therefore numerical predictors must be binned and converted to categorical predictors. *The Bayesian classifier is the only classification or prediction method presented in this book that is especially suited for (and limited to) categorical predictor variables.*

Example 1: Predicting Fraudulent Financial Reporting

An accounting firm has many large companies as customers. Each customer submits an annual financial report to the firm, which is then audited by the accounting firm. For simplicity, we will designate the outcome of the audit as "fraudulent" or "truthful," referring to the accounting firm's assessment of the customer's financial report. The accounting firm has a strong incentive to be accurate in identifying fraudulent reports—if it passes a fraudulent report as truthful, it would be in legal trouble.

The accounting firm notes that, in addition to all the financial records, it has information on whether or not the customer has had prior legal trouble (criminal or civil charges of any nature filed against it). This information has not been used in previous audits, but the accounting firm is wondering whether it could be used in the future to identify reports that merit more intensive review. Specifically, it wants to know whether having had prior legal trouble is predictive of fraudulent reporting.

In this case, each customer is a record, and the response of interest, $Y = \{$fraudulent, truthful$\}$, has two classes into which a company can be classified: $C_1 = $ fraudulent and $C_2 = $ truthful. The predictor variable—"prior legal trouble"—has two values: 0 (no prior legal trouble) and 1 (prior legal trouble).

The accounting firm has data on 1500 companies that it has investigated in the past. For each company, it has information on whether the financial report was judged fraudulent or truthful and whether the company had prior legal trouble. The data were partitioned into a training set (1000 firms) and a validation set (500 firms), and the counts in the training set are shown in Table 8.1.

TABLE 8.1   **PIVOT TABLE FOR FINANCIAL REPORTING EXAMPLE**

	Prior Legal ($X = 1$)	No Prior Legal ($X = 0$)	Total
Fraudulent (C_1)	50	50	100
Truthful (C_2)	180	720	900
Total	230	770	1000

8.2 APPLYING THE FULL (EXACT) BAYESIAN CLASSIFIER

Now consider the financial report from a new company, which we wish to classify as either fraudulent or truthful by using these data. To do this, we compute the probabilities, as above, of belonging to each of the two classes.

If the new company had had prior legal trouble, the probability of belonging to the fraudulent class would be P(fraudulent | prior legal) = 50/230 (there were 230 companies with prior legal trouble in the training set, and 50 of them had fraudulent financial reports). The probability of belonging to the other class, "truthful," is, of course, the remainder = 180/230.

Using the "Assign to the Most Probable Class" Method

If a company had prior legal trouble, we assign it to the "truthful" class. Similar calculations for the case of no prior legal trouble are left as an exercise to the reader. In this example, using the rule "assign to the most probable class," all records are assigned to the "truthful" class. This is the same result as the naive rule of "assign all records to the majority class."

Using the Cutoff Probability Method

In this example, we are more interested in identifying the fraudulent reports—those are the ones that can land the auditor in jail. We recognize that, in order to identify the fraudulent reports, some truthful reports will be misidentified as fraudulent, and the overall classification accuracy may decline. Our approach is therefore to establish a cutoff value for the probability of being fraudulent, and classify all records above that value as fraudulent. The Bayesian formula for the calculation of this probability that a record belongs to class C_i is as follows:

$$P(C_i|x_1,\ldots,x_p) = \frac{P(x_1,\ldots,x_p|C_i)P(C_i)}{P(x_1,\ldots,x_p|C_1)P(C_1) + \cdots + P(x_1,\ldots,x_p|C_m)P(C_m)}.$$

$$(8.2)$$

In this example (where frauds are rarer), if the cutoff were established at 0.20, we would classify a prior legal trouble record as fraudulent because P(fraudulent | prior legal) = 50/230 = 0.22. The user can treat this cutoff as a "slider" to be adjusted to optimize performance, like other parameters in any classification model.

Practical Difficulty with the Complete (Exact) Bayes Procedure

The approach outlined above amounts to finding all the records in the sample that are exactly like the new record to be classified in the sense that all the predictor values are all identical. This was easy in the small example presented above, where there was just one predictor.

When the number of predictors gets larger (even to a modest number like 20), many of the records to be classified will be without exact matches. This can be understood in the context of a model to predict voting on the basis of demographic variables. Even a sizable sample may not contain even a single match for a new record who is a male Hispanic with high income from the US Midwest who voted in the last election, did not vote in the prior election, has three daughters and one son, and is divorced. And this is just eight variables, a small number for most data mining exercises. The addition of just a single new variable with five equally frequent categories reduces the probability of a match by a factor of 5.

Solution: Naive Bayes

In the naive Bayes solution, we no longer restrict the probability calculation to those records that match the record to be classified. Instead, we use the entire dataset.

Returning to our original basic classification procedure outlined at the beginning of the chapter, we recall that this procedure for classifying a new record was:

1. Find all the other records with the same predictor profile (i.e., where the predictor values are the same).

2. Determine what classes the records belong to and which class is most prevalent.

3. Assign that class to the new record.

The naive Bayes modification (for the basic classification procedure) is as follows:

1. For class C_1, estimate the individual conditional probabilities for each predictor $P(x_j|C_1)$—these are the probabilities that the predictor value in the record to be classified occurs in class C_1. For example, for X_1 this probability is estimated by the proportion of x_1 values among the C_1 records in the training set.

2. Multiply these probabilities by each other, then by the proportion of records belonging to class C_1.

3. Repeat steps 1 and 2 for all the classes.

4. Estimate a probability for class i by taking the value calculated in step 2 for class i and dividing it by the sum of such values for all classes.

5. Assign the record to the class with the highest probability for this set of predictor values.

The preceding steps lead to the naive Bayes formula for calculating the probability that a record with a given set of predictor values $x_1, \ldots, x_p$ belongs to class C_1 among m classes. The formula can be written as follows:

$$P_{nb}(C_1 \mid x_1, \ldots x_p) =$$

$$\frac{P(C_1)[P(x_1 \mid C_1)P(x_2 \mid C_1) \cdots P(x_p \mid C_1)]}{P(C_1)[P(x_1 \mid C_1)P(x_2 \mid C_1) \cdots P(x_p \mid C_1)] + \cdots + P(C_m)[P(x_1 \mid C_m)P(x_2 \mid C_m) \cdots P(x_p \mid C_m)]}.$$

$$(8.3)$$

This is a somewhat formidable formula; see Example 2 for a simpler numerical version. Note that all the needed quantities can be obtained from pivot tables of Y vs. each of the categorical predictors.

The Naive Bayes Assumption of Conditional Independence In probability terms, we have made a simplifying assumption that the exact *conditional probability* of seeing a record with predictor profile $x_1, x_2, \ldots, x_p$ within a certain class, $P(x_1, x_2, \ldots, x_p | C_i)$, is well approximated by the product of the individual conditional probabilities $P(x_1 | C_i) \times P(x_2 | C_i) \times \cdots \times P(x_p | C_i)$. These two quantities are identical when the predictors are independent within each class.

For example, suppose that "lost money last year" is an additional variable in the accounting fraud example. The simplifying assumption we make with naive Bayes is that, within a given class, we no longer need to look for the records characterized both by "prior legal trouble" and "lost money last year." Rather, assuming that the two are independent, we can simply multiply the probability of "prior legal trouble" by the probability of "lost money last year." Of course, complete independence is unlikely in practice, where some correlation between predictors is expected.

In practice, despite the assumption violation, the procedure works quite well—primarily because what is usually needed is not a propensity for each record that is accurate in absolute terms but just a reasonably accurate *rank ordering* of propensities. Even when the assumption is violated, the rank ordering of the records' propensities is typically preserved.

Note that if all we are interested in is a rank ordering, and the denominator remains the same for all classes, it is sufficient to concentrate only on the numerator. The disadvantage of this approach is that the probability values it yields (the propensities), while ordered correctly, are not on the same scale as the exact values that the user would anticipate.

Using the Cutoff Probability Method The procedure above is for the basic case where we seek maximum classification accuracy for all classes. In the case of the *relatively rare class of special interest*, the procedure is:

1. Establish a cutoff probability for the class of interest above which we consider that a record belongs to that class.
2. For the class of interest, compute the probability that each individual predictor value in the record to be classified occurs in the training data.
3. Multiply these probabilities times each other, then times the proportion of records belonging to the class of interest.

4. Estimate the probability for the class of interest by taking the value calculated in step 3 for the class of interest and dividing it by the sum of the similar values for all classes.

5. If this value falls above the cutoff, assign the new record to the class of interest; otherwise not.

6. Adjust the cutoff value as needed, as a parameter of the model.

Example 2: Predicting Fraudulent Financial Reports, Two Predictors

Let us expand the financial reports example to two predictors, and, using a small subset of data, compare the complete (exact) Bayes calculations to the naive Bayes calculations.

TABLE 8.2   **INFORMATION ON 10 COMPANIES**

Company	Prior Legal Trouble	Company Size	Status
1	Yes	Small	Truthful
2	No	Small	Truthful
3	No	Large	Truthful
4	No	Large	Truthful
5	No	Small	Truthful
6	No	Small	Truthful
7	Yes	Small	Fraudulent
8	Yes	Large	Fraudulent
9	No	Large	Fraudulent
10	Yes	Large	Fraudulent

Consider the 10 customers of the accounting firm listed in Table 8.2. For each customer, we have information on whether it had prior legal trouble, whether it is a small or large company, and whether the financial report was found to be fraudulent or truthful. Using this information, we will calculate the conditional probability of fraud, given each of the four possible combinations $\{y, \text{small}\}, \{y, \text{large}\}, \{n, \text{small}\}, \{n, \text{large}\}$.

Complete (Exact) Bayes Calculations   The probabilities are computed as

$$P(\text{fraudulent}|\text{PriorLegal} = y, \text{Size} = \text{small}) = 1/2 = 0.5$$

$$P(\text{fraudulent}|\text{PriorLegal} = y, \text{Size} = \text{large}) = 2/2 = 1$$

$$P(\text{fraudulent}|\text{PriorLegal} = n, \text{Size} = \text{small}) = 0/3 = 0$$

$$P(\text{fraudulent}|\text{PriorLegal} = n, \text{Size} = \text{large}) = 1/3 = 0.33$$

Naive Bayes Calculations Now we compute the naive Bayes probabilities. For the conditional probability of fraudulent behaviors given {PriorLegal = y, Size = small}, the numerator is a multiplication of the proportion of {PriorLegal = y} instances among the fraudulent companies, times the proportion of {Size = small} instances among the fraudulent companies, times the proportion of fraudulent companies: $(3/4)(1/4)(4/10) = 0.075$. To get the actual probabilities, we must also compute the numerator for the conditional probability of truthful behaviors given {PriorLegal = y, Size = small}: $(1/6)(4/6)(6/10) = 0.067$. The denominator is then the sum of these two conditional probabilities $(0.075 + 0.067 = 0.14)$. The conditional probability of fraudulent behaviors given {PriorLegal = y, Size = small} is therefore $0.075/0.14 = 0.53$. In a similar fashion, we compute all four conditional probabilities:

$$P_{nb}(\text{fraudulent}|\text{PriorLegal} = y, \text{Size} = \text{small}) = \frac{(3/4)(1/4)(4/10)}{(3/4)(1/4)(4/10) + (1/6)(4/6)(6/10)}$$

$$= 0.53$$

$$P_{nb}(\text{fraudulent}|\text{PriorLegal} = y, \text{Size} = \text{large}) = 0.87$$

$$P_{nb}(\text{fraudulent}|\text{PriorLegal} = n, \text{Size} = \text{small}) = 0.07$$

$$P_{nb}(\text{fraudulent}|\text{PriorLegal} = n, \text{Size} = \text{large}) = 0.31$$

Note how close these naive Bayes probabilities are to the exact Bayes probabilities. Although they are not equal, both would lead to exactly the same classification for a cutoff of 0.5 (and many other values). It is often the case that the rank ordering of probabilities is even closer to the exact Bayes method than the probabilities themselves, and for classification purposes it is the rank orderings that matter.

We now consider a larger numerical example, where information on flights is used to predict flight delays.

Example 3: Predicting Delayed Flights

Predicting flight delays can be useful to a variety of organizations: airport authorities, airlines, and aviation authorities. At times, joint task forces have been formed to address the problem. If such an organization were to provide ongoing real-time assistance with flight delays, it would benefit from some advance notice about flights that are likely to be delayed.

In this simplified illustration, we look at six predictors (see Table 8.3). The outcome of interest is whether or not the flight is delayed (*delayed* here means arrived more than 15 minutes late). Our data consist of all flights from the Washington, DC, area into the New York City area during January 2004. The percentage of delayed flights among these 2201 flights is 19.5%. The data were obtained from the Bureau of Transportation Statistics (available on the Web at

TABLE 8.3	DESCRIPTION OF VARIABLES FOR FLIGHT DELAYS EXAMPLE

Day of Week	Coded as: 1 = Monday, 2 = Tuesday ,..., 7 = Sunday
Departure Time	Broken down into 18 intervals between 6:00 am and 10:00 pm
Origin	Three airport codes: DCA (Reagan National), IAD (Dulles), BWI (Baltimore–Washington Int'l)
Destination	Three airport codes: JFK (Kennedy), LGA (LaGuardia), EWR (Newark)
Carrier	Eight airline codes: CO (Continental), DH (Atlantic Coast), DL (Delta), MQ (American Eagle), OH (Comair), RU (Continental Express), UA (United), and US (USAirways)
Weather	Coded as 1 if inclement, 0 otherwise

www.transtats.bts.gov). The goal is to accurately predict whether or not a new flight (not in this dataset), will be delayed.

A record is a particular flight. The response is whether the flight was delayed, and thus it has two classes (1 = Delayed and 0 = On Time). In addition, information is collected on the predictors listed in Table 8.3.

The data were first partitioned into training (60%) and validation (40%) sets, and then a naive Bayes classifier was applied to the training set.

The top table in Figure 8.1 shows the ratios of delayed flights and on time flights in the training set (called prior class probabilities). The bottom table shows the conditional probabilities for each class, as a function of the predictor values. Note that the conditional probabilities in the output can be computed simply by using pivot tables in Excel, looking at the percentage of records in a cell relative to the entire class. This is illustrated in Table 8.4, which displays the percent of delayed (or on time) flights by destination airport as a percentage of the total delayed (or on time) flights.

Note that in this example there are no predictor values that were not represented in the training data except for on time flights (Class = 0) when the weather was bad (Weather = 1). When the weather was bad, all flights in the training set were delayed.

To classify a new flight, we compute the probability that it will be delayed and the probability that it will be on time. Recall that since both will have

TABLE 8.4	PIVOT TABLE OF DELAYED AND ON TIME FLIGHTS BY DESTINATION AIRPORT (ROWS)

	Delayed %	On Time%	Total %
EWR	38.67	28.36	30.36
JFK	18.75	17.65	17.87
LGA	42.58	53.99	51.78
Total	100.00	100.00	100.00

Prior class probabilities

According to relative occurrences in training data		

Class	Prob.	
1	0.193792581	<-- Success Class
0	0.806207419	

Conditional probabilities

Classes-->				
Input Variables	**1**		**0**	
	Value	Prob	Value	Prob
CARRIER	CO	0.06640625	CO	0.038497653
	DH	0.33984375	DH	0.243192488
	DL	0.109375	DL	0.2
	MQ	0.1796875	MQ	0.112676056
	OH	0.01171875	OH	0.017840376
	RU	0.21484375	RU	0.170892019
	UA	0.0078125	UA	0.016901408
	US	0.0703125	US	0.2
DAY_OF_WEEK	1	0.203125	1	0.128638498
	2	0.16015625	2	0.139906103
	3	0.12890625	3	0.152112676
	4	0.12890625	4	0.159624413
	5	0.1640625	5	0.181220657
	6	0.0703125	6	0.131455399
	7	0.14453125	7	0.107042254
DEP_TIME_BLK	0600–0659	0.03515625	0600–0659	0.061971831
	0700–0759	0.05078125	0700–0759	0.060093897
	0800–0859	0.0546875	0800–0859	0.071361502
	0900–0959	0.0234375	0900–0959	0.053521127
	1000–1059	0.01953125	1000–1059	0.057276995
	1100–1159	0.01953125	1100–1159	0.038497653
	1200–1259	0.0546875	1200–1259	0.062910798
	1300–1359	0.05078125	1300–1359	0.068544601
	1400–1459	0.15234375	1400–1459	0.110798122
	1500–1559	0.08203125	1500–1559	0.064788732
	1600–1659	0.07421875	1600–1659	0.078873239
	1700–1759	0.15625	1700–1759	0.094835681
	1800–1859	0.03125	1800–1859	0.043192488
	1900–1959	0.08984375	1900–1959	0.040375587
	2000–2059	0.01953125	2000–2059	0.030985915
	2100–2159	0.0859375	2100–2159	0.061971831
DEST	EWR	0.38671875	EWR	0.283568075
	JFK	0.1875	JFK	0.176525822
	LGA	0.42578125	LGA	0.539906103
ORIGIN	BWI	0.09375	BWI	0.068544601
	DCA	0.484375	DCA	0.635680751
	IAD	0.421875	IAD	0.295774648
Weather	0	0.92578125	0	1
	1	0.07421875	1	0

FIGURE 8.1   OUTPUT FROM NAIVE BAYES CLASSIFIER APPLIED TO FLIGHT DELAYS (TRAINING) DATA

the same denominator, we can just compare the numerators. Each numerator is computed by multiplying all the conditional probabilities of the relevant predictor values and, finally, multiplying by the proportion of that class [in this case $\hat{P}$(delayed) = 0.19]. For example, to classify a Delta flight from DCA to LGA between 10–11am on a Sunday with good weather, we compute the numerators:

$$\hat{P}(\text{delayed} \mid \text{Carrier} = \text{DL, DayofWeek} = 7, \text{DepartureTime} = 1000 - 1059,$$

$$\text{Destination} = \text{LGA, Origin} = \text{DCA, Weather} = 0)$$

$$\propto (0.11)(0.14)(0.020)(0.43)(0.48)(0.93)(0.19) = 0.000011.$$

$$\hat{P}(\text{ontime} \mid \text{Carrier} = \text{DL, DayofWeek} = 7, \text{DepartureTime} = 1000 - 1059,$$

$$\text{Destination} = \text{LGA, Origin} = \text{DCA, Weather} = 0)$$

$$\propto (0.2)(0.11)(0.06)(0.54)(0.64)(1)(0.81) = 0.00034.$$

The symbol $\propto$ means "is proportional to," reflecting the fact that this calculation deals only with the numerator in the naive Bayes formula (8.3).

It is therefore more likely that the flight will be on time. Note that a record with such a combination of predictor values does not exist in the training set, and therefore we use the naive Bayes rather than the exact Bayes.

To compute the actual probability, we divide each of the numerators by their sum

$$\hat{P}(\text{delayed} \mid \text{Carrier} = \text{DL, DayofWeek} = 7, \text{DepartureTime} = 1000 - 1059,$$

$$\text{Destination} = \text{LGA, Origin} = \text{DCA, Weather} = 0)$$

$$= \frac{0.000011}{0.000011 + 0.00034} = 0.03.$$

$$\hat{P}(\text{ontime} \mid \text{Carrier} = \text{DL, DayofWeek} = 7, \text{DepartureTime} = 1000 - 1059,$$

$$\text{Destination} = \text{LGA, Origin} = \text{DCA, Weather} = 0)$$

$$= \frac{0.00034}{0.000011 + 0.00034} = 0.97.$$

Of course, we rely on software to compute these probabilities for any records of interest (in the training set, the validation set, or for scoring new data). Figure 8.2 shows the estimated probabilities and classifications for a sample of flights in the validation set.

Finally, to evaluate the performance of the naive Bayes classifier for our data, we use the classification matrix, lift charts, and all the measures that were described in Chapter 5. For our example, the classification matrices for the training and validation sets are shown in Figure 8.3. We see that the overall error level is around 18% for both the training and validation data. In comparison, a naive rule that would classify all 880 flights in the validation set as on time

XLMiner: Naive Bayes—Classification of Validation Data

Data range	['Flight Delays.xls']'Data_Partition1'!$C$1340:$H$2219		Back to Navigator

Cutoff Prob.Val. for Success (Updatable)	0.5	(Updating the value here will NOT update value in summary report)

Row ID	Predicted Class	Actual Class	Prob. for 1 (success)	CARRIER	DAY_OF_WEEK	DEP_TIME_BLK	DEST	ORIGIN	Weather
2	0	0	0.160552079	DH	4	1600–1659	JFK	DCA	0
3	0	0	0.197147877	DH	4	1200–1259	LGA	IAD	0
7	0	0	0.248536067	DH	4	1200–1259	JFK	IAD	0
8	0	0	0.263631618	DH	4	1600–1659	JFK	IAD	0
11	0	0	0.281467602	DH	4	2100–2159	LGA	IAD	0
13	0	0	0.025209812	DL	4	0900–0959	LGA	DCA	0
14	0	0	0.048830719	DL	4	1200–1259	LGA	DCA	0
15	0	0	0.07510312	DL	4	1400–1459	LGA	DCA	0
16	0	0	0.088673655	DL	4	1700–1759	LGA	DCA	0
22	0	0	0.113149152	MQ	4	1300–1359	LGA	DCA	0
24	0	0	0.179013723	MQ	4	1500–1559	LGA	DCA	0
25	0	0	0.277045255	MQ	4	1900–1959	LGA	DCA	0
28	0	0	0.018897189	US	4	1100–1159	LGA	DCA	0
33	0	0	0.366861013	RU	4	1400–1459	EWR	BWI	0
34	0	0	0.409792076	RU	4	1700–1759	EWR	BWI	0
40	0	0	0.44593539	DH	4	1700–1759	EWR	IAD	0
42	0	0	0.403842858	DH	4	2100–2159	EWR	IAD	0
46	0	0	0.343153176	RU	4	1900–1959	EWR	DCA	0
47	0	0	0.244033602	RU	4	1400–1459	EWR	DCA	0
50	0	0	0.18094682	RU	4	1600–1659	EWR	DCA	0
57	0	1	0.097462126	DH	5	1000–1059	LGA	IAD	0

FIGURE 8.2 ESTIMATED PROBABILITY OF DELAY FOR A SAMPLE OF THE VALIDATION SET

would have missed the 172 delayed flights, resulting in a 20% error level. In other words, the naive rule is only slightly less accurate. However, examining the lift chart (Figure 8.4) shows the strength of the naive Bayes in capturing the delayed flights well.

8.3 ADVANTAGES AND SHORTCOMINGS OF THE NAIVE BAYES CLASSIFIER

The naive Bayes classifier's beauty is in its simplicity, computational efficiency, good classification performance, and ability to handle categorical variables directly. In fact it often outperforms more sophisticated classifiers even when the underlying assumption of independent predictors is far from true. This advantage is especially pronounced when the number of predictors is very large.

Three main issues should be kept in mind, however. First, the naive Bayes classifier requires a very large number of records to obtain good results. Second, where a predictor category is not present in the training data, naive Bayes assumes that a new record with that category of the predictor has zero probability. This can be a problem if this rare predictor value is important. For example, assume

Training Data Scoring—Summary Report

Cutoff Prob.Val. for Success (Updatable)	0.5

Classification Confusion Matrix

	Predicted Class	
Actual Class	1	0
1	43	213
0	35	1030

Error Report

Class	#Cases	#Errors	% Error
1	256	213	83.20
0	1065	35	3.29
Overall	1321	248	18.77

Validation Data Scoring—Summary Report

Cutoff Prob.Val. for Success (Updatable)	0.5

Classification Confusion Matrix

	Predicted Class	
Actual Class	1	0
1	30	142
0	15	693

Error Report

Class	#Cases	#Errors	% Error
1	172	142	82.56
0	708	15	2.12
Overall	880	157	17.84

FIGURE 8.3 CLASSIFICATION MATRICES FOR FLIGHT DELAYS USING A NAIVE BAYES CLASSIFIER

Lift chart (validation dataset)

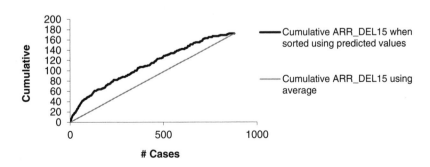

FIGURE 8.4 LIFT CHART OF NAIVE BAYES CLASSIFIER APPLIED TO FLIGHT DELAY DATA.

that the target variable is *bought high-value life insurance* and a predictor category is *owns yacht*. If the training data have no records with *owns yacht* = 1, for any new records where *owns yacht* = 1, naive Bayes will assign a probability of 0 to the target variable *bought high-value life insurance*. With no training records with *owns yacht* = 1, of course, no data mining technique will be able to incorporate this potentially important variable into the classification model—it will be ignored. With naive Bayes, however, the absence of this predictor actively "outvotes" any other information in the record to assign a 0 to the target value (when, in this case, it has a relatively good chance of being a 1). The presence of a large training set (and judicious binning of continuous variables, if required) helps mitigate this effect.

Finally, good performance is obtained when the goal is *classification* or *ranking* of records according to their probability of belonging to a certain class. However, when the goal is actually to *estimate* the probability of class membership (propensity), this method provides very biased results. For this reason the naive Bayes method is rarely used in credit scoring (Larsen, 2005).

SPAM FILTERING

Filtering spam in email has long been a widely familiar application of data mining. Spam filtering, which is based in large part on natural language vocabulary, is a natural fit for a naive Bayesian classifier, which uses exclusively categorical variables. Most spam filters are based on this method, which works as follows:

1. Humans review a large number of emails, classify them as "spam" or "not spam," from these select an equal (also large) number of spam emails and nonspam emails. This is the training data.

2. These emails will contain thousands of words; for each word compute the frequency with which it occurs in the spam dataset, and the frequency with which it occurs in the nonspam dataset. Convert these frequencies into estimated probabilities (i.e., if the word "free" occurs in 500 out of 1000 spam emails, and only 100 out of 1000 nonspam emails, the probability that a spam email will contain the word "free" is 0.5, and the probability that a nonspam email will contain the word "free" is 0.1).

3. If the only word in a new message that needs to be classified as spam or not spam is "free," we would classify the message as spam, since the Bayesian posterior probability is 0.5/(0.5 + 01) or 5/6 that, given the appearance of "free," the message is spam.

4. Of course, we will have many more words to consider. For each such word, the probabilities described in step 2 are calculated, and multiplied together, and formula (8.3) is applied to determine the naive Bayes probability of belonging to the classes. In the simple version, class membership (spam or not spam) is determined by the higher probability.

5. In a more flexible interpretation, the ratio between the "spam" and "not spam" probabilities is treated as a score for which the operator can establish (and change) a cutoff threshold—anything above that level is classified as spam.

6. Users have the option of building a personalized training database by classifying incoming messages as spam or not spam, and adding them to the training database. One person's spam may be another person's substance.

It is clear that, even with the "naive" simplification, this is an enormous computational burden. Spam filters now typically operate at two levels—at servers (intercepting some spam that never makes it to your computer) and on individual computers (where you have the option of reviewing it). Spammers have also found ways to "poison" the vocabulary-based Bayesian approach, by including sequences of randomly selected irrelevant words. Since these words are randomly selected, they are unlikely to be systematically more prevalent in spam than in nonspam, and they dilute the effect of key spam terms such as "Viagra" and "free." For this reason sophisticated spam classifiers also include variables based on elements other than vocabulary, such as the number of links in the message, the vocabulary in the subject line, determination of whether the "From:" email address is the real originator (anti-spoofing), use of HTML and images, and origination at a dynamic or static IP address (the latter are more expensive and cannot be set up quickly).

PROBLEMS

8.1 **Personal Loan Acceptance.** The file `UniversalBank.xls` contains data on 5000 customers of Universal Bank. The data include customer demographic information (age, income, etc.), the customer's relationship with the bank (mortgage, securities account, etc.), and the customer response to the last personal loan campaign (Personal Loan). Among these 5000 customers, only 480 (= 9.6%) accepted the personal loan that was offered to them in the earlier campaign. In this exercise we focus on two predictors: Online (whether or not the customer is an active user of online banking services) and Credit Card (abbreviated CC below) (does the customer hold a credit card issued by the bank), and the outcome Personal Loan (abbreviated Loan below).

Partition the data into training (60%) and validation (40%) sets.

   a. Create a pivot table for the training data with Online as a column variable, CC as a row variable, and Loan as a secondary row variable. The values inside the cells should convey the count (how many records are in that cell).

   b. Consider the task of classifying a customer who owns a bank credit card and is actively using online banking services. Looking at the pivot table, what is the probability that this customer will accept the loan offer? (This is the probability of loan acceptance (Loan = 1) conditional on having a bank credit card (CC = 1) and being an active user of online banking services (Online = 1)).

   c. Create two separate pivot tables for the training data. One will have Loan (rows) as a function of Online (columns) and the other will have Loan (rows) as a function of CC.

   d. Compute the following quantities [$P(A \mid B)$ means "the probability of A given B"]:

      i. $P(CC = 1 \mid Loan = 1)$ (the proportion of credit card holders among the loan acceptors)

      ii. $P(Online = 1 \mid Loan = 1)$

      iii. $P(Loan = 1)$ (the proportion of loan acceptors)

      iv. $P(CC = 1 \mid Loan = 0)$

      v. $P(Online = 1 \mid Loan = 0)$

      vi. $P(Loan = 0)$

   e. Use the quantities computed above to compute the naive Bayes probability $P(Loan = 1 \mid CC = 1, Online = 1)$.

   f. Compare this value with the one obtained from the crossed pivot table in (b). Which is a more accurate estimate?

   g. Which of the entries in this table are needed for computing $P(Loan = 1 \mid CC = 1, Online = 1)$? In XLMiner, run naive Bayes on the data. Examine the "Detailed report on training data," and find the entry that corresponds to $P(Loan = 1 \mid CC = 1, Online = 1)$. Compare this to the number you obtained in (e).

8.2 **Automobile Accidents.** The file `Accidents.xls` contains information on 42,183 actual automobile accidents in 2001 in the United States that involved one of three levels of injury: NO INJURY, INJURY, or FATALITY. For each accident, additional information is recorded, such as day of week, weather conditions, and road type. A firm might be interested in developing a system for quickly classifying the severity of an accident based on initial reports and associated data in the system (some of which rely on GPS-assisted reporting).

Our goal here is to predict whether an accident just reported will involve an injury (MAX_SEV_IR = 1 or 2) or will not (MAX_SEV_IR = 0). For this purpose, create a dummy variable called INJURY that takes the value "yes" if MAX_SEV_IR = 1 or 2, and otherwise "no."

a. Using the information in this dataset, if an accident has just been reported and no further information is available, what should the prediction be? (INJURY = Yes or No?) Why?

b. Select the first 12 records in the dataset and look only at the response (INJURY) and the two predictors WEATHER_R and TRAF_CON_R.

    i. Create a pivot table that examines INJURY as a function of the two predictors for these 12 records. Use all three variables in the pivot table as rows/columns, and use counts for the cells.

    ii. Compute the exact Bayes conditional probabilities of an injury (INJURY = Yes) given the six possible combinations of the predictors.

    iii. Classify the 12 accidents using these probabilities and a cutoff of 0.5.

    iv. Compute manually the naive Bayes conditional probability of an injury given WEATHER_R = 1 and TRAF_CON_R = 1.

    v. Run a naive Bayes classifier on the 12 records and two predictors using XLMiner. Check *detailed report* to obtain probabilities and classifications for all 12 records. Compare this to the exact Bayes classification. Are the resulting classifications equivalent? Is the ranking (= ordering) of observations equivalent?

c. Let us now return to the entire dataset. Partition the data into training/validation sets (use XLMiner's "automatic" option for partitioning percentages).

    i. Assuming that no information or initial reports about the accident itself are available at the time of prediction (only location characteristics, weather conditions, etc.), which predictors can we include in the analysis? (Use the Data_Codes sheet.)

    ii. Run a naive Bayes classifier on the complete training set with the relevant predictors (and INJURY as the response). Note that all predictors are categorical. Show the classification matrix.

    iii. What is the overall error for the validation set?

    iv. What is the percent improvement relative to the naive rule (using the validation set)?

    v. Examine the conditional probabilities output. Why do we get a probability of zero for $P(\text{INJURY} = \text{No} \mid \text{SPD\_LIM} = 5)$?

## CHAPTER 9

Classification and Regression Trees

This chapter describes a flexible data-driven method that can be used for both classification (called *classification tree*) and prediction (called *regression tree*). Among the data-driven methods, trees are the most transparent and easy to interpret. Trees are based on separating observations into subgroups by creating splits on predictors. These splits create logical rules that are transparent and easily understandable, such as "IF Age < 55 AND Education > 12 THEN class = 1." The resulting subgroups should be more homogeneous in terms of the outcome variable, thereby creating useful prediction or classification rules. We discuss the two key ideas underlying trees: *recursive partitioning* (for constructing the tree) and *pruning* (for cutting the tree back). In the context of tree construction, we also describe a few metrics of homogeneity that are popular in tree algorithms, for determining the homogeneity of the resulting subgroups of observations. We explain that pruning is a useful strategy for avoiding overfitting and show how it is done. We also describe alternative strategies for avoiding overfitting. As with other data-driven methods, trees require large amounts of data. However, once constructed, they are computationally cheap to deploy even on large samples. They also have other advantages such as being highly automated, robust to outliers, and able to handle missing values. In addition to prediction and classification, we describe how trees can be used for dimension reduction. Finally, we introduce *random forests* and *boosted trees*, which combine results from multiple trees to improve predictive power.

Data Mining for Business Analytics: Concepts, Techniques, and Applications with XLMiner®, Third Edition.
Galit Shmueli, Peter C. Bruce, and Nitin R. Patel.
© 2016 John Wiley & Sons Inc. Published 2016 by John Wiley & Sons, Inc.

9.1 INTRODUCTION

If one had to choose a classification technique that performs well across a wide range of situations without requiring much effort from the analyst while being readily understandable by the consumer of the analysis, a strong contender would be the tree methodology developed by Breiman et al. (1984). We discuss this classification procedure first, then in later sections we show how the procedure can be extended to prediction of a numerical outcome. The program that Breiman et al. created to implement these procedures was called CART (Classification And Regression Trees). A related procedure is called C4.5.

What is a classification tree? Figure 9.1 shows a tree for classifying bank customers who receive a loan offer as either acceptors or nonacceptors, based on information such as their income, education level, and average credit card expenditure. One of the reasons that tree classifiers are very popular is that they provide easily understandable classification rules (at least if the trees are not too

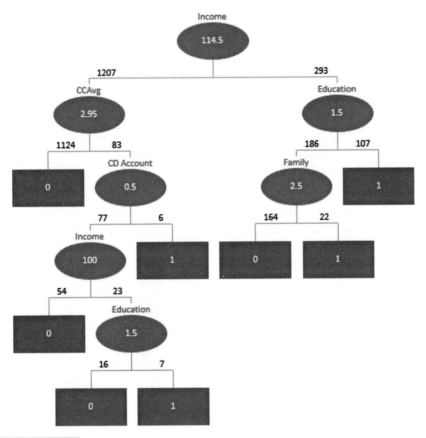

FIGURE 9.1   BEST-PRUNED TREE OBTAINED BY FITTING A FULL TREE TO THE TRAINING DATA AND PRUNING IT USING THE VALIDATION DATA.

large). Consider the tree in the example. The rectangle *terminal nodes* are marked with 0 or 1 corresponding to a nonacceptor (0) or acceptor (1). The values in the circle nodes give the splitting value on a predictor. This tree can easily be translated into a set of rules for classifying a bank customer. For example, the bottom left rectangle node under the "Family" circle in this tree gives us the following rule:

$$\text{IF}(Income \geq 114.5) \text{ AND } (Education < 1.5) \text{ AND } (Family < 2.5),$$
$$\text{THEN } Class = 0 \text{ (nonacceptor)}.$$

In the following we show how trees are constructed and evaluated.

9.2 CLASSIFICATION TREES

Two key ideas underlie classification trees. The first is the idea of *recursive partitioning* of the space of the predictor variables. The second is the idea of *pruning* using validation data. In the next few sections we describe recursive partitioning and in subsequent sections explain the pruning methodology.

Recursive Partitioning

Let us denote the outcome (response) variable by y and the input (predictor) variables by $x_1, x_2, x_3, \ldots, x_p$. In classification, the outcome variable will be a categorical variable. Recursive partitioning divides up the p-dimensional space of the X variables into non-overlapping multidimensional rectangles. The X variables here are considered to be continuous, binary, or ordinal. This division is accomplished recursively (i.e., operating on the results of prior divisions). First, one of the predictor variables is selected, say X_i, and a value of X_i, say s_i, is chosen to split the p-dimensional space into two parts: one part that contains all the points with $X_i < s_i$ and the other with all the points with $X_i \geq s_i$. Then one of these two parts is divided in a similar manner by again choosing a predictor variable (it could be X_i or another variable) and a split value for that variable. This results in three (multidimensional) rectangular regions. This process is continued so that we get smaller and smaller rectangular regions. The idea is to divide the entire X-space up into rectangles such that each rectangle is as homogeneous or "pure" as possible. By *pure*, we mean containing records that belong to just one class. (Of course, this is not always possible, as there may be records that belong to different classes but have exactly the same values for every one of the predictor variables.)

Let us illustrate recursive partitioning with an example.

Example 1: Riding Mowers

We again use the riding-mower example presented in Chapter 3. A riding-mower manufacturer would like to find a way of classifying families in a city into those likely to purchase a riding mower and those not likely to buy one. A pilot random sample of 12 owners and 12 nonowners in the city is undertaken. The data are shown and plotted in Table 9.1 and Figure 9.2.

If we apply the classification tree procedure to these data, the procedure will choose *Income* for the first split with a splitting value of 59.7. The (X_1, X_2)-space is now divided into two rectangles, one with Income $<$ 59.7 and the other with Income $\geq$ 59.7. This is illustrated in Figure 9.3.

Notice how the split has created two rectangles, each of which is much more homogeneous than the rectangle before the split. The left rectangle contains points that are mostly nonowners (seven nonowners and one owner) and the right rectangle contains mostly owners (11 owners and five nonowners).

How was this particular split selected? The algorithm examined each predictor variable (in this case, Income and Lot Size) and all possible split values for

TABLE 9.1	LOT SIZE, INCOME, AND OWNERSHIP OF A RIDING MOWER FOR 24 HOUSEHOLDS		
Household Number	Income ($000s)	Lot Size (000s ft^2)	Ownership of Riding Mower
1	60.0	18.4	Owner
2	85.5	16.8	Owner
3	64.8	21.6	Owner
4	61.5	20.8	Owner
5	87.0	23.6	Owner
6	110.1	19.2	Owner
7	108.0	17.6	Owner
8	82.8	22.4	Owner
9	69.0	20.0	Owner
10	93.0	20.8	Owner
11	51.0	22.0	Owner
12	81.0	20.0	Owner
13	75.0	19.6	Nonowner
14	52.8	20.8	Nonowner
15	64.8	17.2	Nonowner
16	43.2	20.4	Nonowner
17	84.0	17.6	Nonowner
18	49.2	17.6	Nonowner
19	59.4	16.0	Nonowner
20	66.0	18.4	Nonowner
21	47.4	16.4	Nonowner
22	33.0	18.8	Nonowner
23	51.0	14.0	Nonowner
24	63.0	14.8	Nonowner

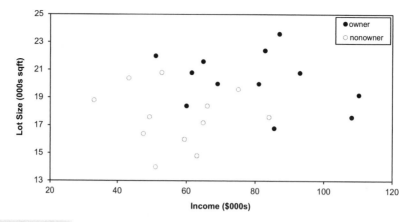

FIGURE 9.2 SCATTER PLOT OF LOT SIZE VS. INCOME FOR 24 OWNERS AND NONOWNERS OF RIDING MOWERS.

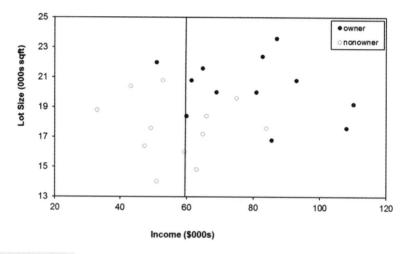

FIGURE 9.3 SPLITTING THE 24 OBSERVATIONS BY INCOME VALUE OF 59.7.

each variable to find the best split. What are the possible split values for a variable? They are simply the midpoints between pairs of consecutive values for the variable. The possible split points for Income are $\{38.1, 45.3, 50.1, \dots, 109.5\}$ and those for Lot Size are $\{14.4, 15.4, 16.2, \dots, 23\}$. These split points are ranked according to how much they reduce impurity (heterogeneity) in the resulting rectangle. A pure rectangle is one that is composed of a single class (e.g., owners). The reduction in impurity is defined as overall impurity before the split minus the sum of the impurities for the two rectangles that result from a split.

Categorical Predictors The previous description used numerical predictors; however, categorical predictors can also be used in the recursive partitioning context. To handle categorical predictors, the split choices for a categorical predictor are all ways in which the set of categories can be divided into two subsets. For example, a categorical variable with four categories, say {a, b, c, d}, can be split in seven ways into two subsets: {a} and {b, c, d}; {b} and {a, c, d}; {c} and {a, b, d}; {d} and {a, b, c}; {a, b} and {c, d}; {a, c} and {b, d}; and finally {a, d} and {b, c}. When the number of categories is large, the number of splits becomes very large. XLMiner supports only binary categorical variables (coded as 0/1). If you have a categorical predictor that takes more than two values, you will need to replace the variable with several dummy variables, each of which is binary in a manner that is identical to the use of dummy variables in regression.[1]

Measures of Impurity

There are a number of ways to measure impurity. The two most popular measures are the *Gini index* and an *entropy measure*. We describe both next. Denote the m classes of the response variable by $k = 1, 2, \ldots, m$.

The Gini impurity index for a rectangle A is defined by

$$I(A) = 1 - \sum_{k=1}^{m} p_k^2,$$

where p_k is the proportion of observations in rectangle A that belong to class k. This measure takes values between 0 (when all the observations belong to the same class) and $(m-1)/m$ (when all m classes are equally represented). Figure 9.4 shows the values of the Gini index for a two-class case as a function of p_k. It can be seen that the impurity measure is at its peak when $p_k = 0.5$ (i.e., when the rectangle contains 50% of each of the two classes).

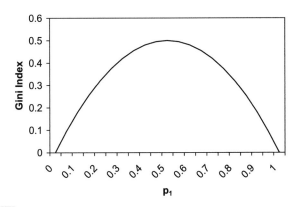

| FIGURE 9.4 | VALUES OF THE GINI INDEX FOR A TWO-CLASS CASE AS A FUNCTION OF THE PROPORTION OF OBSERVATIONS IN CLASS 1 (p_1). |

---

[1] This is a difference between CART and C4.5; the former performs only binary splits, leading to binary trees, whereas the latter performs splits that are as large as the number of categories, leading to "bushlike" structures.

A second impurity measure is the entropy measure. The entropy for a rectangle A is defined by

$$\text{entropy}(A) = -\sum_{k=1}^{m} p_k \log_2(p_k)$$

[to compute $\log_2(x)$ in Excel, use the function $= log(x, 2)$]. This measure ranges between 0 (most pure, all observations belong to the same class) and $\log_2(m)$ (when all m classes are represented equally). In the two-class case, the entropy measure is maximized (like the Gini index) at $p_k = 0.5$.

Let us compute the impurity in the riding-mower example before and after the first split (using Income with the value of 59.7). The unsplit dataset contains 12 owners and 12 nonowners. This is a two-class case with an equal number of observations from each class. Both impurity measures are therefore at their maximum value: Gini = 0.5 and entropy = $\log_2(2) = 1$. After the split, the left rectangle contains seven nonowners and one nonowner. The impurity measures for this rectangle are

$$\text{Gini\_left} = 1 - (7/8)^2 - (1/8)^2 = 0.219.$$
$$\text{entropy\_left} = -(7/8)\log_2(7/8) - (1/8)\log_2(1/8) = 0.544.$$

The right rectangle contains 11 owners and five nonowners. The impurity measures of the right rectangle are therefore

$$\text{Gini\_right} = 1 - (11/16)^2 - (5/16)^2 = 0.430.$$
$$\text{entropy\_right} = -(11/16)\log_2(11/16) - (5/16)\log_2(5/16) = 0.896.$$

The combined impurity of the two rectangles that were created by the split is a weighted average of the two impurity measures, weighted by the number of observations in each:

$$\text{Gini} = (8/24)(0.219) + (16/24)(0.430) = 0.359.$$
$$\text{entropy} = (8/24)(0.544) + (16/24)(0.896) = 0.779.$$

Thus the Gini impurity index decreased from 0.5 before the split to 0.359 after the split. Similarly, the entropy impurity measure decreased from 1 before the split to 0.779 after the split.

By comparing the reduction in impurity across all possible splits in all possible predictors, the next split is chosen. If we continue splitting the mower data, the next split is on the Lot Size variable at the value 21.4. Figure 9.5 shows that once again the tree procedure has astutely chosen to split a rectangle to increase the purity of the resulting rectangles. The lower left rectangle, which contains data points with Income < 59.7 and Lot Size < 21.4, has all points that are nonowners, whereas the upper left rectangle, which contains data points with Income < 59.7 and Lot Size ≥ 21.4, consists exclusively of a single owner.

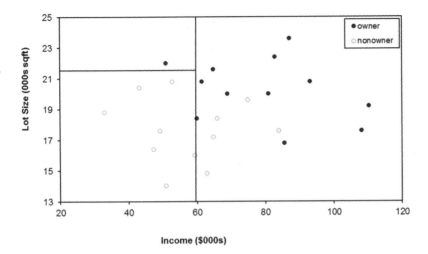

FIGURE 9.5   SPLITTING THE 24 OBSERVATIONS FIRST BY INCOME VALUE OF 59.7 AND THEN LOT SIZE VALUE OF 21.4.

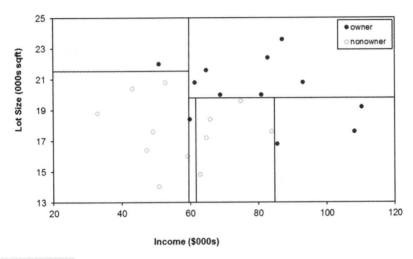

FIGURE 9.6   FINAL STAGE OF RECURSIVE PARTITIONING; EACH RECTANGLE CONSISTING OF A SINGLE CLASS (OWNERS OR NONOWNERS).

In other words, the two left rectangles are now "pure." We can see how the recursive partitioning is refining the set of constituent rectangles to become purer as the algorithm proceeds. The final stage of the recursive partitioning is shown in Figure 9.6. Notice that each rectangle is now pure: it contains data points from just one of the two classes.

The reason the method is called a *classification tree algorithm* is that each split can be depicted as a split of a node into two successor nodes. The first split is shown as a branching of the root node of a tree in Figure 9.7. The tree representing the first three splits is shown in Figure 9.8. The full-grown tree is shown in Figure 9.9.

FIGURE 9.7 TREE REPRESENTATION OF FIRST SPLIT (CORRESPONDS TO FIGURE 9.3).

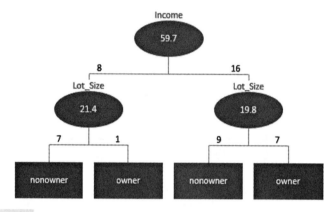

FIGURE 9.8 TREE REPRESENTATION OF FIRST THREE SPLITS.

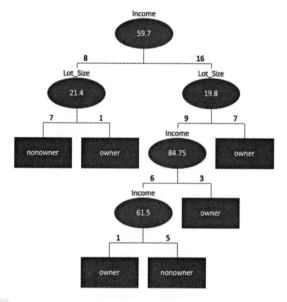

FIGURE 9.9 TREE REPRESENTATION AFTER ALL SPLITS (CORRESPONDS TO FIGURE 9.6). THIS IS THE FULL-GROWN TREE.

Tree Structure

We represent the nodes that have successors by circles. The numbers inside the circles are the splitting values and the name of the variable chosen for splitting at that node is shown above the node. The numbers on the left fork at a decision node are the number of records in the decision node that had values less than the splitting value, and the numbers on the right fork show the number that had a greater or equal value. These are called *decision nodes* because if we were to use a tree to classify a new observation for which we knew only the values of the predictor variables, we would "drop" the observation down the tree so that at each decision node the appropriate branch is taken until we get to a node that has no successors. Such terminal nodes are called the *leaves* of the tree. Each terminal node is depicted with a rectangle rather than a circle, and corresponds to one of the final rectangles into which the predictor space is partitioned.

Classifying a New Observation

To classify a new observation, it is "dropped" down the tree. When it has dropped all the way down to a terminal node, we can assign its class simply by taking a "vote" of all the training data that belonged to the terminal node when the tree was grown. The class with the highest vote is assigned to the new observation. For instance, a new observation reaching the rightmost terminal node in Figure 9.9, which has a majority of observations that belong to the owner class, would be classified as "owner." Alternatively, if a single class is of interest, the algorithm counts the number of "votes" for this class, converts it to a proportion (propensity), then compares it to a user-specified cutoff value. See Chapter 5 for further discussion of the use of a cutoff value in classification, for cases where a single class is of interest.

In a binary classification situation (typically, with a success class that is relatively rare and of particular interest), we can also establish a lower cutoff to better capture those rare successes (at the cost of lumping in more failures as successes). With a lower cutoff, the votes for the *success* class only need attain that lower cutoff level for the entire terminal node to be classified as a *success*. The cutoff therefore determines the proportion of votes needed for determining the terminal node class. It is useful to note that the type of trees grown by CART (called *binary trees*) have the property that the number of terminal nodes is exactly one more than the number of decision nodes.

9.3 EVALUATING THE PERFORMANCE OF A CLASSIFICATION TREE

To assess the accuracy of a tree in classifying new cases, we use the tools and criteria discussed in Chapter 5. We start by partitioning the data into training and validation sets. The training set is used to grow the tree, and the validation set is used to assess its performance. In the next section we discuss an important step in constructing trees that involves using the validation data. In that case, a third set of test data is preferable for assessing the accuracy of the final tree.

Each observation in the validation (or test) data is "dropped down" the tree and classified according to the terminal node it reaches. These predicted classes can then be compared to the actual memberships via a classification matrix. When a particular class is of interest, a lift chart is useful for assessing the model's ability to capture those members. We use the following example to illustrate this.

Example 2: Acceptance of Personal Loan

Universal Bank is a relatively young bank that is growing rapidly in terms of overall customer acquisition. The majority of these customers are liability customers with varying sizes of relationship with the bank. The customer base of asset customers is quite small, and the bank is interested in growing this base rapidly to bring in more loan business. In particular, it wants to explore ways of converting its liability (deposit) customers to personal loan customers.

A campaign the bank ran for liability customers showed a healthy conversion rate of over 9% successes. This has encouraged the retail marketing department to devise smarter campaigns with better target marketing. The goal of our analysis is to model the previous campaign's customer behavior to find what combination of factors make a customer more likely to accept a personal loan. This will serve as the basis for the design of a new campaign.

The bank's dataset includes data on 5000 customers. The data include customer demographic information (age, income, etc.), customer response to the last personal loan campaign (*Personal Loan*), and the customer's relationship with the bank (mortgage, securities account, etc.). Among these 5000 customers, only 480 (= 9.6%) accepted the personal loan that was offered to them in the earlier campaign. Table 9.2 contains a sample of the bank's customer database for 20 customers, to illustrate the structure of the data.

After randomly partitioning the data into training (2500 observations), validation (1500 observations), and test (1000 observations) sets, we use the training data to construct a full-grown tree.[2] The first four levels of the tree are shown in Figure 9.10. Each of the dark rectangle nodes at the bottom level indicate that they further branch into subtrees.

---

[2] In XLMiner, we set the maximum number of tree levels to 100 and the minimum number of records in a terminal node to 1.

TABLE 9.2   SAMPLE OF DATA FOR 20 CUSTOMERS OF UNIVERSAL BANK

ID	Age	Professional Experience	Income	Family Size	CC Avg	Education	Mortgage	Personal Loan	Securities Account	CD Account	Online Banking	Credit Card
1	25	1	49	4	1.60	UG	0	No	Yes	No	No	No
2	45	19	34	3	1.50	UG	0	No	Yes	No	No	No
3	39	15	11	1	1.00	UG	0	No	No	No	No	No
4	35	9	100	1	2.70	Grad	0	No	No	No	No	No
5	35	8	45	4	1.00	Grad	0	No	No	No	No	Yes
6	37	13	29	4	0.40	Grad	155	No	No	No	Yes	No
7	53	27	72	2	1.50	Grad	0	No	No	No	Yes	No
8	50	24	22	1	0.30	Prof	0	No	No	No	No	Yes
9	35	10	81	3	0.60	Grad	104	No	No	No	Yes	No
10	34	9	180	1	8.90	Prof	0	Yes	No	No	No	No
11	65	39	105	4	2.40	Prof	0	No	No	No	Yes	No
12	29	5	45	3	0.10	Grad	0	No	Yes	No	No	No
13	48	23	114	2	3.80	Prof	0	No	No	No	Yes	No
14	59	32	40	4	2.50	Grad	0	No	Yes	No	No	No
15	67	41	112	1	2.00	UG	0	No	No	No	Yes	Yes
16	60	30	22	1	1.50	Prof	0	Yes	No	No	No	No
17	38	14	130	4	4.70	Prof	134	No	No	No	No	No
18	42	18	81	4	2.40	UG	0	No	No	No	No	No
19	46	21	193	2	8.10	Prof	0	Yes	No	No	No	No
20	55	28	21	1	0.50	Grad	0	No	Yes	No	No	Yes

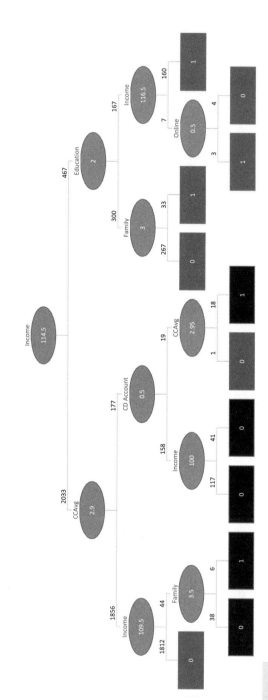

FIGURE 9.10 FIRST FOUR LEVELS OF THE FULL-GROWN TREE FOR THE LOAN ACCEPTANCE DATA USING THE TRAINING SET (2500 OBSERVATIONS). DARK RECTANGLE (TERMINAL) NODES INDICATE AN UNSEEN SUBTREE.

Even with just four levels, it is difficult to see the complete picture. A look at the top tree node or the first row of the table reveals that the first predictor that is chosen to split the data is *Income*, with a value of 114.5 ($000s).

Since the full-grown tree leads to completely pure terminal leaves, it is 100% accurate in classifying the training data. This can be seen in Figure 9.11. In contrast, the confusion matrix for the validation and test data[3] (which were not used to construct the full-grown tree) show lower classification accuracy. The main reason is that the full-grown tree overfits the training data (to complete accuracy!). This motivates the next section, where we describe ways to avoid overfitting by either stopping the growth of the tree before it is fully grown or by pruning the full-grown tree.

9.4 AVOIDING OVERFITTING

As the last example illustrated, using a full-grown tree (based on the training data) leads to complete overfitting of the data. As discussed in Chapter 5, overfitting will lead to poor performance on new data. If we look at the overall error at the various levels of the tree, it is expected to decrease as the number of levels grows until the point of overfitting. Of course, for the training data the overall error decreases more and more until it is zero at the maximum level of the tree. However, for new data, the overall error is expected to decrease until the point where the tree models the relationship between class and the predictors. After that, the tree starts to model the noise in the training set, and we expect the overall error for the validation set to start increasing. This is depicted in Figure 9.12. One intuitive reason for the overfitting at the high levels of the tree is that these splits are based on very small numbers of observations. In such cases, class difference is likely to be attributed to noise rather than predictor information.

Two ways to try and avoid exceeding this level, thereby limiting overfitting, are by setting rules to stop tree growth, or alternatively, by pruning the full-grown tree back to a level where it does not overfit. These solutions are discussed next.

Stopping Tree Growth: CHAID

One can think of different criteria for stopping the tree growth before it starts overfitting the data. Examples are tree depth (i.e., number of splits), minimum number of records in a terminal node, and minimum reduction in impurity. The problem is that it is not simple to determine what is a good stopping point using such rules.

---

[3] The confusion matrices for the validation and test sets were obtained manually: For each set, we first scored the data using the Score function. We then used Excel's Pivot Table to get a confusion matrix.

Training Data Scoring—Summary Report (Using Full-Grown Tree)

Cutoff probability value for success (UPDATABLE)	0.5

Confusion Matrix

Actual Class	Predicted Class	
	1	0
1	244	0
0	0	2256

Error Report

Class	#Cases	#Errors	% Error
1	244	0	0
0	2256	0	0
Overall	2500	0	0

Validation Data Scoring (Using Full-Grown Tree)

Confusion Matrix

Actual Class	Predicted Class	
	1	0
1	134	21
0	10	1335

Error Report

Class	#Cases	#Errors	% Error
1	155	21	13.55
0	1345	10	0.74
Overall	1500	31	2.07

Test Data Scoring (Using Full-Grown Tree)

Confusion Matrix

Actual Class	Predicted Class	
	1	0
1	76	5
0	13	906

Error Report

Class	#Cases	#Errors	% Error
1	81	5	6.17
0	919	13	1.41
Overall	1000	18	1.80

FIGURE 9.11   LOAN ACCEPTANCE DATA: CLASSIFICATION MATRIX AND ERROR RATES FOR THE TRAINING, VALIDATION, AND TEST DATA USING THE FULL TREE (RESULTS FOR VALIDATION AND TEST SETS WERE OBTAINED MANUALLY).

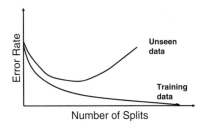

FIGURE 9.12	**ERROR RATE AS A FUNCTION OF THE NUMBER OF SPLITS FOR TRAINING VS. VALIDATION DATA: OVERFITTING.**

Previous methods developed were based on the idea of recursive partitioning, using rules to prevent the tree from growing excessively and overfitting the training data. One popular method called *CHAID* (chi-squared automatic interaction detection) is a recursive partitioning method that predates classification and regression tree (CART) procedures by several years and is widely used in database marketing applications to this day. It uses a well-known statistical test (the chi-square test for independence) to assess whether splitting a node improves the purity by a statistically significant amount. In particular, at each node we split on the predictor that has the strongest association with the response variable. The strength of association is measured by the p-value of a chi-squared test of independence. If for the best predictor the test does not show a significant improvement, the split is not carried out, and the tree is terminated. This method is more suitable for categorical predictors, but it can be adapted to continuous predictors by binning the continuous values into categorical bins.

Pruning the Tree

An alternative solution that has proved to be more successful than stopping tree growth is pruning the full-grown tree. This is the basis of methods such as CART (developed by Breiman et al., implemented in multiple data mining software packages such as SAS Enterprise Miner, CART, MARS, and in XLMiner) and C4.5 (developed by Quinlan and implemented in packages such as IBM SPSS Modeler). In C4.5 the training data are used both for growing and pruning the tree. In CART the innovation is to use the validation data to prune back the tree that is grown from training data. CART and CART-like procedures use validation data to prune back the tree that has deliberately been overgrown using the training data. This approach is also used by XLMiner.

The idea behind pruning is to recognize that a very large tree is likely to be overfitting the training data, and that the weakest branches, which hardly reduce the error rate, should be removed. In the mower example the last few splits resulted in rectangles with very few points (four rectangles in the full tree had just one point). We can see intuitively that these last splits are likely just capturing noise in the training set rather than reflecting patterns that would occur in future data, such as the validation data. Pruning consists of successively selecting a decision node and re-designating it as a terminal node [lopping off the branches extending beyond that decision node (its *subtree*) and thereby reducing the size of the tree]. The pruning process trades off misclassification error in the validation dataset against the number of decision nodes in the pruned tree to arrive at a tree that captures the patterns—but not the noise—in the training data. Returning to Figure 9.12, we would like to find the point where the curve for the unseen data begins to increase.

To find this point, the CART algorithm uses a criterion called the *cost complexity* of a tree to generate a sequence of trees that are successively smaller to the point of having a tree with just the root node. (What is the classification rule for a tree with just one node?) This means that the first step is to find the best subtree of each size $(1, 2, 3, \ldots)$. Then we choose among these trees the tree that minimizes the misclassification error rate of the validation set. This is called the *minimum error tree*. Let us look at this process in a little more detail next.

Constructing the best tree of each size is based on the CC criterion, which is equal to the misclassification error of a tree (based on the training data) plus a penalty factor for the size of the tree. For a tree T that has $L(T)$ terminal nodes, the cost complexity can be written as

$$CC(T) = \text{err}(T) + \alpha L(T),$$

where $\text{err}(T)$ is the fraction of training data observations that are misclassified by tree T and α is a penalty factor for tree size. When $\alpha = 0$, there is no penalty for having too many nodes in a tree, and the best tree using the cost complexity criterion is the full-grown unpruned tree. When we increase α to a very large value the penalty cost component swamps the misclassification error component of the cost complexity criterion function, and the best tree is simply the tree with the fewest terminal nodes: namely the tree with simply one node. The idea is therefore to start with the full-grown tree and then increase the penalty factor α gradually until the cost complexity of the full tree exceeds that of a subtree. Then the same procedure is repeated using the subtree. Continuing in this manner, we generate a succession of trees with a diminishing number of nodes all the way to a trivial tree consisting of just one terminal node.

From this sequence of trees it seems natural to choose the one that gave the lowest misclassification error on the validation dataset. We call this the *minimum error tree*. To illustrate this, Figure 9.13 shows the error rate for both the training and validation data as a function of the tree size. It can be seen that the training set error steadily decreases as the tree grows, with a noticeable drop in error rate between two and three nodes. The validation set error rate, however, reaches

#Decision Nodes	% Training Error	% Validation Error
44	0.00	2.07
43	0.04	2.07
42	0.08	2.07
41	0.08	2.07
40	0.08	2.07
39	0.12	2.07
38	0.16	2.00
37	0.24	2.00
36	0.24	2.07
35	0.32	2.07
34	0.36	2.07
33	0.36	2.07
32	0.36	2.07
31	0.4	2.07
30	0.44	2.07
29	0.48	2.07
28	0.48	2.07
27	0.48	2.07
26	0.52	2.07
25	0.52	2.07
24	0.56	2.07
23	0.6	2.07
22	0.68	2.13
21	0.76	2.13
20	0.88	2.13
19	0.88	2.13
18	1	2.13
17	1.04	2.13
16	1.08	1.93
15	1.08	1.93
14	1.36	1.93
13	1.36	1.93
12	1.44	1.93
11	1.44	1.93
10	1.56	1.73
9	1.56	1.53
8	1.56	1.53
7	1.56	1.53 <-- Best-Pruned & Min. Error Tree	Std. Error	0.003173
6	1.6	1.87
5	2.92	1.87
4	3.4	2.00
3	3.4	2.00
2	9.76	3.47
1	9.76	10.33
0	9.76	10.33

FIGURE 9.13 ERROR RATE AS A FUNCTION OF THE NUMBER OF SPLITS FOR TRAINING VS. VALIDATION DATA FOR THE LOAN EXAMPLE.

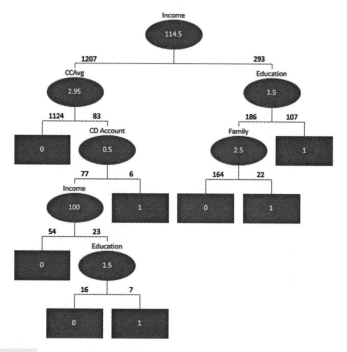

FIGURE 9.14   BEST-PRUNED TREE OBTAINED BY FITTING A FULL TREE TO THE TRAINING DATA, PRUNING IT USING THE VALIDATION DATA, AND CHOOSING THE SMALLEST TREE WITHIN ONE STANDARD ERROR OF THE MINIMUM ERROR TREE.

a minimum at seven nodes and then starts to increase as the tree grows. At this point the tree is pruned, and we obtain the *minimum error tree*.

A further enhancement is to incorporate the sampling error that might cause this minimum to vary if we had a different sample. The enhancement uses the estimated standard error of the error rate to prune the tree even further; we add one standard error to the minimum validation error. The *best-pruned tree* is the smallest tree in the pruning sequence with error within one standard error of the minimum error tree. The best pruned tree for the loan acceptance example is shown in Figure 9.14. In this case it coincides with the minimum error tree. It is obtained by adding one standard error[4] (0.3173%) to the validation error of the minimum error tree (1.53%). The smallest tree with validation error below the total of 1.8473% is the best-pruned tree (see Figure 9.13).

Returning to the loan acceptance example, we expect that the classification accuracy of the validation set using the pruned tree would be higher than using the full-grown tree (compare Figure 9.11 with Figure 9.15). However, the performance of the pruned tree on the validation data is not fully reflective of

---

[4] Note: The version of XLMiner that produced this figure expresses error as a percent, but standard error as a proportion, so a conversion of one of them is required for comparability.

Training Data Scoring—Summary Report (Using Full-Grown Tree)

Cutoff probability value for success (UPDATABLE)	0.5

Confusion Matrix

Actual Class	Predicted Class	
	1	0
1	244	0
0	0	2256

Error Report

Class	#Cases	#Errors	% Error
1	244	0	0
0	2256	0	0
Overall	2500	0	0

Validation Data Scoring—Summary Report (Using Best-Pruned Tree)

Cutoff probability value for success (UPDATABLE)	0.5

Confusion Matrix

Actual Class	Predicted Class	
	1	0
1	137	18
0	5	1340

Error Report

Class	#Cases	#Errors	% Error
1	155	18	11.6129
0	1345	5	0.371747
Overall	1500	23	1.533333

Test Data Scoring—Summary Report (Using Best-Pruned Tree)

Cutoff probability value for success (UPDATABLE)	0.5

Confusion Matrix

Actual Class	Predicted Class	
	1	0
1	72	9
0	3	916

Error Report

Class	#Cases	#Errors	% Error
1	81	9	11.11111
0	919	3	0.326442
Overall	1000	12	1.2

FIGURE 9.15   LOAN ACCEPTANCE DATA: CONFUSION MATRIX AND ERROR RATES FOR THE TRAINING, VALIDATION, AND TEST DATA BASED ON THE PRUNED TREE.

the performance on completely new data, since the validation data were actually used for the pruning. This is a situation where it is particularly useful to evaluate the performance of the chosen model—whatever it may be—on a third set of data: the test set, which has not been used at all. In our example, the pruned tree applied to the test data yields an overall error rate of 1.2% (compared to 1.53% for the validation data). This is a highly accurate model, yielding only 12 errors in the test set, and thus small differences in error rates between test and validation data are in the range of chance fluctuation. With much larger datasets and less accurate classifiers, the typical tendency will be for the test data to show higher error rates than the validation data, since the latter are in effect part of the model-building process.

9.5 CLASSIFICATION RULES FROM TREES

As described in Section 9.1, classification trees provide easily understandable *classification rules* (if the trees are not too large). Each terminal node is equivalent to a classification rule. Returning to the example, the left terminal node in the best pruned tree (Figure 9.14) gives us the rule

IF ($Income \geq 114.5$) AND ($Education < 1.5$) AND ($Family < 2.5$), THEN $Class = 0$.

However, in many cases the number of rules can be reduced by removing redundancies. For example, consider the rule from the second-from-bottom left-most terminal node in Figure 9.10:

$$\text{IF } (Income < 114.5) \text{ AND } (CCAvg \geq 2.95)$$
$$\text{AND } (CD\ Account < 0.5) \text{ AND } (Income < 109.5),$$
$$\text{THEN } Class = 0.$$

This rule can be simplified to

$$\text{IF } (Income < 109.5) \text{ AND } (CCAvg \geq 2.9) \text{ AND } (CD\ Account < 0.5),$$
$$\text{THEN } Class = 0.$$

This transparency in the process and understandability of the algorithm that leads to classifying a record as belonging to a certain class is very advantageous in settings where the final classification is not solely of interest. Berry and Linoff (2000) give the example of health insurance underwriting, where the insurer is required to show that coverage denial is not based on discrimination. By showing rules that led to denial (e.g., income < $20K AND low credit history), the company can avoid law suits. Compared to the output of other classifiers, such as discriminant functions, tree-based classification rules are easily explained to managers and operating staff. Their logic is certainly far more transparent than that of weights in neural networks!

9.6 CLASSIFICATION TREES FOR MORE THAN TWO CLASSES

Classification trees can be used with an outcome that has more than two classes. In terms of measuring impurity, the two measures that were presented earlier (the Gini impurity index and the entropy measure) were defined for m classes and hence can be used for any number of classes. The tree itself would have the same structure, except that its terminal nodes would take one of the m-class labels.

9.7 REGRESSION TREES

The tree method can also be used for numerical response variables. Regression trees for prediction operate in much the same fashion as classification trees. The output variable, Y, is a numerical variable in this case, but both the principle and the procedure are the same: Many splits are attempted, and for each, we measure "impurity" in each branch of the resulting tree. The tree procedure then selects the split that minimizes the sum of such measures. To illustrate a regression tree, consider the example of predicting prices of Toyota Corolla automobiles (from Chapter 5). The dataset includes information on 1000 sold Toyota Corolla cars (we use the first 1000 cars from the dataset `ToyotoCorolla.xls`). The goal is to find a predictive model of price as a function of 10 predictors (mileage, horsepower, number of doors, etc.). A regression tree for these data was built using a training set of 600. The best-pruned tree is shown in Figure 9.16.

We see that from the 12 input variables (including dummies), only six predictors show up as useful for predicting price: the age of the car, its weight, mileage, quarterly tax, horsepower, and CC. Three details differ between regression trees and classification trees: prediction, impurity measures, and evaluating performance. We describe these next.

Prediction

Predicting the value of the response Y for an observation is performed in a fashion similar to the classification case: The predictor information is used for "dropping" the observation down the tree until reaching a terminal node. For instance, to predict the price of a Toyota Corolla with Age $= 60$, Mileage $= 160,000$, Horse_Power $= 100$, Weight $= 1200$, Quarterly_Tax $= 50$, and CC $= 1000$, we drop it down the tree and reach the node that has the value $7823.53. This is the price prediction for this car according to the tree. In classification trees the value of the terminal node (which is one of the categories) is determined by the "voting" of the training observations that were in that terminal node. In regression trees the value of the terminal node is determined by the average

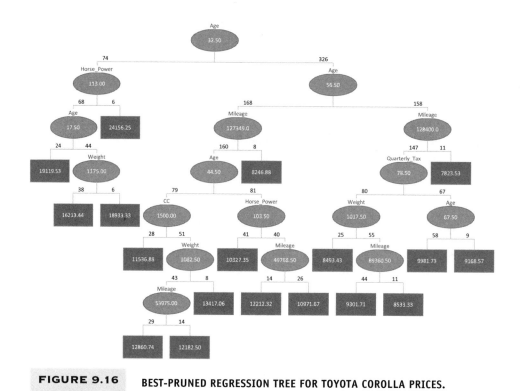

FIGURE 9.16   **BEST-PRUNED REGRESSION TREE FOR TOYOTA COROLLA PRICES.**

output value of the training observations in that terminal node. In the example above, the value \$7823.53 is the average of the 11 cars in the training set that fall in the category of Age $\geq$ 56.5 AND Mileage $\geq$ 128,400.

Measuring Impurity

We described two types of impurity measures for nodes in classification trees: the Gini index and the entropy-based measure. In both cases the index is a function of the ratio between the categories of the observations in that node. In regression trees a typical impurity measure is the sum of the squared deviations from the mean of the terminal node. This is equivalent to the squared errors, since the mean of the terminal node is exactly the prediction. In the example above, the impurity of the node with the value \$7823.53 is computed by subtracting 7823.53 from the price of each of the 11 cars in the training set that fell in that terminal node, then squaring these deviations and summing them up. The lowest impurity possible is zero, when all values in the node are equal.

Evaluating Performance

As stated above, predictions are obtained by averaging the values of the responses in the nodes. We therefore have the usual definition of predictions and errors. The predictive performance of regression trees can be measured in the same way that other predictive methods are evaluated, using summary measures such as RMSE and charts such as lift charts.

9.8   ADVANTAGES, WEAKNESSES, AND EXTENSIONS

Tree methods are good off-the-shelf classifiers and predictors. They are also useful for variable selection, with the most important predictors usually showing up at the top of the tree. Trees require relatively little effort from users in the following senses: First, there is no need for transformation of variables (any monotone transformation of the variables will give the same trees). Second, variable subset selection is automatic because it is part of the split selection. In the loan example, note that the best-pruned tree has automatically selected just three variables (Income, Education, and Family) out of the set of 14 variables available.

Trees are also intrinsically robust to outliers, since the choice of a split depends on the *ordering* of observation values and not on the absolute *magnitudes* of these values. However, they are sensitive to changes in the data, and even a slight change can cause very different splits!

Unlike models that assume a particular relationship between the response and predictors (e.g., a linear relationship such as in linear regression and linear discriminant analysis), classification and regression trees are nonlinear and non-parametric. This allows for a wide range of relationships between the predictors and the response. However, this can also be a weakness: because the splits are done on single predictors rather than on combinations of predictors, the tree is likely to miss relationships between predictors, in particular linear structures like those in linear or logistic regression models. Classification trees are useful classifiers in cases where horizontal and vertical splitting of the predictor space adequately divides the classes. But consider, for instance, a dataset with two predictors and two classes, where separation between the two classes is obviously achieved by using a diagonal line (as shown in Figure 9.17). In such cases a classification tree is expected to have lower performance than methods such as discriminant analysis. One way to improve performance is to create new predictors that are derived from existing predictors, which can capture hypothesized relationships between predictors (similar to interactions in regression models).

Another performance issue with classification trees is that they require a large dataset in order to construct a good classifier. From a computational point of view, trees can be relatively expensive to grow because of the multiple sorting involved in computing all possible splits on every variable. Pruning the data using the validation set adds further computation time.

Although trees are useful for variable selection, one challenge is that they "favor" predictors with many potential split points. This includes categorical

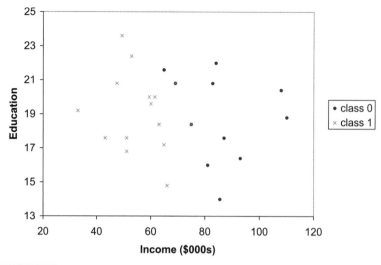

FIGURE 9.17 SCATTERPLOT DESCRIBING A TWO-PREDICTOR CASE WITH TWO CLASSES. THE BEST SEPARATION IS ACHIEVED WITH A DIAGONAL LINE, WHICH CLASSIFICATION TREES CANNOT DO.

predictors with many categories and numerical predictors with many different values. Such predictors have a higher chance of appearing in a tree. One simplistic solution is to combine multiple categories into a smaller set and bin numerical predictors with many values. Alternatively, there are special algorithms that avoid this problem by using a different splitting criterion (e.g., QUEST classification trees; see Loh and Shih, 1997, and www.math.ccu.edu.tw/yshih/quest.html)

An appealing feature of trees is that they handle missing data without having to impute values or delete observations with missing values. Finally, a very important practical advantage of trees is the transparent rules that they generate. Such transparency is often useful in managerial applications.

9.9 IMPROVING PREDICTION: MULTIPLE TREES

To address the shortcomings of a single tree—especially poor predictive power—researchers developed several extensions to trees that combine results from multiple trees. These are examples of *ensembles* (see Chapter 13). Two popular multi-tree approaches are *random forests* and *boosted trees*.

Breiman and Cutler introduced *random forests*.[5] Random forests are a special case of *bagging*, a method for improving predictive power by combining multiple

---

[5] For further details on random forests, see www.stat.berkeley.edu/users/breiman/RandomForests/cc_home.htm

classifiers or prediction algorithms. See Chapter 13 for further details on bagging. The basic idea in random forests is to:

1. Draw multiple random samples, with replacement, from the data (this sampling approach is called the *bootstrap*).
2. Fit a classification (or regression) tree to each sample (and thus obtain a "forest").
3. Combine the predictions/classifications from the individual trees to obtain improved predictions. Use voting for classification and averaging for prediction.

To run a random forest in XLMiner, choose "Random Trees" in Classify > Classification Tree or in Predict > Regression Tree.

Unlike a single tree, results from a random forest cannot be displayed in a tree-like diagram, thereby losing the interpretability that a single tree provides. However, random forests can produce "predictor importance" scores, which measure the relative contribution of the different predictors. The importance score for a particular predictor is computed by summing up the decrease in the Gini index for that predictor over all the trees in the forest.

The second type of multi-tree improvement is *boosted trees*. Here a sequence of trees is fitted, so that each tree concentrates on misclassified records from the previous tree.[6] The general steps are:

1. Fit a single tree.
2. Draw a sample that gives higher selection probabilities to misclassified records.
3. Fit a tree to the new sample.
4. Repeat steps 2 and 3 multiple times.
5. Use weighted voting to classify records, where heavier weight is given to later trees.

Figure 9.18 shows the result of running a boosted tree on the loan acceptance example that we saw earlier. In XLMiner, we choose "Boosting" from the Classification Tree menu (a similar option is available for predicting a numerical outcome in the Regression Tree menu), and set the number of trees (called "weak learners") to 20. We can see that compared to the performance of the single best-pruned tree (Figure 9.15), the boosted tree has better performance on the validation and test sets in terms of lower overall error rate and especially in

---

[6] In "boosted trees" algorithms, predictors are randomly drawn for each tree; in contrast, in random forests, the number of predictors is fixed.

Training Data Scoring—Summary Report

Cutoff probability value for success (UPDATABLE)	0.5

Confusion Matrix

Actual Class	Predicted Class	
	1	0
1	244	0
0	0	2256

Error Report

Class	#Cases	#Errors	% Error
1	244	0	0
0	2256	0	0
Overall	2500	0	0

Validation Data Scoring—Summary Report

Cutoff probability value for success (UPDATABLE)	0.5

Confusion Matrix

Actual Class	Predicted Class	
	1	0
1	137	18
0	6	1339

Error Report

Class	#Cases	#Errors	% Error
1	155	18	11.6129
0	1345	6	0.446097
Overall	1500	24	1.6

Test Data Scoring Scoring—Summary

Cutoff probability value for success (UPDATABLE)	0.5

Confusion Matrix

Actual Class	Predicted Class	
	1	0
1	74	7
0	5	914

Error Report

Class	#Cases	#Errors	% Error
1	81	7	8.641975
0	919	5	0.54407
Overall	1000	12	1.2

FIGURE 9.18 LOAN ACCEPTANCE DATA: CONFUSION MATRIX AND ERROR RATES FOR THE TRAINING, VALIDATION, AND TEST DATA BASED ON BOOSTED TREE.

terms of correct classification of 1's—the rare class of special interest. Where does boosting's special talent for finding 1's come from? When one class is dominant (0's constitute over 90% of the data here), basic classifiers are tempted to classify cases as belonging to the dominant class, and the 1's in this case constitute most of the misclassifications with the single best-pruned tree. The boosting algorithm concentrates on the misclassifications (which are mostly 1's), so it is naturally going to do well in reducing the misclassification of 1's (from 9 in the single tree to 7 in the boosted tree, in the test sets).

PROBLEMS

9.1   **Competitive Auctions on eBay.com.** The file `eBayAuctions.xls` contains information on 1972 auctions that transacted on eBay.com during May–June 2004. The goal is to use these data to build a model that will classify auctions as competitive or noncompetitive. A *competitive auction* is defined as an auction with at least two bids placed on the item auctioned. The data include variables that describe the item (auction category), the seller (his/her eBay rating), and the auction terms that the seller selected (auction duration, opening price, currency, day-of-week of auction close). In addition, we have the price at which the auction closed. The task is to predict whether or not the auction will be competitive.

Data Preprocessing. Create dummy variables for the categorical predictors. These include *Category* (18 categories), *Currency* (USD, GBP, euro), *EndDay* (Monday–Sunday), and *Duration* (1, 3, 5, 7, or 10 days). Split the data into training and validation datasets using a 60 : 40% ratio.

   a. Fit a classification tree using all predictors, using the best-pruned tree. To avoid overfitting, set the minimum number of records in a terminal node to 50. Also set the maximum number of levels to be displayed at seven (the maximum allowed in XLMiner). Write down the results in terms of rules. (*N*ote: If you had to slightly reduce the number of predictors due to software limitations, or for clarity of presentation, which would be a good variable to choose?)

   b. Is this model practical for predicting the outcome of a new auction?

   c. Describe the interesting and uninteresting information that these rules provide.

   d. Fit another classification tree (using the best-pruned tree, with a minimum number of records per terminal node = 50 and maximum allowed number of displayed levels), this time only with predictors that can be used for predicting the outcome of a new auction. Describe the resulting tree in terms of rules. Make sure to report the smallest set of rules required for classification.

   e. Plot the resulting tree on a scatter plot: Use the two axes for the two best (quantitative) predictors. Each auction will appear as a point, with coordinates corresponding to its values on those two predictors. Use different colors or symbols to separate competitive and noncompetitive auctions. Draw lines (you can sketch these by hand or use Excel) at the values that create splits. Does this splitting seem reasonable with respect to the meaning of the two predictors? Does it seem to do a good job of separating the two classes?

   f. Examine the lift chart and the classification table for the tree. What can you say about the predictive performance of this model?

   g. Based on this last tree, what can you conclude from these data about the chances of an auction obtaining at least two bids and its relationship to the auction settings set by the seller (duration, opening price, ending day, currency)? What would you recommend for a seller as the strategy that will most likely lead to a competitive auction?

9.2   **Predicting Delayed Flights.** The file `FlightDelays.xls` contains information on all commercial flights departing the Washington, DC, area and arriving at New York during January 2004. For each flight there is information on the departure and arrival airports, the distance of the route, the scheduled time and date of the flight, and so on. The variable that we are trying to predict is whether or not a flight is delayed. A delay is defined as an arrival that is at least 15 minutes later than scheduled.

Data Preprocessing. Create dummies for day of week, carrier, departure airport, and arrival airport. This will give you 17 dummies. Bin the scheduled departure time into eight bins (in XLMiner use *Transform* > *Bin Continuous Data* and select equal width). After binning CRS_DEP_TIME into the 8 bins, this new variable should be broken down into dummies (because the effect will not be linear, due to the morning and afternoon rush hours). This will avoid treating the departure time as a continuous predictor, since it is reasonable that delays are related to rush-hour times. Partition the data into training and validation sets.

a. Fit a classification tree to the flight delay variable using all the relevant predictors. Do not include DEP_TIME (actual departure time) in the model because it is unknown at the time of prediction (unless we are generating our predictions of delays after the plane takes off, which is unlikely). In the third step of the Classification Tree menu, choose "Maximum # levels to be displayed = 6." Use the best-pruned tree, setting the minimum number of observations in the final nodes to 1. Express the resulting tree as a set of rules.

b. If you needed to fly between DCA and EWR on a Monday at 7 AM, would you be able to use this tree? What other information would you need? Is it available in practice? What information is redundant?

c. Fit another tree, this time excluding the Weather predictor. (Why?) Select the option of seeing both the full tree and the best-pruned tree. You will find that the best-pruned tree contains a single terminal node.

   i. How is this tree used for classification? (What is the rule for classifying?)

   ii. To what is this rule equivalent?

   iii. Examine the full tree. What are the top three predictors according to this tree?

   iv. Why, technically, does the pruned tree result in a tree with a single node?

   v. What is the disadvantage of using the top levels of the full tree as opposed to the best-pruned tree?

   vi. Compare this general result to that from logistic regression in the example in Chapter 10. What are possible reasons for the classification tree's failure to find a good predictive model?

9.3 **Predicting Prices of Used Cars (Regression Trees).** The file `Toyota-Corolla.xls` contains the data on used cars (Toyota Corolla) on sale during late summer of 2004 in the Netherlands. It has 1436 records containing details on 38 attributes, including Price, Age, Kilometers, HP, and other specifications. The goal is to predict the price of a used Toyota Corolla based on its specifications. (The example in Section 9.7 is a subset of this dataset.)

   Data Preprocessing. Create dummy variables for the categorical predictors (Fuel Type and Color). Split the data into training (50%), validation (30%), and test (20%) datasets.

   a. Run a regression tree (RT) using the Prediction menu in XLMiner with the output variable Price and input variables Age_08_04, KM, Fuel_Type, HP, Automatic, Doors, Quarterly_Tax, Mfg_Guarantee, Guarantee_Period, Airco, Automatic_Airco, CD_Player, Powered_Windows, Sport_Model, and Tow_Bar. Keep the minimum number of records in a terminal node to 1, maximum number of tree levels to 100, and the scoring option to Full Tree, to make the run least restrictive.

    i. Which appear to be the three or four most important car specifications for predicting the car's price?

    ii. Compare the prediction errors of the training, validation, and test sets by examining their RMS error and by plotting the three boxplots. What is happening with the training set predictions? How does the predictive performance of the test set compare to the other two? Why does this occur?

    iii. How can we achieve predictions for the training set that are not equal to the actual prices?

    iv. If we used the full tree instead of the best-pruned tree to score the validation set, how would this affect the predictive performance for the validation set? (**Hint**: Does the full tree use the validation data?)

b. Let us see the effect of turning the price variable into a categorical variable. First, create a new variable that categorizes price into 20 bins. Use *Transform → Bin continuous data* to categorize Price into 20 bins of equal counts (leave all other options at their default). Next, repartition the data keeping Binned_Price instead of Price. Run a classification tree (CT) using the *Classification* menu of XLMiner with the same set of input variables as in the RT, and with Binned_Price as the output variable. Keep the minimum number of records in a terminal node to 1 and uncheck the *Prune Tree* option, to make the run least restrictive.

    i. Compare the tree generated by the CT with the one generated by the RT. Are they different? (Look at structure, the top predictors, size of tree, etc.) Why?

    ii. Predict the price, using the RT and the CT, of a used Toyota Corolla with the specifications listed in Table 9.3.

TABLE 9.3	SPECIFICATIONS FOR A PARTICULAR TOYOTA COROLLA

Variable	Value
Age_-08_-04	77
KM	117,000
Fuel_Type	Petrol
HP	110
Automatic	No
Doors	5
Quarterly_Tax	100
Mfg_Guarantee	No
Guarantee_Period	3
Airco	Yes
Automatic_Airco	No
CD_Player	No
Powered_Windows	No
Sport_Model	No
Tow_Bar	Yes

    iii. Compare the predictions in terms of the predictors that were used, the magnitude of the difference between the two predictions, and the advantages and disadvantages of the two methods.

9.4 **Predicting Delayed Flights (Boosting).** Return to the flight delays problem above and run it again, only this time choose a boosted classification tree. Leave the default number of weak learners, and select re-sampling. Set maximum levels to display at 6, and minimum number of records in a terminal node to 1.

 a. Compared with the single tree, how does the boosted tree behave in terms of overall error rate?

 b. Compared with the single tree, how does the boosted tree behave in terms of error rate in identifying delayed flights?

 c. Explain why this model might have the best performance over the other models you fit.

CHAPTER 10

Logistic Regression

In this chapter we describe the highly popular and powerful classification method called logistic regression. Like linear regression, it relies on a specific model relating the predictors with the outcome. The user must specify the predictors to include as well as their form (e.g., including any interaction terms). This means that even small datasets can be used for building logistic regression classifiers, and that once the model is estimated, it is computationally fast and cheap to classify even large samples of new observations. We describe the logistic regression model formulation and its estimation from data. We also explain the concepts of "logit," "odds," and "probability" of an event that arise in the logistic model context and the relations among the three. We discuss variable importance using coefficient and statistical significance and also mention variable selection algorithms for dimension reduction. All this is illustrated on an authentic dataset of flight information where the goal is to predict flight delays. Our presentation is strictly from a data mining perspective, where classification is the goal and performance is evaluated on a separate validation set. However, because logistic regression is also heavily used in statistical analyses for purposes of inference, we give a brief review of key concepts related to coefficient interpretation, goodness-of-fit evaluation, inference, and multiclass models in the Appendix at the end of this chapter.

Data Mining for Business Analytics: Concepts, Techniques, and Applications with XLMiner®, Third Edition.
Galit Shmueli, Peter C. Bruce, and Nitin R. Patel.
© 2016 John Wiley & Sons Inc. Published 2016 by John Wiley & Sons, Inc.

10.1   INTRODUCTION

Logistic regression extends the ideas of linear regression to the situation where the dependent variable Y is categorical. We can think of a categorical variable as dividing the observations into classes. For example, if Y denotes a recommendation on holding/selling/buying a stock, we have a categorical variable with three categories. We can think of each of the stocks in the dataset (the observations) as belonging to one of three classes: the *hold* class, the *sell* class, and the *buy* class. Logistic regression can be used for classifying a new observation, where its class is unknown, into one of the classes, based on the values of its predictor variables (called *classification*). It can also be used in data where the class is known, to find factors distinguishing between observations in different classes in terms of their predictor variables, or "predictor profile" (called *profiling*). Logistic regression is used in applications such as:

1. Classifying customers as returning or nonreturning (classification).
2. Finding factors that differentiate between male and female top executives (profiling).
3. Predicting the approval or disapproval of a loan based on information such as credit scores (classification).

The logistic regression model is used in a variety of fields: whenever a structured model is needed to explain or predict categorical (in particular, binary) outcomes. One such application is in describing choice behavior in econometrics (see the box on Consumer Choice Theory).

LOGISTIC REGRESSION AND CONSUMER CHOICE THEORY

In the context of choice behavior, the logistic model can be shown to follow from the *random utility theory* developed by Manski (1977) as an extension of the standard economic theory of consumer behavior. In essence, the consumer theory states that when faced with a set of choices, a consumer makes the choice that has the highest utility (a numerical measure of worth with arbitrary zero and scale). It assumes that the consumer has a preference order on the list of choices that satisfies reasonable criteria such as transitivity. The preference order can depend on the person (e.g., socioeconomic characteristics) as well as attributes of the choice. The random utility model considers the utility of a choice to incorporate a random element. When we model the random element as coming from a "reasonable" distribution, we can logically derive the logistic model for predicting choice behavior.

In this chapter we focus on the use of logistic regression for classification. We deal only with a binary dependent variable having two possible classes. In the Appendix we show how the results can be extended to the case where Y assumes more than two possible outcomes. Popular examples of binary response outcomes are success/failure, yes/no, buy/don't buy, default/don't default, and survive/die. For convenience, we often code the values of a binary response Y as 0 and 1.

Note that in some cases we may choose to convert continuous data or data with multiple outcomes into binary data for purposes of simplification, reflecting the fact that decision making may be binary (approve the loan/don't approve, make an offer/don't make an offer). As with multiple linear regression, the independent variables $X_1, X_2, \ldots, X_k$ may be categorical variables, continuous variables, or a mixture of these two types. While in multiple linear regression the aim is to predict the value of the continuous Y for a new observation, in logistic regression the goal is to predict which class a new observation will belong to, or simply to *classify* the observation into one of the classes. In the stock example, we would want to classify a new stock into one of the three recommendation classes: sell, hold, or buy. Or, we might want to compute for a new observation its *propensity* (= the probability) to belong to each class, and then possibly rank a set of new observations from highest to lowest propensity in order to act on those with the highest propensity.

In logistic regression we take two steps: the first step yields estimates of the *propensities* or *probabilities* of belonging to each class. In the binary case we get an estimate of $p = P(Y = 1)$, the probability of belonging to class 1 (which also tells us the probability of belonging to class 0). In the next step we use a cutoff value on these probabilities in order to classify each case into one of the classes. For example, in a binary case, a cutoff of 0.5 means that cases with an estimated probability of $P(Y = 1) \geq 0.5$ are classified as belonging to class 1, whereas cases with $P(Y = 1) < 0.5$ are classified as belonging to class 0. This cutoff does not need to be set at 0.5. When the event in question is a low-probability, but notable or important event (e.g., 1= fraudulent transaction), a lower cutoff may be used to classify more cases as belonging to class 1.

10.2 THE LOGISTIC REGRESSION MODEL

The idea behind logistic regression is straightforward: instead of using Y as the dependent variable, we use a function of it, which is called the *logit*. The logit, it turns out, can be modeled as a linear function of the predictors. Once the logit has been predicted, it can be mapped back to a probability.

To understand the logit, we take several intermediate steps: First, we look at $p = P(Y = 1)$, the probability of belonging to class 1 (as opposed to class 0).

In contrast to Y, the class label, which only takes the values 0 and 1, p can take any value in the interval $[0, 1]$. However, if we express p as a linear function of the q predictors[1] in the form

$$p = \beta_0 + \beta_1 x_1 + \beta_2 x_2 + \cdots + \beta_q x_q, \tag{10.1}$$

it is not guaranteed that the right-hand side will lead to values within the interval $[0, 1]$. The solution is to use a nonlinear function of the predictors in the form

$$p = \frac{1}{1 + e^{-(\beta_0 + \beta_1 x_1 + \beta_2 x_2 + \cdots + \beta_q x_q)}}. \tag{10.2}$$

This is called the *logistic response function*. For any values of $x_1, \ldots, x_q$, the right-hand side will always lead to values in the interval $[0, 1]$. Next, we look at a different measure of belonging to a certain class, known as *odds*. The odds of belonging to class 1 ($Y = 1$) is defined as *the ratio of the probability of belonging to class 1 to the probability of belonging to class 0*:

$$\text{Odds}(Y = 1) = \frac{p}{1 - p}. \tag{10.3}$$

This metric is very popular in horse races, sports, gambling, epidemiology, and many other areas. Instead of talking about the *probability* of winning or contacting a disease, people talk about the *odds* of winning or contacting a disease. How are these two different? If, for example, the probability of winning is 0.5, the odds of winning are $0.5/0.5 = 1$. We can also perform the reverse calculation: Given the odds of an event, we can compute its probability by manipulating equation (10.3):

$$p = \frac{\text{Odds}}{1 + \text{Odds}}. \tag{10.4}$$

Substituting (10.2) into (10.4), we can write the relationship between the odds and the predictors as

$$\text{Odds}(Y = 1) = e^{\beta_0 + \beta_1 x_1 + \beta_2 x_2 + \cdots + \beta_q x_q}. \tag{10.5}$$

This last equation describes a multiplicative (proportional) relationship between the predictors and the odds. Such a relationship is interpretable in terms of percentages: for example, a unit increase in predictor x_j is associated with an average increase of $\beta_j \times 100\%$ in the odds (holding all other predictors constant).

Now, if we take a natural logarithm[2] on both sides, we get the standard formulation of a logistic model:

$$\log(\text{Odds}) = \beta_0 + \beta_1 x_1 + \beta_2 x_2 + \cdots + \beta_q x_q. \tag{10.6}$$

---

[1] Unlike elsewhere in the book, where p denotes the number of predictors, in this chapter we indicate predictors by q, to avoid confusion with the probability p.

[2] The natural logarithm function is typically denoted ln() or log(). In this book, we use log().

FIGURE 10.1 (a) ODDS AND (b) LOGIT AS A FUNCTION OF p.

The log(Odds), called the *logit*, takes values from $-\infty$ (very low odds) to ∞ (very high odds).[3] A logit of 0 corresponds to even odds of 1 (probability = 0.5). Thus our final formulation of the relation between the response and the predictors uses the logit as the dependent variable and models it as a *linear function* of the q predictors.

To see the relation between the probability, odds, and logit of belonging to class 1, look at Figure 10.1, which shows the odds (top) and logit (bottom) as a function of p. Notice that the odds can take any nonnegative value, and that the logit can take any real value. Let us examine some data to illustrate the use of logistic regression.

Example: Acceptance of Personal Loan

Recall the example described in Chapter 9 of acceptance of a personal loan by Universal Bank. The bank's dataset includes data on 5000 customers. The

---

[3] We use the terms *Odds* and *Odds(Y = 1)* interchangeably.

data include customer demographic information (*Age, Income*, etc.), customer response to the last personal loan campaign (*Personal Loan*), and the customer's relationship with the bank (mortgage, securities account, etc.). Among these 5000 customers, only 480 (= 9.6%) accepted the personal loan that was offered to them in a previous campaign. The goal is to build a model that identifies customers who are most likely to accept the loan offer in future mailings.

Data Preprocessing We start by creating dummy variables for each of the categorical predictors. Except for Education, which has three categories, the remaining four categorical variables have two categories. We therefore need $6 = 2 + 1 + 1 + 1 + 1$ dummy variables to describe these five categorical predictors. In XLMiner's classification functions, the response can remain in text form (Yes, No, etc.), but the predictor variables must be coded into dummy variables. We use the following coding:

$$\text{EducProf} = \begin{cases} 1 & \text{if education is } \textit{Professional} \\ 0 & \text{otherwise} \end{cases}$$

$$\text{EducGrad} = \begin{cases} 1 & \text{if education is at } \textit{Graduate} \text{ level} \\ 0 & \text{otherwise} \end{cases}$$

$$\text{Securities} = \begin{cases} 1 & \text{if customer has securities account in bank} \\ 0 & \text{otherwise} \end{cases}$$

$$\text{CD} = \begin{cases} 1 & \text{if customer has CD account in bank} \\ 0 & \text{otherwise} \end{cases}$$

$$\text{Online} = \begin{cases} 1 & \text{if customer uses online banking} \\ 0 & \text{otherwise} \end{cases}$$

$$\text{CreditCard} = \begin{cases} 1 & \text{if customer holds Universal Bank credit card} \\ 0 & \text{otherwise} \end{cases}$$

Next, we partition the data randomly into training (60%) and validation (40%) sets. We use the training set to fit a model and the validation set to assess the model's performance.

Model with a Single Predictor

Consider first a simple logistic regression model with just one predictor. This is conceptually analogous to the simple linear regression model in which we fit a straight line to relate the outcome, Y, to a single predictor, X.

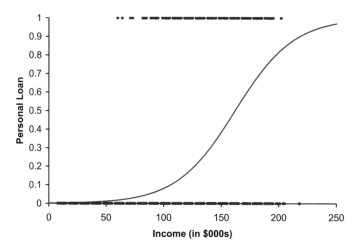

FIGURE 10.2   PLOT OF DATA POINTS (PERSONAL LOAN AS A FUNCTION OF INCOME.) AND THE FITTED LOGISTIC CURVE

Let us construct a simple logistic regression model for classification of customers using the single predictor *Income*. The equation relating the outcome variable to the predictor in terms of probabilities is

$$P(Personal\ Loan = Yes \mid Income = x) = \frac{1}{1 + e^{-(\beta_0 + \beta_1 x)}},$$

or equivalently, in terms of odds,

$$\text{Odds}(Personal\ Loan = Yes) = e^{\beta_0 + \beta_1 x}. \tag{10.7}$$

The estimated coefficients for the model are $b_0 = -6.3525$ and $b_1 = 0.0392$. So the fitted model is

$$P(Personal\ Loan = Yes \mid Income = x) = \frac{1}{1 + e^{6.3525 - 0.0392x}}. \tag{10.8}$$

Although logistic regression can be used for prediction in the sense that we predict the *probability* of a categorical outcome, it is most often used for classification. To see the difference between the two, consider predicting the probability of a customer accepting the loan offer as opposed to classifying the customer as an acceptor/nonacceptor. From Figure 10.2 it can be seen that the loan acceptance probabilities that are produced by the logistic regression model (the S-shaped curve in Figure 10.2) can yield values between 0 and 1. To end up with classifications into either 1 or 0 (e.g., a customer either accepts the loan offer or not), we need a threshold, or cutoff value (see section on "Propensities and Cutoff for Classification" in Chapter 5). This is true in the case of multiple predictor variables as well.

In the Universal Bank example, in order to classify a new customer as an acceptor/nonacceptor of the loan offer, we use the information on his/her income by plugging it into the fitted equation in (10.8). This yields an estimated probability of accepting the loan offer. We then compare it to the cutoff value. The customer is classified as an acceptor if the probability of his/her accepting the offer is above the cutoff.[4]

Estimating the Logistic Model from Data: Computing Parameter Estimates

In logistic regression, the relation between Y and the β parameters is nonlinear. For this reason the β parameters are not estimated using the method of least squares (as in multiple linear regression). Instead, a method called *maximum likelihood* is used. The idea, in brief, is to find the estimates that maximize the chance of obtaining the data that we have. This requires iterations using a computer program.[5]

Algorithms to compute the coefficient estimates are less robust than algorithms for linear regression. Computed estimates are generally reliable for well-behaved datasets where the number of observations with outcome variable values of both 0 and 1 are large; their ratio is "not too close" to either 0 or 1; and when the number of coefficients in the logistic regression model is small relative to the sample size (e.g., no more than 10%). As with linear regression, collinearity (strong correlation among the predictors) can lead to computational difficulties. Computationally intensive algorithms have been developed recently that circumvent some of these difficulties. For technical details on the maximum likelihood estimation in logistic regression, see Hosmer and Lemeshow (2000).

To illustrate a typical output from such a procedure, look at the output in Figure 10.3 for the logistic model fitted to the training set of 3000 Universal Bank customers. The outcome variable is *Personal Loan*, with Yes defined as the *success* (this is equivalent to setting the outcome variable to 1 for an acceptor

---

[4] Here we compared the probability to a cutoff c. If we prefer to look at *odds* of accepting rather than the probability, an equivalent method is to use the equation in (10.7) and compare the odds to $c/(1 - c)$. If the odds are higher than this number, the customer is classified as an acceptor. If the odds are lower, we classify the customer as a nonacceptor.

[5] The method of maximum likelihood ensures good asymptotic (large sample) properties for the estimates. Under very general conditions, maximum likelihood estimators are: (1) *consistent*—the probability of the estimator differing from the true value approaches zero with increasing sample size; (2) *asymptotically efficient*—the variance is the smallest possible among consistent estimators; and (3) *asymptotically normally distributed*—this allows us to compute confidence intervals and perform statistical tests in a manner analogous to the analysis of multiple linear regression models, provided that the sample size is *large*.

Input Variables	Coefficient	Std. Error	χ^2-Statistic	P-Value	Odds
Intercept	−13.2012	2.4677	28.6189	0	0
Age	−0.0445	0.091	0.2397	0.6244	0.9564
Experience	0.0566	0.0901	0.3946	0.5299	1.0582
Income	0.0658	0.0042	242.6842	0	1.068
Family	0.5715	0.1012	31.9029	0	1.771
CCAvg	0.1872	0.0615	9.2581	0.0023	1.2059
Mortgage	0.0018	0.0008	4.757	0.0292	1.0018
Securities Account	−0.8548	0.4186	4.1695	0.0412	0.4254
CD Account	3.4689	0.4489	59.7096	0	32.1008
Online	−0.8435	0.2283	13.649	0.0002	0.4302
CreditCard	−0.964	0.2825	11.6419	0.0006	0.3814
EducGrad	4.5889	0.3871	140.5552	0	98.3867
EducProf	4.5225	0.3842	138.5354	0	92.0689

FIGURE 10.3   LOGISTIC REGRESSION COEFFICIENT TABLE FOR PERSONAL LOAN ACCEPTANCE AS A FUNCTION OF 12 PREDICTORS.

and 0 for a nonacceptor). Here we use all 12 predictors, but note that for each categorical predictor with m categories we use only $m − 1$ dummies.

Ignoring p-values for the coefficients, a model based on all 12 predictors would have the estimated logistic equation

$$\text{Logit}(\textit{Personal Loan} = \textit{Yes}) = -13.201 - 0.045 \text{ Age} + 0.057 \text{ Experience}$$

$$+ 0.066 \textit{ Income} + 0.572 \textit{ Family} + 0.187 \textit{ CCAvg}$$

$$+ 0.002 \textit{ Mortgage} - 0.855 \textit{ Securities} + 3.469 \textit{ CD}$$

$$- 0.844 \textit{ Online} - 0.964 \textit{ Credit Card}$$

$$+ 4.589 \textit{ EducGrad} + 4.523 \textit{ EducProf}. \tag{10.9}$$

The positive coefficients for the dummy variables *CD, EducGrad,* and *EducProf* mean that holding a CD account and having graduate or professional education (all marked by 1 in the dummy variables) are associated with higher probabilities of accepting the loan offer. In contrast, having a securities account, using online banking, and owning a Universal Bank credit card are associated with lower acceptance rates. For the continuous predictors, positive coefficients indicate that a higher value on that predictor is associated with a higher probability of accepting the loan offer (e.g., income: higher income customers tend more to accept the offer). Similarly, negative coefficients indicate that a higher value on that predictor is associated with a lower probability of accepting the loan offer (e.g., *Age*: older customers are less likely to accept the offer).

If we want to talk about the *odds* of offer acceptance, we can use the last column (entitled "odds") to obtain the equation:

$$
\begin{aligned}
\text{Odds}(\textit{Personal Loan} = \text{Yes}) \ = \ & e^{-13.201}(0.956)^{\textit{Age}}\,(1.058)^{\textit{Experience}}\,(1.068)^{\textit{Income}} \\
& \times (1.771)^{\textit{Family}}\,(1.206)^{\textit{CCAvg}}\,(1.002)^{\textit{Mortgage}} \\
& \times (0.425)^{\textit{Securities}}\,(32.105)^{\textit{CD}}\,(0.430)^{\textit{Online}} \\
& \times (0.381)^{\textit{CreditCard}}(98.405)^{\textit{EducGrad}}\,(92.086)^{\textit{EducProf}}.
\end{aligned}
$$

$$(10.10)$$

Notice how positive coefficients in the logit model translate into coefficients larger than 1 in the odds model, and negative coefficients in the logit translate into coefficients smaller than 1 in the odds.

A third option is to look directly at an equation for the probability of acceptance, using equation (10.2). This is the probability (the *propensity*) of accepting the offer for a customer with given values of the 12 predictors.[6]

Interpreting Results in Terms of Odds (for a Profiling Goal)

Logistic models, when they are appropriate for the data, can give useful information about the roles played by different predictor variables. Say we want to know how increasing family income by one unit will affect the probability of loan acceptance. This can be found straightforwardly if we consider not probabilities, but odds.

Recall that the odds are given by

$$
\text{Odds} = e^{\beta_0 + \beta_1 x_1 + \beta_2 x_2 + \cdots + \beta_k x_q}.
$$

At first let us return to the single predictor example, where we model a customer's acceptance of a personal loan offer as a function of his/her income:

$$
\text{Odds}(\textit{Personal Loan} = \textit{Yes}) = e^{\beta_0 + \beta_1 \cdot \textit{Income}}.
$$

We can think of the model as a multiplicative model of odds. The odds that a customer with income zero will accept the loan is estimated by $e^{-6.353+(0.039)(0)} = 0.0017$. These are the *base-case odds*. In this example, it is obviously economically

---

[6] If all q predictors are categorical, each having m_q categories, we need not compute probabilities/odds for each of the n observations. The number of different probabilities/odds is exactly $m_1 \times m_2 \times \cdots \times m_q$.

meaningless to talk about a zero income; the value zero and the corresponding base-case odds could be meaningful, however, in the context of other predictors. The odds of accepting the loan with an income of \$100K will increase by a multiplicative factor of $e^{(0.039)(100)} = 50.5$ over the base case, so the odds that such a customer will accept the offer are $e^{-6.353+(0.039)(100)} = 0.088$.

To generalize this to the multiple-predictor case, consider the 12 predictors in the personal loan offer example. The odds of a customer accepting the offer as a function of the 12 predictors are given in (10.10).

Suppose that the value of Income, or in general x_1, is increased by one unit from x_1 to $x_1 + 1$, while the other predictors (denoted $x_2, \ldots, x_{12}$) are held at their current value. We get the odds ratio

$$\frac{\text{Odds}(x_1 + 1, x_2, \ldots, x_{12})}{\text{Odds}(x_1, \ldots x_{12})} = \frac{e^{\beta_0 + \beta_1(x_1+1) + \beta_2 x_2 + \cdots + \beta_{12} x_{12}}}{e^{\beta_0 + \beta_1 x_1 + \beta_2 x_2 + \cdots + \beta_{12} x_{12}}} = e^{\beta_1}.$$

This tells us that a single unit increase in x_1, holding $x_2, \ldots, x_{12}$ constant, is associated with an increase in the odds that a customer accepts the offer by a factor of e^{β_1}. In other words, e^{β_1} is the multiplicative factor by which the odds (of belonging to class 1) increase when the value of x_1 is increased by 1 unit, *holding all other predictors constant*. If $\beta_1 < 0$, an increase in x_1 is associated with a decrease in the odds of belonging to class 1, whereas a positive value of β_1 is associated with an increase in the odds.

When a predictor is a dummy variable, the interpretation is technically the same but has a different practical meaning. For instance, the coefficient for **CD** was estimated from the data to be 3.469. Recall that the reference group is customers not holding CD account. We interpret this coefficient as follows: $e^{3.469} = 32.10$ are the odds that a customer who has a CD account will accept the offer relative to a customer who does not have a CD account, holding all other factors constant. This means that customers who have CD accounts in Universal Bank are more likely to accept the offer than customers without a CD account (holding all other variables constant).

The advantage of reporting results in odds as opposed to probabilities is that statements such as those above are true for any value of x_1. Unless x_1 is a dummy variable, we cannot apply such statements about the effect of increasing x_1 by a single unit to probabilities. This is because the result depends on the actual value of x_1. So, if we increase x_1 from, say, 3 to 4, the effect on p, the probability of belonging to class 1, will be different than if we increase x_1 from 30 to 31. In short, the change in the probability, p, for a unit increase in a particular predictor variable, while holding all other predictors constant, is not a constant—it depends on the specific values of the predictor variables. We therefore talk about probabilities only in the context of specific observations.

10.3   EVALUATING CLASSIFICATION PERFORMANCE

The general measures of performance that were described in Chapter 5 are used to assess how well the logistic model does. Recall that there are several performance measures, the most popular being those based on the classification matrix (accuracy alone or combined with costs) and the lift chart. As in other classification methods, the goal is to find a model that accurately classifies observations to their class, using only the predictor information. A variant of this goal is to find a model that does a superior job of identifying the members of a particular class of interest (which might come at some cost to overall accuracy). Since the training data are used for selecting the model, we expect the model to perform quite well for those data, and therefore prefer to test its performance on the validation set. Recall that the data in the validation set were not involved in the model building, and thus we can use them to test the model's ability to classify data that it has not "seen" before.

To obtain the classification matrix from a logistic regression analysis, we use the estimated equation to predict the probability of class membership (the *propensities*) for each observation in the validation set, and use the cutoff value to decide on the class assignment of these observations. We then compare these classifications to the actual class memberships of these observations. In the Universal Bank case we use the estimated model in equation (10.9) to predict the probability of offer acceptance in a validation set that contains 2000 customers (these data were not used in the modeling step). Technically, this is done by predicting the logit using the estimated model in equation (10.9) and then obtaining the probabilities p through the relation $p = e^{logit}/(1 + e^{logit})$. We next compare these probabilities to our chosen cutoff value in order to classify each of the 2000 validation observations as acceptors or nonacceptors. XLMiner created the validation classification matrix automatically, and it is possible to obtain the detailed probabilities and classification for each observation. For example, Figure 10.4 shows a partial XLMiner output of scoring the validation set. We see that the first four customers have a probability of accepting the offer that is lower than the cutoff of 0.5, and therefore they are classified as nonacceptors (0). The fifth customer's probability of acceptance is estimated by the model to exceed 0.5, and he or she is therefore classified as an acceptor (1), which in fact is a misclassification.

Predicted Class	Actual Class	Success Probability	Log Odds	Age	Experience	Income	Family	CCAvg
0	0	2.14E-05	−10.754	45	19	34	3	1.5
0	0	3.35E-06	−12.6074	39	15	11	1	1
0	0	0.015825	−4.1302	53	27	72	2	1.5
0	0	0.000217	−8.4374	50	24	22	1	0.3
1	0	0.567811	0.2729	65	39	105	4	2.4

FIGURE 10.4　SCORING THE VALIDATION DATA: XLMINER'S OUTPUT FOR THE FIRST FIVE CUSTOMERS OF UNIVERSAL BANK (BASED ON 12 PREDICTORS).

(a)

(b)

FIGURE 10.5　(a) LIFT AND (b) DECILE CHARTS OF VALIDATION DATA FOR UNIVERSAL BANK LOAN OFFER: COMPARING LOGISTIC MODEL CLASSIFICATION WITH CLASSIFICATION BY NAIVE MODEL.

Another useful tool for assessing model classification performance is the lift (gains) chart (see Chapter 5). Figure 10.5 (top) illustrates the lift chart obtained for the personal loan offer model using the validation set. The "lift" over the base curve indicates for a given number of cases (read on the x-axis), the additional responders that you can identify by using the model. The same information is portrayed in in Figure 10.5 (bottom): Taking the 10% of the records that are ranked by the model as "most probable 1's" yields 7.7 times as many 1's as would simply selecting 10% of the records at random.

Variable Selection

The next step includes searching for alternative models. As with multiple linear regression, each categorical variable must be turned into dummy variables, resulting in additional columns. In addition, we can build more complex models that reflect interactions among predictors by including new variables that are derived from the predictors. For example, if we hypothesize that there is an interactive effect between income and family size, we should add an interaction term of the form Income × Family. The choice among the set of alternative models is guided primarily by performance on the validation data. For models that perform roughly equally well, simpler models are generally preferred over more complex models. Note also that performance on validation data may be overly optimistic when it comes to predicting performance on data that have not been exposed to the model at all. This is because when the validation data are used to select a final model, we are selecting based on how well the model performs with those data and therefore may be incorporating some of the random idiosyncrasies of those data into the judgment about the best model. The model still may be the best among those considered, but it will probably not do as well with the unseen data. Therefore it is useful to evaluate the chosen model on a new test set to get a sense of how well it will perform on new data. In addition, one must consider practical issues such as costs of collecting variables, error-proneness, and model complexity in the selection of the final model.

As in linear regression, in logistic regression we can use automated variable selection heuristics such as stepwise selection, forward selection, and backward elimination (see Section 6.4 in Chapter 6). If the dataset is not too large, we can even try an exhaustive search (Best Subset in XLMiner) over all possible models.

10.4 EXAMPLE OF COMPLETE ANALYSIS: PREDICTING DELAYED FLIGHTS

Predicting flight delays would be useful to a variety of organizations: airport authorities, airlines, aviation authorities. At times, joint task forces have been formed to address the problem. Such an organization, if it were to provide ongoing real-time assistance with flight delays, would benefit from some advance notice about flights that are likely to be delayed.

TABLE 10.1	**DESCRIPTION OF PREDICTORS FOR FLIGHT DELAYS EXAMPLE**
Day of Week	Coded as: 1 = Monday, 2 = Tuesday,. . . , 7 = Sunday
Departure Time	Broken down into 18 intervals between 6 AM and 10 PM
Origin	Three airport codes: DCA (Reagan National), IAD (Dulles), BWI (Baltimore–Washington Int'l)
Destination	Three airport codes: JFK (Kennedy), LGA (LaGuardia), EWR (Newark)
Carrier	Eight airline codes: CO (Continental), DH (Atlantic Coast), DL (Delta), MQ (American Eagle), OH (Comair), RU (Continental Express), UA (United), and US (USAirways)
Weather	Coded as 1 if there was inclement weather

In this simplified illustration, we look at six predictors (see Table 10.1). The outcome of interest is whether the flight is delayed or not (*delayed* means more than 15 minutes late). Our data consist of all flights from the Washington, DC, area into the New York City area during January 2004. The percent of delayed flights among these 2201 flights is 19.5%. The data were obtained from the Bureau of Transportation Statistics website (www.transtats.bts.gov).

The goal is to predict accurately whether or not a new flight, not in this dataset, will be delayed. Our dependent variable is a binary variable called *Delayed*, coded as 1 for a delayed flight and 0 otherwise.

Other information that is available on the website, such as distance and arrival time, is irrelevant because we are looking at a certain route (distance, flight time, etc., should be approximately equal). A sample of the data for 20 flights is shown in Table 10.2. Figures 10.6 and 10.7 show visualizations of the relationships between flight delays and different predictors or combinations of predictors. From Figure 10.6 we see that Sundays and Mondays saw the largest proportion of delays. Delay rates also seem to differ by carrier, by time of day, as well as by origin and destination airports. For Weather, we see a strong distinction between delays when Weather = 1 (in that case there is always a delay) and Weather = 0. The heatmap in Figure 10.7 reveals some specific combinations with high rates of delays, such as Sunday flights by carrier RU, departing from BWI, or Sunday flights by MQ departing from DCA. We can also see combinations with very low delay rates.

Our main goal is to find a model that can obtain accurate classifications of new flights based on their predictor information. An alternative goal is finding a certain percentage of flights that are most/least likely to get delayed (*ranking*). And a third different goal is profiling flights: finding out which factors are associated with a delay (not only in this sample but in the entire population of flights on this route), and for those factors we would like to quantify these

TABLE 10.2	SAMPLE OF 20 FLIGHTS

Delayed	Carrier	Day of Week	Departure Time	Destination	Origin	Weather
0	DL	2	728	LGA	DCA	0
1	US	3	1600	LGA	DCA	0
0	DH	5	1242	EWR	IAD	0
0	US	2	2057	LGA	DCA	0
0	DH	3	1603	JFK	IAD	0
0	CO	6	1252	EWR	DCA	0
0	RU	6	1728	EWR	DCA	0
0	DL	5	1031	LGA	DCA	0
0	RU	6	1722	EWR	IAD	0
1	US	1	627	LGA	DCA	0
1	DH	2	1756	JFK	IAD	0
0	MQ	6	1529	JFK	DCA	0
0	US	6	1259	LGA	DCA	0
0	DL	2	1329	LGA	DCA	0
0	RU	2	1453	EWR	BWI	0
0	RU	5	1356	EWR	DCA	0
1	DH	7	2244	LGA	IAD	0
0	US	7	1053	LGA	DCA	0
0	US	2	1057	LGA	DCA	0
0	US	4	632	LGA	DCA	0

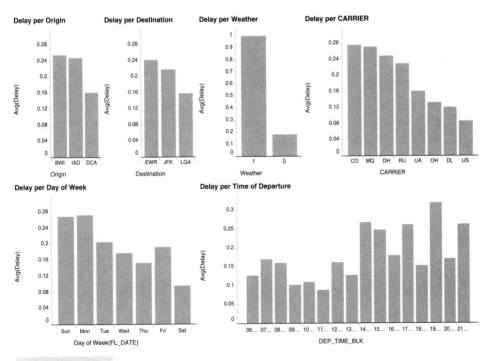

FIGURE 10.6	PROPORTION OF DELAYED FLIGHTS BY EACH OF THE FIVE PREDICTORS. TIME OF DAY IS DIVIDED INTO HOURLY BINS.

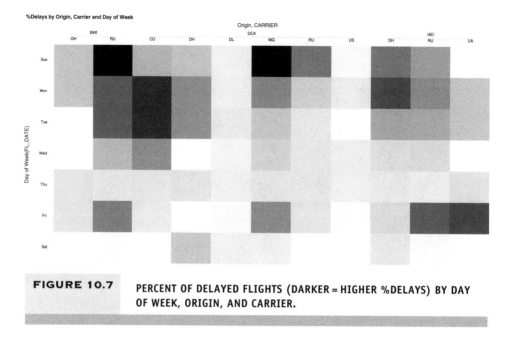

FIGURE 10.7   PERCENT OF DELAYED FLIGHTS (DARKER = HIGHER %DELAYS) BY DAY OF WEEK, ORIGIN, AND CARRIER.

effects. A logistic regression model can be used for all these goals, albeit in different ways.

Data Preprocessing

We first create dummy variables for each of the categorical predictors: two dummies for the departure airport (with IAD as the reference airport), two for the arrival airport (with LGA as the reference), seven for the carrier (with USAirways as the reference carrier), six for the day (with Wednesday as the reference group), 15 for the departure hour (hourly intervals between 6 AM and 10 PM). This yields a total of 33 dummies. In addition, we have a single dummy for Weather. While this is a large number of predictors, we start by using all of them, look at performance, and then explore reducing the dimensions.

We then partition the data into training set (60%) and validation set (40%). We use the training set to fit a model and the validation set to assess the model's performance.

Model Fitting and Estimation

The estimated model with 33 predictors is given in Figure 10.8. Notice how negative coefficients in the logit model (the "Coefficient" column) translate into odds coefficients lower than 1, and positive logit coefficients translate into odds coefficients larger than 1.

Input Variables	Coefficient	Std. Error	Chi2-Statistic	p-Value	Odds
Intercept	−2.767405	0.635035	18.99110849	1.31329E-05	0.062825
Weather	8.7011206	8.025675	1.175404312	0.278293576	6009.643
CARRIER_CO	1.8085832	0.536808	11.35114797	0.000754013	6.101796
CARRIER_DH	1.2697338	0.523548	5.881818042	0.015298018	3.559905
CARRIER_DL	0.4193065	0.335389	1.563018961	0.211223733	1.520906
CARRIER_MQ	1.7176629	0.330507	27.00937692	2.02471E-07	5.571492
CARRIER_OH	0.1281504	0.854056	0.022514732	0.880725879	1.136724
CARRIER_RU	1.2950371	0.513515	6.360026089	0.011671983	3.651131
CARRIER_UA	−0.536269	1.175101	0.208264596	0.648131009	0.584926
DEP_TIME_BLK_0700-0759	0.1990101	0.522474	0.145084515	0.70327774	1.220194
DEP_TIME_BLK_0800-0859	0.3922193	0.488976	0.643402861	0.422481251	1.480262
DEP_TIME_BLK_0900-0959	−0.231681	0.612402	0.143122276	0.705196568	0.793199
DEP_TIME_BLK_1000-1059	−0.845045	0.659046	1.644097725	0.19976417	0.429538
DEP_TIME_BLK_1100-1159	0.2729511	0.622993	0.191956873	0.661292895	1.313836
DEP_TIME_BLK_1200-1259	0.4466879	0.478279	0.872261462	0.35033	1.563126
DEP_TIME_BLK_1300-1359	0.2935675	0.498882	0.346273948	0.556229975	1.341204
DEP_TIME_BLK_1400-1459	0.9615407	0.426268	5.08825999	0.024088349	2.615723
DEP_TIME_BLK_1500-1559	0.8522894	0.475057	3.218722578	0.072800422	2.345009
DEP_TIME_BLK_1600-1659	0.7592518	0.463078	2.688208509	0.101093417	2.136677
DEP_TIME_BLK_1700-1759	0.8462042	0.421629	4.028006867	0.044750781	2.330783
DEP_TIME_BLK_1800-1859	0.7256879	0.573265	1.602467794	0.205553839	2.066152
DEP_TIME_BLK_1900-1959	1.421361	0.48125	8.723012387	0.003142184	4.142755
DEP_TIME_BLK_2000-2059	0.963043	0.63341	2.311647197	0.128407911	2.619656
DEP_TIME_BLK_2100-2159	0.7957765	0.452274	3.095836563	0.078492798	2.216161
Destination_EWR	−0.348665	0.319026	1.194436212	0.274436543	0.70563
Destination_JFK	−0.64583	0.267018	5.849979223	0.015577235	0.524227
DOW_Fri	0.0512637	0.268224	0.036528019	0.848429191	1.0526
DOW_Mon	0.5244692	0.272285	3.710168231	0.05408194	1.689562
DOW_Sat	−0.407275	0.329605	1.526818805	0.216590514	0.665461
DOW_Sun	0.4665922	0.287453	2.634767196	0.104547385	1.594551
DOW_Thu	−0.097089	0.281637	0.118840009	0.730296014	0.907475
DOW_Tue	0.1333263	0.285531	0.21803466	0.640541507	1.142623
Origin_BWI	0.3546605	0.409812	0.748956871	0.386806693	1.425697
Origin_DCA	−0.524996	0.380317	1.905557044	0.167457626	0.591557

FIGURE 10.8   ESTIMATED LOGISTIC REGRESSION MODEL FOR DELAYED FLIGHTS (BASED ON THE TRAINING SET)

Model Interpretation

The coefficient for Arrival Airport JFK is estimated from the data to be −0.65. Recall that the reference group is LGA. We interpret this coefficient as follows: $e^{-0.65} = 0.52$ are the odds of a flight arriving at JFK being delayed relative to a flight to LGA being delayed (= the base-case odds), holding all other factors constant. This means that flights to LGA are more likely to be delayed than those to JFK (holding everything else constant). If we take into account statistical significance of the coefficients, we see that in general the departure airport is not associated with the chance of delays. For carriers, it appears that four carriers are significantly different from the base carrier (USAirways), with odds of delay ranging between 3.5 to 6.1 relative to the other airlines. Weather has an enormous coefficient, which is not statistically significant. Flights leaving on Sunday or Monday have, on average, odds of 1.6 of delays relative to other days of the week (the other days seem statistically similar to the reference group Wednesday). Odds of delays also appear to change over the course of the day, with the most noticeable difference from the reference category (6–7 AM) being 7–8 PM.

Model Performance

How should we measure the performance of models? One possible measure is "percent of flights correctly classified." Accurate classification can be obtained from the classification matrix for the validation data. The classification matrix gives a sense of the classification accuracy and what type of misclassification is more frequent. From the classification matrix and error rates in Figure 10.9, it can be seen that the model more accurately classifies nondelayed flights and is less accurate in classifying flights that were delayed. (*Note:* The same pattern appears in the classification matrix for the training data, so it is not surprising to see it emerge for new data.) If there is an asymmetric cost structure so that one type of misclassification is more costly than the other, the cutoff value can be selected to minimize the cost. Of course, this tweaking should be carried out on the training data and assessed only using the validation data.

In most conceivable situations, the purpose of the model would be to identify those flights most likely to be delayed so that resources can be directed toward

| Cutoff probability value for success (UPDATABLE) | 0.5 | Updating the value here will NOT update value in detailed report |

Confusion Matrix

Actual Class	Predicted Class	
	1	0
1	19	153
0	3	705

Error Report

Class	# Cases	# Errors	% Error
1	172	153	88.95
0	708	3	0.42
Overall	880	156	17.73

Lift chart (validation dataset)

Cumulative Delay when sorted using predicted values

Cumulative Delay using average

FIGURE 10.9   CLASSIFICATION MATRIX AND ERROR RATES, AND LIFT CHART FOR THE FLIGHT DELAY VALIDATION DATA either reducing the delay or mitigating its effects. Air traffic controllers might work to open up additional air routes or allocate more controllers to a specific area for a short time. Airlines might bring on personnel to rebook passengers and to activate standby flight crews and aircraft. Hotels might allocate space for stranded travelers. In all cases, the resources available are going to be limited and might vary over time and from organization to organization. In this situation, the most useful model would provide an ordering of flights by their probability of delay, letting the model users decide how far down that list to go in taking action. Therefore model lift is a useful measure of performance—as you move down that list of flights, ordered by their delay probability, how much better does the model do in predicting delay than would a naive model which is simply the average delay rate for all flights? From the lift curve for the validation data (Figure 10.9) we see that our model is superior to the baseline (simple random selection of flights).

Variable Selection

From the data exploration charts (Figures 10.6 and 10.7) and from the coefficient table for the flights delay model, it appears that several of the variables might be dropped or coded differently. Additionally, we look at the number of flights in different categories to identify categories with very few or no flights—such categories are candidates for removal or merger (see Table 10.3).

First, we find that most carriers depart from a single airport (DCA). For those that depart from all three airports, the delay rates are similar regardless of airport. We therefore drop the departure airport distinction by excluding Origin dummies and find that the model performance and fit is not harmed. We also

TABLE 10.3   NUMBER OF FLIGHTS BY CARRIER AND ORIGIN

	BWI	DCA	IAD	Total
CO		94		94
DH		27	524	551
DL		388		388
MQ		295		295
OH	30			30
RU	115	162	131	408
UA			31	31
US		404		404
Total	145	1370	686	2201

FIGURE 10.10   XLMINER'S REDUCE CATEGORIES UTILITY (*TRANSFORM > TRANSFORM CATEGORICAL DATA*). HERE WE USE IT MANUALLY TO CREATE TWO CATEGORIES OF CARRIERS FROM THE ORIGINAL 8 CATEGORIES.

drop the destination airport for a practical reason: not all carriers fly to all airports. Our model would then be invalid for prediction in nonexistent combinations of carrier and destination airport. We also try grouping carriers, day of week, and hour of day into fewer categories that are more distinguishable with respect to delays (this can be done using XLMiner's Reduce Categories utility; see Figure 10.10). For example, Sundays and Mondays seem to have a similar rate of delays, which differs from the lower rate on Tuesday to Saturday. We therefore group the days of week into Sunday + Monday and Other, resulting in a single dummy variable.

Figure 10.11 displays the estimated model, with its training and validation classification matrices and error rates, as well as the lift charts. It can be seen that this model competes well with the larger model in terms of classification accuracy and lift, while using much less information.

Regression Model

Input Variables	Coefficient	Std. Error	Chi2-Statistic	p-Value	Odds	CI Lower	CI Upper
Intercept	-2.783912	0.205258	183.9549438	6.64E-42	0.061796	0.041328	0.092401325
Sun-Mon	0.530707	0.16187	10.74921936	0.001043	1.700134	1.237939	2.334893348
Weather	8.731349	8.046588	1.177440838	0.277878	6194.076	0.000876	43776887785
CARRIER_CO_MQ_DH_RU	1.197526	0.178343	45.08749955	1.88E-11	3.311912	2.334924	4.69769579
DEP_TIME_BLK_9-11am	-0.745774	0.375615	3.942101728	0.047092	0.474367	0.227191	0.990464306
DEP_TIME_BLK_After2pm	0.501026	0.167951	8.899255268	0.002853	1.650413	1.187496	2.293786096
DEP_TIME_BLK_7-8pm	1.210624	0.306554	15.59569907	7.84E-05	3.355577	1.840044	6.11936398

Residual DF	1314
Residual Dev.	1142.615
# Iterations Used	5
Multiple R²	0.120394

Training Data Scoring—Summary Report

Validation Data Scoring—Summary Report

FIGURE 10.11   OUTPUT FOR LOGISTIC REGRESSION MODEL WITH FEWER PREDICTORS

We therefore conclude with a six-predictor model that requires only knowledge of the carrier, the day of week, the hour of the day, and whether the day had inclement weather. However, this weather variable refers to actual weather at flight time, not a forecast, and is not known in advance! If the aim is to predict in advance whether a particular flight will be delayed, a model without *Weather* must be used. In contrast, if the goal is profiling delayed vs. nondelayed flights, we can keep *Weather* in the model to allow evaluating the impact of the other factors while holding weather constant [i.e., (approximately) comparing days with inclement weather to days without inclement weather].

To conclude, based on the model built from the January 2004 data, the highest chance of a nondelayed flight from DC to New York is on Tuesday to Saturday between 9 and 11am, on Delta, Comair, United or USAirways. And clearly, good weather is advantageous!

10.5 APPENDIX: LOGISTIC REGRESSION FOR PROFILING

The presentation of logistic regression in this chapter has been primarily from a data mining perspective where classification or ranking is the goal, and performance is evaluated by reviewing results with a validation sample. For reference, some key concepts of a classical statistical perspective are included below.

Appendix A: Why Linear Regression Is Problematic for a Categorical Response

Now that you have seen how logistic regression works, we explain why it is considered preferable over linear regression for a binary outcome. Technically, one can apply a multiple linear regression model to this problem, treating the outcome variable Y as continuous. This is called a *Linear Probability Model*. Of course, Y must be coded numerically (e.g., 1 for customers who did accept the loan offer and 0 for customers who did not accept it). Although software will yield an output that at first glance may seem standard (e.g., Figure 10.12), a closer look will reveal several anomalies:

1. Using the model to predict Y for each of the observations (or classify them) yields predictions that are not necessarily 0 or 1.

2. A look at the histogram or probability plot of the residuals reveals that the assumption that the dependent variable (or residuals) follows a normal distribution is violated. Clearly, if Y takes only the values 0 and 1, then it cannot be normally distributed. In fact, a more appropriate distribution

Regression Model

Input Variables	Coefficient	Std. Error	t-Statistic	p-Value	CI Lower	CI Upper	RSS Reduction
Intercept	−0.23463	0.01329	−17.65839	0.00000	−0.26068	−0.20858	27.26533
Income	0.00319	0.00010	32.25651	0.00000	0.00300	0.00338	67.95862
Family	0.03294	0.00384	8.58057	0.00000	0.02541	0.04047	4.53180
CD Account	0.27016	0.01789	15.10542	0.00000	0.23510	0.30523	13.18045

ANOVA

Source	DF	SS	MS	F-Statistic	p-Value
Regression	3	85.6709	28.557	494.3648	0
Error	2996	173.0638	0.0578		
Total	2999	258.7347	28.6147		

FIGURE 10.12   OUTPUT FOR MULTIPLE LINEAR REGRESSION MODEL OF *PERSONAL LOAN* ON THREE PREDICTORS

for the number of 1's in the dataset is the binomial distribution with $p = P(Y = 1)$.

3. The assumption that the variance of Y is constant across all classes is violated. Since Y follows a binomial distribution, its variance is $np(1 - p)$. This means that the variance will be higher for classes where the probability of adoption, p, is near 0.5 than where it is near 0 or 1.

The first anomaly is the main challenge when the goal is classification, especially if we are interested in propensities ($P(Y = 1)$). The second and third anomalies are relevant to profiling, where we use statistical inference that relies on standard errors.

Below you will find partial output from running a multiple linear regression of Personal Loan (PL, coded as PL = 1 for customers who accepted the loan offer and PL = 0 otherwise) on three of the predictors.

The estimated model is

$$\widehat{PL} = -0.2346 + 0.0032 \text{ Income} + 0.0329 \text{ Family} + 0.27016363 \text{ CD}.$$

To predict whether a new customer will accept the personal loan offer (PL = 1) or not (PL = 0), we input the information on its values for these three predictors. For example, we would predict the loan offer acceptance of a customer with an annual income of \$50K with two family members who does not hold CD accounts in Universal Bank to be $-0.2346 + (0.0032)(50) + (0.0329)(2) = -0.009$. Clearly, this is not a valid "loan acceptance" value. Furthermore the histogram of the residuals (Figure 10.13) reveals that the residuals are probably not normally distributed. Therefore our estimated model is based on violated assumptions and cannot be used for inference.

Appendix B: Evaluating Explanatory Power

When the purpose of the analysis is profiling (identifying predictor profiles that distinguish the two classes, or explaining the differences between the classes in terms of predictor values), we are less interested in how well the model classifies new data than in how well the model fits the data it was trained on. For example, if we are interested in characterizing the average loan offer acceptor vs. nonacceptor in terms of income, education, and so on, we want to find a model that fits the data best. We therefore mention popular measures used to assess how well the model fits the data. Clearly, we look at the training set in order to evaluate goodness of fit.

Overall Strength of Fit   As in multiple linear regression, we first evaluate the overall explanatory power of the model before looking at single predictors.

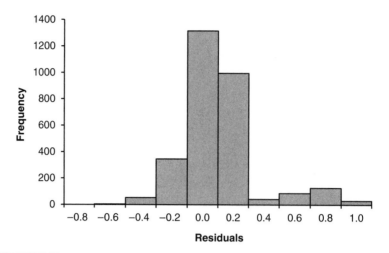

FIGURE 10.13 HISTOGRAM OF RESIDUALS FROM A MULTIPLE LINEAR REGRESSION MODEL OF LOAN ACCEPTANCE ON THE THREE PREDICTORS. THIS SHOWS THAT THE RESIDUALS DO NOT FOLLOW THE NORMAL DISTRIBUTION THAT THE MODEL ASSUMES.

We ask: Is this set of predictors better than a simple naive model for explaining the difference between classes?[7]

The deviance D is a statistic that measures overall goodness of fit. It is similar to the concept of sum of squared errors (SSE) in the case of least squares estimation (used in linear regression). We compare the deviance of our model, D (called *Std Dev Estimate* in XLMiner, e.g., in Figure 10.14), to the deviance of the naive model, D_0. If the reduction in deviance is statistically significant (as indicated by a low p-value[8] or in XLMiner by a high *multiple R^2*), we consider our model to provide a good overall fit. XLMiner's *Multiple-R^2* measure is computed as $(D_0 - D)/D_0$. Given the model deviance and the multiple R^2, we can compute the null deviance by $D_0 = D/(1 - R^2)$.

Finally, the classification matrix and lift chart for the *training data* (Figure 10.15) give a sense of how accurately the model classifies the data. If the model fits the data well, we expect it to classify these data accurately into their actual classes.

Impact of Single Predictors As in multiple linear regression, the output from a logistic regression procedure typically yields a coefficient table, where for each predictor X_i we have an estimated coefficient b_i and an associated standard error σ_i. The associated p-value indicates the statistical significance of the pre-

---

[7] In a naive model, no explanatory variables exist and each observation is classified as belonging to the majority class.
[8] The difference between the deviance of a naive model and deviance of the model at hand approximately follows a chi-squared distribution with k degrees of freedom, where k is the number of predictors in the model at hand. Therefore, to get the p-value, compute the difference between the deviances (d) and then look up the probability that a chi-squared variable with k degrees of freedom is larger than d. This can be done using =CHIDIST(d, k) in Excel.

Residual DF	2987
Residual Dev.	652.5186
# Iterations Used	6
Multiple R^2	0.65443

FIGURE 10.14 MEASURES OF EXPLANATORY POWER FOR UNIVERSAL BANK TRAINING DATA WITH A 12 PREDICTOR MODEL

Training Data Scoring—Summary Report

Confusion Matrix

	Predicted Class	
Actual Class	1	0
1	201	85
0	25	2689

Error Report

Class	# Cases	# Errors	% Error
1	286	85	29.720
0	2714	25	0.921
Overall	3000	110	3.667

Lift chart (training dataset)

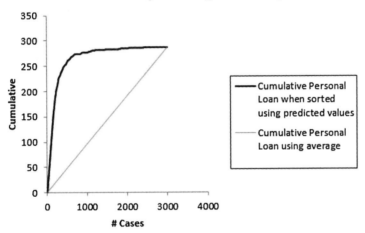

FIGURE 10.15 CLASSIFICATION MATRIX (TOP) AND LIFT CHART (BOTTOM) FOR TRAINING DATA FOR UNIVERSAL BANK TRAINING DATA WITH 12 PREDICTORS

dictor X_i, with very low p-values indicating a statistically significant relationship between the predictor and the outcome (given that the other predictors are accounted for), a relationship that is most likely not a result of chance. Three important points to remember are:

1. A statistically significant relationship is not necessarily a *practically significant* one, in which the predictor has great impact. If the sample is very large, the *p*-value will be very small simply because the chance uncertainty associated with a small sample is negligible.

2. Comparing the coefficient magnitudes, or equivalently the odds magnitudes, is meaningless unless all predictors have the same scale. Recall that each coefficient is multiplied by the predictor value, so that different predictor scales would lead to different coefficient scales.

3. A statistically significant predictor means that *on average*, a unit increase in that predictor is associated with a certain effect on the outcome (holding all other predictors constant). It does not, however, indicate predictive power. Statistical significance is of major importance in explanatory modeling, or *profiling*, but of secondary importance in predictive modeling (*classification*). In predictive modeling, statistically significant predictors might give hints as to more and less important predictors, but the eventual choice of predictors should be based on predictive measures, such as the validation set confusion matrix (for classification) or the validation set lift chart (for ranking).

Appendix C: Logistic Regression for More Than Two Classes

The logistic model for a binary response can be extended for more than two classes. Suppose that there are m classes. Using a logistic regression model, for each observation we would have m probabilities of belonging to each of the m classes. Since the m probabilities must add up to 1, we need estimate only $m - 1$ probabilities.

Ordinal Classes   Ordinal classes are classes that have a meaningful order. For example, in stock recommendations, the three classes *buy*, *hold*, and *sell* can be treated as ordered. As a simple rule, if classes can be numbered in a meaningful way, we consider them ordinal. When the number of classes is large (typically, more than 5), we can treat the dependent variable as continuous and perform multiple linear regression. When $m = 2$, the logistic model described above is used. We therefore need an extension of the logistic regression for a small number of ordinal classes ($3 \le m \le 5$). There are several ways to extend the binary-class case. Here we describe the *proportional odds* or *cumulative logit method*. For other methods, see Hosmer and Lemeshow (2000).

For simplicity of interpretation and computation, we look at *cumulative* probabilities of class membership. For example, in the stock recommendations we have m = 3 classes. Let us denote them by 1 = *buy*, 2 = *hold*, and 3 = *sell*. The probabilities that are estimated by the model are $P(Y \le 1)$, (the probability of a *buy* recommendation) and $P(Y \le 2)$ (the probability of a *buy* or *hold* recommendation). The three noncumulative probabilities of class membership can easily be recovered from the two cumulative probabilities:

$$P(Y = 1) = P(Y \leq 1),$$

$$P(Y = 2) = P(Y \leq 2) - P(Y \leq 1),$$

$$P(Y = 3) = 1 - P(Y \leq 2).$$

Next, we want to model each logit as a function of the predictors. Corresponding to each of the $m - 1$ cumulative probabilities is a logit. In our example we would have

$$\text{logit(buy)} = \log \frac{P(Y \leq 1)}{1 - P(Y \leq 1)},$$

$$\text{logit (buy or hold)} = \log \frac{P(Y \leq 2)}{1 - P(Y \leq 2)}.$$

Each of the logits is then modeled as a linear function of the predictors (as in the two-class case). If in the stock recommendations we have a single predictor x, we have two equations:

$$\text{logit(buy)} = \alpha_0 + \beta_1 x,$$

$$\text{logit(buy or hold)} = \beta_0 + \beta_1 x.$$

This means that both lines have the same slope (β_1) but different intercepts. Once the coefficients $\alpha_0, \beta_0, \beta_1$ are estimated, we can compute the class membership probabilities by rewriting the logit equations in terms of probabilities. For the three-class case, for example, we would have

$$P(Y = 1) = P(Y \leq 1) = \frac{1}{1 + e^{-(a_0 + b_1 x)}},$$

$$P(Y = 2) = P(Y \leq 2) - P(Y \leq 1) = \frac{1}{1 + e^{-(b_0 + b_1 x)}} - \frac{1}{1 + e^{-(a_0 + b_1 x)}},$$

$$P(Y = 3) = 1 - P(Y \leq 2) = 1 - \frac{1}{1 + e^{-(b_0 + b_1 x)}},$$

where a_0, b_0, and b_1 are the estimates obtained from the training set.

For each observation we now have the estimated probabilities that it belongs to each of the classes. In our example, each stock would have three probabilities: for a *buy* recommendation, a *hold* recommendation, and a *sell* recommendation. The last step is to classify the observation into one of the classes. This is done by assigning it to the class with the highest membership probability. So, if a stock had estimated probabilities $P(Y = 1)=0.2$, $P(Y = 2) = 0.3$, and $P(Y = 3) = 0.5$, we would classify it as getting a *sell* recommendation.

This procedure is currently not implemented in XLMiner. Other non–Excel-based packages that do have such an implementation are Minitab and SAS.

Nominal Classes When the classes cannot be ordered and are simply different from one another, we are in the case of nominal classes. An example is the choice between several brands of cereal. A simple way to verify that the classes are nominal is when it makes sense to tag them as $A, B, C, \ldots$, and the assignment of letters to classes does not matter. For simplicity, let us assume that there are $m = 3$ brands of cereal that consumers can choose from (assuming that each consumer chooses one). Then we estimate the probabilities $P(Y = A)$, $P(Y = B)$, and $P(Y = C)$. As before, if we know two of the probabilities, the third probability is determined. We therefore use one of the classes as the reference class. Let us use C as the reference brand.

The goal, once again, is to model the class membership as a function of predictors. So in the cereals example we might want to predict which cereal will be chosen if we know the cereal's price, x.

Next, we form $m - 1$ pseudologit equations that are linear in the predictors. In our example we would have

$$\text{logit}(A) = \log \frac{P(Y = A)}{P(Y = C)} = \alpha_0 + \alpha_1 x,$$

$$\text{logit}(B) = \log \frac{P(Y = B)}{P(Y = C)} = \beta_0 + \beta_1 x.$$

Once the four coefficients are estimated from the training set, we can estimate the class membership probabilities[9]

$$P(Y = A) = \frac{e^{a_0 + a_1 x}}{1 + e^{a_0 + a_1 x} + e^{b_0 + b_1 x}},$$

$$P(Y = B) = \frac{e^{b_0 + b_1 x}}{1 + e^{a_0 + a_1 x} + e^{b_0 + b_1 x}},$$

$$P(Y = C) = 1 - P(Y = A) - P(Y = B),$$

where a_0, a_1, b_0, and b_1 are the coefficient estimates obtained from the training set. Finally, an observation is assigned to the class that has the highest probability.

---

[9] From the two logit equations we see that

$$P(Y = A) = P(Y = C)e^{\alpha_0 + \alpha_1 x}$$
$$P(Y = B) = P(Y = C)e^{\beta_0 + \beta_1 x}$$

Since $P(Y = A) + P(Y = B) + P(Y = C) = 1$, we get

$$P(Y = C) = 1 - P(Y = C)e^{\alpha_0 + \alpha_1 x} - P(Y = C)e^{\beta_0 + \beta_1 x}$$
$$= \frac{1}{e^{\alpha_0 + \alpha_1 x} + e^{\beta_0 + \beta_1 x}}.$$

By plugging this form into the two equations above it, we also obtain the membership probabilities in classes A and B.

PROBLEMS

10.1 **Financial Condition of Banks.** The file `Banks.xls` includes data on a sample of 20 banks. The "Financial Condition" column records the judgment of an expert on the financial condition of each bank. This dependent variable takes one of two possible values—*weak* or *strong*—according to the financial condition of the bank. The predictors are two ratios used in the financial analysis of banks: TotLns&Lses/Assets is the ratio of total loans and leases to total assets and TotExp/Assets is the ratio of total expenses to total assets. The target is to use the two ratios for classifying the financial condition of a new bank.

Run a logistic regression (on the entire dataset) that models the status of a bank as a function of the two financial measures provided. Specify the *success* class as *weak* (this is similar to creating a dummy that is 1 for financially weak banks and 0 otherwise), and use the default cutoff value of 0.5.

  a. Write the estimated equation that associates the financial condition of a bank with its two predictors in three formats:
    i. The logit as a function of the predictors
    ii. The odds as a function of the predictors
    iii. The probability as a function of the predictors

  b. Consider a new bank whose total loans and leases/assets ratio = 0.6 and total expenses/assets ratio = 0.11. From your logistic regression model, estimate the following four quantities for this bank (use Excel to do all the intermediate calculations; show your final answers to four decimal places): the logit, the odds, the probability of being financially weak, and the classification of the bank (use cutoff = 0.5).

  c. The cutoff value of 0.5 is used in conjunction with the probability of being financially weak. Compute the threshold that should be used if we want to make a classification based on the odds of being financially weak, and the threshold for the corresponding logit.

  d. Interpret the estimated coefficient for the total loans and leases to total assets ratio (TotLns&Lses/Assets) in terms of the odds of being financially weak.

  e. When a bank that is in poor financial condition is misclassified as financially strong, the misclassification cost is much higher than when a financially strong bank is misclassified as weak. To minimize the expected cost of misclassification, should the cutoff value for classification (which is currently at 0.5) be increased or decreased?

10.2 **Identifying Good System Administrators.** A management consultant is studying the roles played by experience and training in a system administrator's ability to complete a set of tasks in a specified amount of time. In particular, she is interested in discriminating between administrators who are able to complete given tasks within a specified time and those who are not. Data are collected on the performance of 75 randomly selected administrators. They are stored in the file `SystemAdministrators.xls`. The variable Experience measures months of full-time system administrator experience, while Training measures the number of relevant training credits. The dependent variable Completed is either Yes or No, according to whether or not the administrator completed the tasks.

  a. Create a scatterplot of *Experience* versus *Training* using color or symbol to differentiate programmers who complete the task from those who did not complete it. Which predictor(s) appear(s) potentially useful for classifying task completion?

b. Run a logistic regression model with both predictors using the entire dataset as training data. Among those who complete the task, what is the percentage of programmers who are incorrectly classified as failing to complete the task?

c. To decrease the percentage in part (b), should the cutoff probability be increased or decreased?

d. How much experience must be accumulated by a programmer with four years of training before his or her estimated probability of completing the task exceeds 0.5?

10.3 **Sales of Riding Mowers.** A company that manufactures riding mowers wants to identify the best sales prospects for an intensive sales campaign. In particular, the manufacturer is interested in classifying households as prospective owners or nonowners on the basis of Income (in $1000s) and Lot Size (in 1000 ft^2). The marketing expert looked at a random sample of 24 households, given in the file RidingMowers.xls. Use all the data to fit a logistic regression of ownership on the two predictors.

a. What percentage of households in the study were owners of a riding mower?

b. Create a scatterplot of Income versus Lot Size using color or symbol to differentiate owners from nonowners. From the scatterplot, which class seems to have the higher average income, owners or nonowners?

c. Among nonowners, what is the percentage of households classified correctly?

d. To increase the percentage of correctly classified nonowners, should the cutoff probability be increased or decreased?

e. What are the odds that a household with a $60K income and a lot size of 20,000 ft^2 is an owner?

f. What is the classification of a household with a $60K income and a lot size of 20,000 ft^2? Use cutoff = 0.5.

g. What is the minimum income that a household with 16,000 ft^2 lot size should have before it is classified as an owner?

10.4 **Competitive Auctions on eBay.com.** The file eBayAuctions.xls contains information on 1972 auctions transacted on eBay.com during May–June 2004. The goal is to use these data to build a model that will distinguish competitive auctions from noncompetitive ones. A competitive auction is defined as an auction with at least two bids placed on the item being auctioned. The data include variables that describe the item (auction category), the seller (his or her eBay rating), and the auction terms that the seller selected (auction duration, opening price, currency, day of week of auction close). In addition, we have the price at which the auction closed. The goal is to predict whether or not the auction will be competitive.

Data preprocessing. Create dummy variables for the categorical predictors. These include Category (18 categories), Currency (USD, GBP, euro), EndDay (Monday–Sunday), and Duration (1, 3, 5, 7, or 10 days). Split the data into training (60%) and validation (40%) datasets.

a. Create pivot tables for the average of the binary dependent variable (Competitive?) as a function of the various categorical variables (use the original variables, not the dummies). Use the information in the tables to reduce the number of dummies that will be used in the model. For example, categories that appear most similar with respect to the distribution of competitive auctions could be combined.

b. Run a logistic model with all predictors with a cutoff of 0.5.

c. If we want to predict at the start of an auction whether it will be competitive, we cannot use the information on the closing price. Run a logistic model with all predictors as above, excluding price. How does this model compare to the full model with respect to accurate prediction?

d. Interpret the meaning of the coefficient for closing price. Does closing price have a practical significance? Is it statistically significant for predicting competitiveness of auctions? (Use a 10% significance level.)

e. Use stepwise selection and an exhaustive search to find the model with the best fit to the training data. Which predictors are used?

f. Use stepwise selection and an exhaustive search to find the model with the lowest predictive error rate (use the validation data). Which predictors are used?

g. What is the danger of using the best-predictive model that you found?

h. Explain why the best-fitting model and the best-predictive models are the same or different.

i. If the major objective is accurate classification, what cutoff value should be used?

j. Based on these data, what auction settings set by the seller (duration, opening price, ending day, currency) would you recommend as being most likely to lead to a competitive auction?

CHAPTER 11

Neural Nets

In this chapter we describe neural networks, a flexible data-driven method that can be used for classification or prediction. Although considered a "black box" in terms of interpretability, neural nets have been highly successful in terms of predictive accuracy. We discuss the concepts of "nodes" and "layers" (input layers, output layers, and hidden layers) and how they connect to form the structure of a network. We then explain how a neural network is fitted to data using a numerical example. Because overfitting is a major danger with neural nets, we present a strategy for avoiding it. We describe the different parameters that a user must specify and explain the effect of each on the process. Finally, we discuss the usefulness of neural nets and their limitations.

11.1 INTRODUCTION

Neural networks, also called *artificial neural networks*, are models for classification and prediction. The neural network is based on a model of biological activity in the brain, where neurons are interconnected and learn from experience. Neural networks mimic the way that human experts learn. The learning and memory properties of neural networks resemble the properties of human learning and memory, and they also have a capacity to generalize from particulars.

A number of successful applications have been reported in financial applications (see Trippi and Turban, 1996) such as bankruptcy predictions, currency

Data Mining for Business Analytics: Concepts, Techniques, and Applications with XLMiner®, Third Edition.
Galit Shmueli, Peter C. Bruce, and Nitin R. Patel.
© 2016 John Wiley & Sons Inc. Published 2016 by John Wiley & Sons, Inc.

market trading, picking stocks and commodity trading, detecting fraud in credit card and monetary transactions, and customer relationship management (CRM). There have also been a number of successful applications of neural nets in engineering applications. One of the best known is ALVINN, an autonomous vehicle-driving application for normal speeds on highways. Using as input a 30×32 grid of pixel intensities from a fixed camera on the vehicle, the classifier provides the direction of steering. The response variable is a categorical one with 30 classes, such as *sharp left, straight ahead*, and *bear right.*

The main strength of neural networks is their high predictive performance. Their structure supports capturing very complex relationships between predictors and a response, which is often not possible with other predictive models.

11.2   CONCEPT AND STRUCTURE OF A NEURAL NETWORK

The idea behind neural networks is to combine the input information in a very flexible way that captures complicated relationships among these variables and between them and the response variable. For instance, recall that in linear regression models the form of the relationship between the response and the predictors is specified directly by the user (see Chapter 6.) In many cases the exact form of the relationship is very complicated, or is generally unknown. In linear regression modeling we might try different transformations of the predictors, interactions between predictors, and so on, but the specified form of the relationship remains linear. In comparison, in neural networks the user is not required to specify the correct form. The network tries instead to learn about such relationships from the data. In fact linear regression and logistic regression can be thought of as special cases of very simple neural networks that have only input and output layers and no hidden layers.

Although researchers have studied numerous different neural network architectures, the most successful applications in data mining of neural networks have been *multilayer feedforward networks*. These are networks in which there is an *input layer* consisting of nodes (sometimes called *neurons*) that simply accept the input values, and successive layers of nodes that receive input from the previous layers. The outputs of nodes in each layer are inputs to nodes in the next layer. The last layer is called the *output layer*. Layers between the input and output layers are known as *hidden layers*. A feedforward network is a fully connected network with a one-way flow and no cycles. Figure 11.1 shows a diagram for this architecture, with two hidden layers and one node in the output layer representing the target value to be predicted. In a classification problem with *m* classes, there would be *m* output nodes.

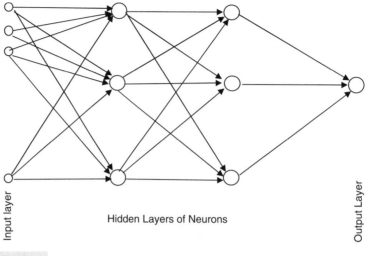

Input layer

Hidden Layers of Neurons

Output Layer

FIGURE 11.1   **MULTILAYER FEEDFORWARD NEURAL NETWORK.**

11.3 FITTING A NETWORK TO DATA

To illustrate how a neural network is fitted to data, we start with a very small illustrative example. Although the method is by no means operational in such a small example, it is useful for explaining the main steps and operations, for showing how computations are done and for integrating all the different aspects of neural network data fitting. We will later discuss a more realistic setting.

Example 1: Tiny Dataset

Consider the following very small dataset. Table 11.1 includes information on a tasting score for a certain processed cheese. The two predictors are scores for fat and salt, indicating the relative presence of fat and salt in the particular cheese sample (where 0 is the minimum amount possible in the manufacturing process,

	TABLE 11.1	**TINY EXAMPLE ON TASTING SCORES FOR SIX CONSUMERS AND TWO PREDICTORS**

Obs.	Fat Score	Salt Score	Acceptance
1	0.2	0.9	like
2	0.1	0.1	dislike
3	0.2	0.4	dislike
4	0.2	0.5	like
5	0.4	0.5	like
6	0.3	0.8	like

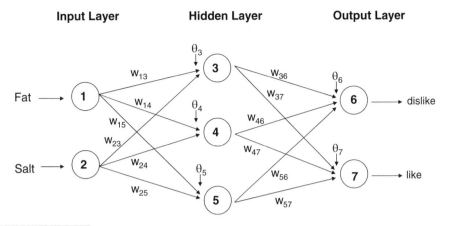

Input Layer   **Hidden Layer**   **Output Layer**

FIGURE 11.2   NEURAL NETWORK FOR THE TINY EXAMPLE. CIRCLES REPRESENT NODES ("NEURONS"), $w_{i,j}$ ON ARROWS ARE WEIGHTS, AND θ_j ARE NODE BIAS VALUES.

and 1 the maximum). The output variable is the cheese sample's consumer taste preference, where "like" or "dislike" indicate whether or not the consumer likes the cheese.

Figure 11.2 describes an example of a typical neural net that could be used for predicting cheese preference (like/dislike) by new consumers, based on these data. We numbered the nodes in the example from 1 to 7. Nodes 1 and 2 belong to the input layer, nodes 3 to 5 belong to the hidden layer, and nodes 6 and 7 belong to the output layer. The values on the connecting arrows are called *weights*, and the weight on the arrow from node i to node j is denoted by $w_{i,j}$. The additional *bias* nodes, denoted by θ_j, serve as an intercept for the output from node j. These are all explained in more detail below.

Computing Output of Nodes

We discuss the input and output of the nodes separately for each of the three types of layers (input, hidden, and output). The main difference is the function used to map from the input to the output of the node.

Input nodes take as input the values of the predictors. Their output is the same as the input. If we have p predictors, the input layer will usually include p nodes. In our example there are two predictors, and therefore the input layer (shown in Figure 11.2) includes two nodes, each feeding into each node of the hidden layer. Consider the first observation: The input into the input layer is $fat = 0.2$ and $salt = 0.9$, and the output of this layer is also $x_1 = 0.2$ and $x_2 = 0.9$.

Hidden layer nodes take as input the output values from the input layer. The hidden layer in this example consists of three nodes, each receiving input from all the input nodes. To compute the output of a hidden layer node, we compute a weighted sum of the inputs and apply a certain function to it. More formally, for a set of input values $x_1, x_2, \dots, x_p$, we compute the output of node j by taking the weighted sum[1] $\theta_j + \sum_{i=1}^{p} w_{ij} x_i$, where $\theta_j, w_{1,j}, \dots, w_{p,j}$ are weights that are initially set randomly, then adjusted as the network "learns." Note that θ_j, also called the *bias* of node j, is a constant that controls the level of contribution of node j. In the next step we take a function g of this sum. The function g, also called a *transfer function* or *activation function*, is some monotone function, and examples are a linear function [$g(s) = bs$], an exponential function [$g(s) = \exp(bs)$], and a logistic/sigmoidal function [$g(s) = 1/1 + e^{-s}$]. This last function is by far the most popular one in neural networks. Its practical value arises from the fact that it has a squashing effect on very small or very large values but is almost linear in the range where the value of the function is between 0.1 and 0.9.

Using a logistic function, we can write the output of node j in the hidden layer as

$$\text{Output}_j = g\left(\theta_j + \sum_{i=1}^{p} w_{ij} x_i\right) = \frac{1}{1 + e^{-(\theta_j + \sum_{i=1}^{p} w_{ij} x_i)}}. \tag{11.1}$$

Initializing the Weights The values of θ_j and w_{ij} are initialized to small, usually random, numbers (typically, but not always, in the range 0.00 ± 0.05). Such values represent a state of no knowledge by the network, similar to a model with no predictors. The initial weights are used in the first round of training.

In our example, suppose that the initial weights for node 3 are $\theta_3 = -0.3$, $w_{1,3} = 0.05$, and $w_{2,3} = 0.01$ (as shown in Figure 11.3). Using the logistic function, we can compute the output of node 3 in the hidden layer (from the

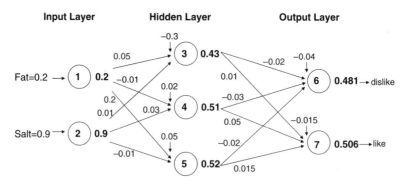

FIGURE 11.3 COMPUTING NODE OUTPUTS (IN BOLDFACE TYPE) USING THE FIRST OBSERVATION IN THE TINY EXAMPLE AND A LOGISTIC FUNCTION.

---

[1] Other options exist for combining inputs, such as taking the maximum or minimum of the weighted inputs rather than their sum, but they are much less popular.

first observation) as

$$\text{Output}_3 = \frac{1}{1 + e^{-[-0.3+(0.05)(0.2)+(0.01)(0.9)]}} = 0.43.$$

Figure 11.3 shows the initial weights, inputs, and outputs for observation 1 in our tiny example. If there is more than one hidden layer, the same calculation applies, except that the input values for the second, third, and subsequent, hidden layers would be the output of the preceding hidden layer. This means that the number of input values into a certain node is equal to the number of nodes in the preceding layer. (If there was an additional hidden layer in our example, its nodes would receive input from the three nodes in the first hidden layer.)

Finally, the output layer obtains input values from the (last) hidden layer. It applies the same function as above to create the output. In other words, it takes a weighted average of its input values and then applies the function g. In our example, output nodes 6 and 7 receive input from the three hidden layer nodes. We can compute the output of these nodes by

$$\text{Output}_6 = \frac{1}{1 + e^{-[-0.04+(-0.02)(0.43)+(-0.03)(0.51)+(0.015)(0.52)]}} = 0.481$$

$$\text{Output}_7 = \frac{1}{1 + e^{-[-0.015+(0.01)(0.430)+(0.05)(0.507)+(0.015)(0.511)]}} = 0.506$$

These two numbers are *almost* the propensities $P(Y = \text{Dislike} \mid \text{Fat} = 0.2, \text{Salt} = 0.9)$ and $P(Y = \text{Like} \mid \text{Fat} = 0.2. \text{Salt} = 0.9)$. The last step involves normalizing these two values so that they add up to 1. In other words,

$$P(Y = \text{Dislike}) = \text{Output}_6/(\text{Output}_6 + \text{Output}_7) = 0.481/(0.481 + 0.506) = 0.49.$$

$$P(Y = \text{Like}) = 1 - P(Y = \text{Dislike}) = 0.506/(0.481 + 0.506) = 0.51.$$

For classification we use a cutoff value (for a binary response) on the propensity. Using a cutoff of 0.5, we would classify this record as "like." For applications with more than two classes, we choose the output node with the largest value.

Relation to Linear and Logistic Regression   Consider a neural network with a single output node and no hidden layers. For a dataset with p predictors, the output node receives $x_1, x_2, \ldots, x_p$, takes a weighted sum of these, and applies the g function. The output of the neural network is therefore $g\left(\theta + \sum_{i=1}^{p} w_i x_i\right)$.

First, consider a numerical output variable y. If g is the identity function [$g(s) = s$], the output is simply

$$\hat{y} = \theta + \sum_{i=1}^{p} w_i x_i.$$

This is exactly equivalent to the formulation of a multiple linear regression! This means that a neural network with no hidden layers, a single output node, and an identity function g searches only for linear relationships between the response and the predictors.

Now consider a binary output variable Y. If g is the logistic function, the output is simply

$$\hat{P}(Y = 1) = \frac{1}{1 + e^{-(\theta + \sum_{i=1}^{p} w_i x_i)}},$$

which is equivalent to the logistic regression formulation!

In both cases, although the formulation is equivalent to the linear and logistic regression models, the resulting estimates for the weights (*coefficients* in linear and logistic regression) can differ because the estimation method is different. The neural net estimation method is different from *least squares*, the method used to calculate coefficients in linear regression, and the *maximum likelihood* method used in logistic regression. We explain below the method by which the neural network learns.

Preprocessing the Data

When using a logistic activation function (called "Standard" in XLMiner), neural networks perform best when the predictors and response variables are on a scale of [0,1]. For this reason all variables should be scaled to a [0,1] interval before entering them into the network. For a numerical variable X that takes values in the range $[a, b]$ where $a < b$, we normalize the measurements by subtracting a and dividing by $b - a$. The normalized measurement is then

$$X_{\text{norm}} = \frac{X - a}{b - a}.$$

Note that if $[a, b]$ is within the [0,1] interval, the original scale will be stretched.

If a and b are unknown, we can estimate them from the minimal and maximal values of X in the data. Even if new data exceed this range by a small amount, yielding normalized values slightly lower than 0 or larger than 1, will not affect the results much.

For binary variables, no adjustment is needed other than creating dummy variables. For categorical variables with m categories, if they are ordinal in nature, a choice of m fractions in [0,1] should reflect their perceived ordering. For example, if four ordinal categories are equally distant from each other, we can map them to [0, 0.25, 0.5, 1]. If the categories are nominal, transforming into $m - 1$ dummies is a good solution.

Another operation that improves the performance of the network is to transform highly skewed predictors. In business applications, there tend to be many highly right-skewed variables (e.g., income). Taking a log transform of a right-skewed variable (before converting to a [0,1] scale) will usually spread out the values more symmetrically.

Another common sigmoidal function is the hyperbolic tangent (tanh, called "Symmetric" in XLMiner). When using this function, it is usually better to scale predictors to a [−1,1] scale.

Training the Model

Training the model means estimating the weights θ_j and w_{ij} that lead to the best predictive results. The process that we described earlier (Section 11.3) for computing the neural network output for an observation is repeated for all the observations in the training set. For each observation the model produces a prediction that is then compared with the actual response value. Their difference is the error for the output node. However, unlike least squares or maximum likelihood, where a global function of the errors (e.g., sum of squared errors) is used for estimating the coefficients, in neural networks, the estimation process uses the errors iteratively to update the estimated weights.

In particular, the error for the output node is distributed across all the hidden nodes that led to it, so that each node is assigned "responsibility" for part of the error. Each of these node-specific errors is then used for updating the weights.

Back Propagation of Error The most popular method for using model errors to update weights ("learning") is an algorithm called *back propagation*. As the name implies, errors are computed from the last layer (the output layer) back to the hidden layers.

Let us denote by $\hat{y}_k$ the output from output node k. The error associated with output node k is computed by

$$\mathrm{err}_k = \hat{y}_k(1 - \hat{y}_k)(y_k - \hat{y}_k).$$

Notice that this is similar to the ordinary definition of an error $(y_k - \hat{y}_k)$ multiplied by a correction factor. The weights are then updated as follows:

$$\theta_j^{\mathrm{new}} = \theta_j^{\mathrm{old}} + l \times \mathrm{err}_j, \tag{11.2}$$

$$w_{i,j}^{\mathrm{new}} = w_{i,j}^{\mathrm{old}} + l \times \mathrm{err}_j,$$

where l is a *learning rate* or *weight decay* parameter, a constant ranging typically between 0 and 1, which controls the amount of change in weights from one iteration to the next.

In our example, the error associated with output node 7 for the first observation is $(0.506)(1 - 0.506)(1 - 0.506) = 0.123$. For output node 6 the error is $0.481(1 - 0.481)(1 - 0.481) = 0.129$. These errors are then used to compute the errors associated with the hidden layer nodes, and those weights are updated accordingly using a formula similar to (11.2).

Two methods for updating the weights are case updating and batch updating. In *case updating*, the weights are updated after each observation is run through the network (called a *trial*). For example, if we used case updating in the tiny example, the weights would first be updated after running observation 1 as follows: for a learning rate of 0.5, the weights $\theta_7, w_{3,7}, w_{4,7}$, and $w_{5,7}$ are updated to

$$\theta_7 = -0.015 + (0.5)(0.123) = 0.047,$$

$$w_{3,7} = 0.01 + (0.5)(0.123) = 0.072,$$

$$w_{4,7} = 0.05 + (0.5)(0.123) = 0.112,$$

$$w_{5,7} = 0.015 + (0.5)(0.123) = 0.077.$$

Similarly, we obtain updated weights $\theta_6 = 0.025, w_{3,6} = 0.045, w_{4,6} = 0.035$, and $w_{5,6} = 0.045$. These new weights are next updated after the second observation is run through the network, the third, and so on, until all observations are used. This is called one *epoch*, *sweep*, or *iteration* through the data. Typically, there are many epochs.

In *batch updating*, the entire training set is run through the network before each updating of weights takes place. In that case, the errors err_k in the updating equation is the sum of the errors from all observations. In practice, case updating tends to yield more accurate results than batch updating but requires a longer runtime. This is a serious consideration, since even in batch updating, hundreds or even thousands of sweeps through the training data are executed.

When does the updating stop? The most common conditions are one of the following:

1. When the new weights are only incrementally different from those of the preceding iteration
2. When the misclassification rate reaches a required threshold
3. When the limit on the number of runs is reached

Let us examine the output from running a neural network on the tiny data. Following Figures 11.2 and 11.3, we used a single hidden layer with three nodes. The weights and classification matrix are shown in Figure 11.4. We can see that the network misclassifies all three "dislike" observations and correctly classifies the three "like" observations. Notice how close all probabilities are to

0.5 (in the bottom-most table), indicating that a small change in the cutoff value would result in different classifications. This unstable result is not surprising, since the number of observations is too small for estimating the 17 weights. However, for purposes of illustration we discuss below the remainder of the output.

Training Data Scoring—Summary Report

Cutoff probability value for success (UPDATABLE)	0.5

Confusion Matrix

	Predicted Class	
Actual Class	like	dislike
like	3	0
dislike	3	0

Error Report

Class	# Cases	# Errors	% Error
like	3	0	0
dislike	3	3	100
Overall	6	3	50

Inter-Layer Connections Weights

	Input Layer		
Hidden Layer 1	fat	salt	Bias
Neuron 1	−0.141514	0.1324437	−0.56303312
Neuron 2	0.3758005	0.4062995	0.27385139
Neuron 3	−0.204786	0.4989756	−0.18658434

	Hidden Layer 1			
Output Layer	Neuron 1	Neuron 2	Neuron 3	Bias
dislike	0.6180442	−0.471879	−0.35678883	0.2269127
like	0.1311993	−0.057994	0.14913854	−0.069866

Predicted Class	Actual Class	Prob. for dislike	Prob. for like(success)	fat	salt
like	like	0.4905665	0.50943352	0.2	0.9
like	dislike	0.4996521	0.50034792	0.1	0.1
like	dislike	0.4960046	0.50399539	0.2	0.4
like	dislike	0.4949069	0.50509311	0.2	0.5
like	like	0.4942435	0.50575647	0.4	0.5
like	like	0.4913139	0.5086861	0.3	0.8

FIGURE 11.4   OUTPUT FOR NEURAL NETWORK WITH A SINGLE HIDDEN LAYER WITH THREE NODES FOR THE TINY DATA EXAMPLE.

To run a neural net in XLMiner, choose the *Neural Network* option in either the *Prediction* or *Classification* menu, depending on whether your response variable is quantitative or categorical. There are several options that you can choose, as described in the software guide. Notice, however, two points to carefully consider:

1. *Normalize input data* applies standard normalization (subtracting the mean and dividing by the standard deviation). However, if the logistic ("Standard") activation function is used, it is preferable not to use standard normalization, but rather to scale the predictors manually to a [0,1] interval. Likewise, if the tanh ("Symmetric") activation function is used, the predictors should be scaled manually to [−1,1].

2. XLMiner employs case updating. The user can specify the number of epochs to run.

The final network weights are shown in two tables in the XLMiner output. Figure 11.5 shows these weights in a format similar to that of our previous diagrams.

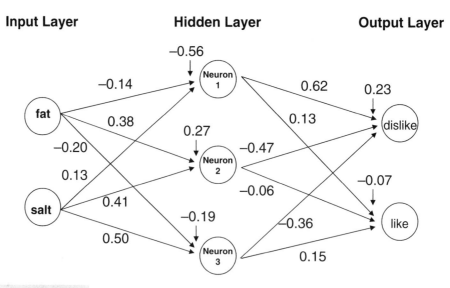

FIGURE 11.5   **NEURAL NETWORK FOR THE TINY EXAMPLE WITH FINAL WEIGHTS FROM XLMINER OUTPUT.**

The first table in the Inter-Layer Connection Weights shows the weights that connect the input layer and the hidden layer. The *Bias nodes* are the weights θ_3, θ_4, and θ_5. The weights in this table are used to compute the output of the hidden layer nodes. They were computed iteratively after choosing a random initial set of weights (like the set we chose in Figure 11.3). We use the weights in the way we described earlier to compute the hidden layer's output. For instance, for the first observation, the output of our previous node 3 (denoted *Neuron 1* in XLMiner) is

$$\text{Output}_3 = \frac{1}{1 + e^{-[-0.56+(-0.14)(0.2)+(0.13)(0.9)]}} = 0.38.$$

Similarly, we can compute the output from the two other hidden nodes for the same observation and get $\text{Output}_4 = 0.67$ and $\text{Output}_5 = 0.56$.

The second table gives the weights connecting the hidden and output layer nodes. This is equivalent to using a single node with a cutoff value. To compute the output for the "dislike" output node for the first observation, we use the outputs from the hidden layer that we computed above, and get

$$\text{Output}_6 = \frac{1}{1 + e^{-[0.23+(0.62)(0.38)+(-0.47)(0.67)+(-0.36)(0.56)]}} = 0.487.$$

Similarly, we can compute the output for the "like" output node, obtaining the value $\text{Output}_7 = 0.506$. Last, we can compute the probability (=propensity) of "like" as $0.506/(0.487 + 0.506) = 0.509$. This is the probability shown in the bottom-most table in Figure 11.4 (for observation 1), which gives the neural network's predicted probabilities and the classifications based on these values. The probabilities for the other five observations are computed equivalently, replacing the input value in the computation of the hidden layer outputs and then plugging these outputs into the computation for the output layer.

Example 2: Classifying Accident Severity

Let's apply the network training process to some real data: US automobile accidents that have been classified by their level of severity as *no injury*, *injury*, or *fatality*. A firm might be interested in developing a system for quickly classifying the severity of an accident, based on initial reports and associated data in the system (some of which rely on GPS-assisted reporting). Such a system could be used to assign emergency response team priorities. Figure 11.6 shows a small extract (999 records, four predictor variables) from a US government database.

The explanation of the four predictor variables and response is given in Table 11.2. For the analysis, we converted ALCHL_I to a 0/1 dummy variable (1 = presence of alcohol) and created four dummies for SUR_COND. This gives us a total of seven predictors.

With the exception of alcohol involvement and a few other variables in the larger database, most of the variables are ones that we might reasonably expect to be available at the time of the initial accident report, before accident details and severity have been determined by first responders. A data mining model that could predict accident severity on the basis of these initial reports would have value in allocating first responder resources.

Row Id.	Selected variables				
	ALCHL_I	PROFIL_I_R	SUR_COND	VEH_INVL	MAX_SEV_IR
1	2	0	1	1	0
4	2	0	2	2	1
5	2	1	1	2	1
6	2	0	1	1	0
9	2	1	1	1	1
10	2	0	1	1	0
12	1	1	1	2	1
17	2	1	1	2	1
18	2	1	4	1	1
19	2	0	1	1	0
20	2	1	1	2	2
21	2	1	1	2	1
23	2	1	3	2	1
26	1	0	1	1	2
27	1	1	1	2	2

FIGURE 11.6   SUBSET FROM THE ACCIDENT DATA, FOR A HIGH-FATALITY REGION.

TABLE 11.2   DESCRIPTION OF VARIABLES FOR AUTOMOBILE ACCIDENT EXAMPLE

ALCHL_I   Presence (1) or absence (2) of alcohol
PROFIL_I_R   Profile of the roadway: level (1), other (0)
SUR_COND   Surface condition of the road: dry (1), wet (2), snow/slush (3), ice (4), unknown (9)
VEH_INVL   Number of vehicles involved
MAX_SEV_IR   Presence of injuries/fatalities: no injuries (0), injury (1), fatality (2)

To use a neural net architecture for this classification problem we put seven nodes in the input layer, one for each of the seven predictors, and three neurons (one for each class) in the output layer. We use a single hidden layer and experiment with the number of nodes. If we increase the number of nodes from four to eight and examine the resulting confusion matrices, we find that five nodes gives a good balance between improving the predictive performance on the training set without deteriorating the performance on the validation set. (Networks with more than five nodes in the hidden layer performed as well as the five-node network but add undesirable complexity.) Some software allow exploring multiple architectures in an automated way. XLMiner's Automatic Network can be used to compare multiple architectures—it tries networks with different numbers of hidden layers and different numbers of nodes in the hidden layer(s).

Note that there are a total of five connections from each node in the input layer to each node in the hidden layer, and a total of $7 \times 5 = 35$ connections between the input layer and the hidden layer. In addition, there is a total of three connections from each node in the hidden layer to each node in the output layer, a total of $5 \times 3 = 15$ connections between the hidden layer and the output layer.

We train the network on the training partition of 600 records. Each iteration in the neural network process consists of a presentation to the input layer of the predictors in a case, followed by successive computations of the outputs of the hidden layer nodes and the output layer nodes using the appropriate weights. The output values of the output layer nodes are used to compute the error. The backward propagation algorithm completes one iteration, using the error to adjust the weights of all the connections in the network. Since the training data has 600 cases, one sweep through the data, termed an *epoch*, consists of 600 iterations. We will train the network using 30 epochs, so there will be a total of 18,000 iterations. The resulting classification results, error rates, and weights following the last epoch of training the neural net on these data are shown in Figure 11.7.

Note that had we stopped after only one pass of the data (600 iterations), the error would have been much worse, and none of the fatal accidents (coded as 2) would have been spotted, as can be seen in Figure 11.8. Our results can depend on how we set the different parameters, and there are a few pitfalls to avoid. We discuss these next.

Avoiding Overfitting

A weakness of the neural network is that it can easily overfit the data, causing the error rate on validation data (and most important, on new data) to be too large. It is therefore important to limit the number of training epochs and not to over-train on the data. As in classification and regression trees, overfitting can be detected by examining the performance on the validation set and seeing when it starts deteriorating, while the training set performance is still improving. This approach is used in some algorithms (but not in XLMiner) to limit the number of training epochs: the error rate on the validation dataset is computed periodically while the network is being trained. The validation error decreases in the early epochs of the training, but after a while, it begins to increase. The point of minimum validation error is a good indicator of the best number of epochs for training, and the weights at that stage are likely to provide the best error rate in new data.

To illustrate the effect of overfitting, compare the confusion matrices of the five-node network (Figure 11.7) with those from a network with four hidden layers, each with 25 nodes, run for 60 epochs (Figure 11.9). Both networks perform similarly on the training set, but the more complex network does worse on the validation set.

Training Data Scoring–Summary Report

Confusion Matrix

Actual Class	Predicted Class		
	0	1	2
0	317	0	12
1	0	173	7
2	21	39	31

Error Report

Class	# Cases	# Errors	% Error
0	329	12	3.647416
1	180	7	3.888889
2	91	60	65.93407
Overall	600	79	13.16667

Validation Data Scoring–Summary Report

Confusion Matrix

Actual Class	Predicted Class		
	0	1	2
0	215	0	7
1	0	114	5
2	17	20	21

Error Report

Class	# Cases	# Errors	% Error
0	222	7	3.153153
1	119	5	4.201681
2	58	37	63.7931
Overall	399	49	12.2807

Inter-Layer Connections Weights

Hidden Layer 1	Input Layer							
	PROFIL_I_R	VEH_INVL	ALCHL_I_1	SUR_COND_1	SUR_COND_2	SUR_COND_3	SUR_COND_4	Bias
Neuron 1	−2.25167	−0.43213	−0.38535	0.931877845	−0.771960503	−0.281095925	−0.745166659	0.202846
Neuron 2	1.90247	1.328339	0.361849	−1.134428133	0.879260049	0.126827105	0.242204068	0.269196
Neuron 3	1.052584	0.441061	0.521507	0.524671291	−1.042115003	−0.24963397	−0.025855574	−0.29391
Neuron 4	−2.51741	−2.40636	−0.16124	1.488628808	−0.755883396	−0.120849569	−0.496020789	−1.2182
Neuron 5	2.151966	0.877305	−1.24761	−0.251986441	1.359928715	0.455704657	1.142010989	−1.7395

Output Layer	Hidden Layer 1					
	Neuron 1	Neuron 2	Neuron 3	Neuron 4	Neuron 5	Bias
0	1.735166	−1.99982	−1.0286	1.918510775	−1.756192337	−0.830315926
1	−2.34624	0.896682	−0.99386	−2.42803985	2.347881234	−0.673641884
2	0.717193	0.888273	1.043032	−2.440551187	−2.445047344	−0.641564506

FIGURE 11.7   XLMINER OUTPUT FOR NEURAL NETWORK FOR ACCIDENT DATA, WITH FIVE NODES IN THE HIDDEN LAYER, AFTER 30 EPOCHS.

Using the Output for Prediction and Classification

When the neural network is used for predicting a numerical response, the resulting output needs to be scaled back to the original units of that response. Recall that numerical variables (both predictor and response variables) are usually rescaled to a [0,1] interval before being used by the network. The output will therefore also be on a [0,1] scale. To transform the prediction back to the original

Training Data Scoring—Summary Report

Confusion Matrix

Actual Class	Predicted Class		
	0	1	2
0	328	1	0
1	0	180	0
2	35	56	0

Error Report

Class	# Cases	# Errors	% Error
0	329	1	0.303951
1	180	0	0
2	91	91	100
Overall	600	92	15.33333

Validation Data Scoring—Summary Report

Confusion Matrix

Actual Class	Predicted Class		
	0	1	2
0	221	1	0
1	0	119	0
2	25	33	0

Error Report

Class	# Cases	# Errors	% Error
0	222	1	0.45045
1	119	0	0
2	58	58	100
Overall	399	59	14.78697

Inter-Layer Connections Weights

Hidden Layer 1	Input Layer							
	PROFIL_I_R	VEH_INVL	ALCHL_I_1	SUR_COND_1	SUR_COND_2	SUR_COND_3	SUR_COND_4	Bias
Neuron 1	−1.191	−0.37133	−0.17645	1.034139886	−0.424482084	0.010851786	−0.379368507	0.185934
Neuron 2	0.745799	0.786711	−0.13268	−0.64494796	0.379113554	0.157017352	0.014123182	−0.08824
Neuron 3	−0.00942	0.501572	0.122154	−0.309988481	0.1206329	0.134147532	−0.220287869	0.067561
Neuron 4	−1.3765	−0.9194	−0.14498	0.977569663	−0.333850918	−0.210656136	−0.463626155	−0.02612
Neuron 5	1.01174	0.549136	−0.44278	−0.632431301	0.62482995	−0.056907479	0.093559284	−0.25899

Output Layer	Hidden Layer 1					
	Neuron 1	Neuron 2	Neuron 3	Neuron 4	Neuron 5	Bias
0	1.285029	−1.36381	−0.61051	1.641970721	−1.267374767	−0.321989822
1	−1.42038	0.681874	−0.00733	−1.612349324	1.296953836	−0.190250726
2	−0.04007	0.133019	0.053135	−0.705875048	−0.537703562	−1.036133649

FIGURE 11.8   **XLMINER OUTPUT FOR NEURAL NETWORK FOR ACCIDENT DATA, WITH FIVE NODES IN THE HIDDEN LAYER, AFTER ONLY ONE EPOCH.**

Y units, which were in the range $[a, b]$, we multiply the network output by $b - a$ and add a.

When the neural net is used for classification and we have m classes, we will obtain an output from each of the m output nodes. How do we translate these m outputs into a classification rule? Usually the output node with the largest value determines the net's classification.

In the case of a binary response ($m = 2$), we can use just one output node with a cutoff value to map a numerical output value to one of the two classes.

Training Data Scoring–Summary Report

Confusion Matrix			
		Predicted Class	
Actual Class	0	1	2
0	316	0	13
1	0	175	5
2	20	45	26

Error Report			
Class	# Cases	# Errors	% Error
0	329	13	3.951368
1	180	5	2.777778
2	91	65	71.42857
Overall	600	83	13.83333

Validation Data Scoring–Summary Report

Confusion Matrix			
		Predicted Class	
Actual Class	0	1	2
0	215	1	6
1	0	115	4
2	17	28	13

Error Report			
Class	# Cases	# Errors	% Error
0	222	7	3.153153
1	119	4	3.361345
2	58	45	77.58621
Overall	399	56	14.03509

FIGURE 11.9   XLMINER OUTPUT FOR NEURAL NETWORK FOR ACCIDENT DATA WITH 25 NODES IN THE HIDDEN LAYER.

Although we typically use a cutoff of 0.5 with other classifiers, in neural networks there is a tendency for values to cluster around 0.5 (from above and below). An alternative is to use the validation set to determine a cutoff that produces reasonable predictive performance.

11.4 REQUIRED USER INPUT

One of the time-consuming and complex aspects of training a model using back propagation is that we first need to decide on a network architecture. This means specifying the number of hidden layers and the number of nodes in each layer. The usual procedure is to make intelligent guesses using past experience and to do several trial-and-error runs on different architectures. Algorithms exist that grow the number of nodes selectively during training or trim them in a manner analogous to what is done in classification and regression trees (see Chapter 9). Research continues on such methods. As of now, no automatic method seems clearly superior to the trial-and-error approach. A few general guidelines for choosing an architecture follow:

Number of hidden layers. The most popular choice for the number of hidden layers is one. A single hidden layer is usually sufficient to capture even very complex relationships between the predictors.

Size of hidden layer. The number of nodes in the hidden layer also determines the level of complexity of the relationship between the predictors that the network captures. The trade-off is between under- and overfitting. On the one hand, using too few nodes might not be sufficient to capture complex relationships (recall the special cases of a linear relationship such as in linear and logistic regression, in the extreme case of zero nodes or no hidden layer). On the other hand, too many nodes might lead to overfitting. A rule of thumb is to start with p (number of predictors) nodes and gradually decrease or increase while checking for overfitting.

Number of output nodes. For a binary response, a single node is sufficient and a cutoff is used for classification. For a categorical response with $m > 2$ classes, the number of nodes should equal the number of classes. Finally, for a numerical response, typically a single output node is used unless we are interested in predicting more than one function.

In addition to the choice of architecture, the user should pay attention to the *choice of predictors*. Since neural networks are highly dependent on the quality of their input, the choice of predictors should be done carefully, using domain knowledge, variable selection, and dimension reduction techniques before using the network. We return to this point in the discussion of advantages and weaknesses below.

Other parameters that the user can control are the *learning rate* (a.k.a. weight decay), l, and the *momentum*. The first is used primarily to avoid overfitting, by down-weighting new information. This helps to tone down the effect of outliers on the weights and avoids getting stuck in local optima. This parameter typically takes a value in the range [0,1]. Berry and Linoff (2000) suggest starting with a large value (moving away from the random initial weights, thereby "learning quickly" from the data) and then slowly decreasing it as the iterations progress and the weights are more reliable. Han and Kamber (2001) suggest the more concrete rule of thumb of setting $l = 1/$(current number of iterations). This means that at the start, $l = 1$, during the second iteration it is $l = 0.5$, and then it keeps decaying toward $l = 0$. Notice that in XLMiner the default is $l = 0$, which means that the weights do not decay at all.

The second parameter, called *momentum*, is used to "keep the ball rolling" (hence the term *momentum*) in the convergence of the weights to the optimum. The idea is to keep the weights changing in the same direction as they did in the preceding iteration. This helps avoid getting stuck in a local optimum. High values of momentum mean that the network will be "reluctant" to learn from data that want to change the direction of the weights, especially when we consider case updating. In general, values in the range 0 to 2 are used.

11.5 EXPLORING THE RELATIONSHIP BETWEEN PREDICTORS AND RESPONSE

Neural networks are known to be black boxes in the sense that their output does not shed light on the patterns in the data that it models (like our brains). That is in fact one of the biggest criticisms of the method. However, in some cases it is possible to learn more about the relationships that the network captures by conducting a sensitivity analysis on the validation set. This is done by setting all predictor values to their mean and obtaining the network's prediction. Next, the process is repeated by setting each predictor sequentially to its minimum, and then maximum, value. By comparing the predictions from different levels of the predictors, we can get a sense of which predictors affect predictions more and in what way.

11.6 ADVANTAGES AND WEAKNESSES OF NEURAL NETWORKS

As mentioned in Section 11.1, the big advantage of neural networks is their good predictive performance. They are known to have high tolerance to noisy data and the ability to capture highly complicated relationships between the predictors and a response. Their weakest point is in providing insight into the structure of the relationship, hence their blackbox reputation.

Several considerations and dangers should be kept in mind when using neural networks. First, although they are capable of generalizing from a set of examples, extrapolation is still a serious danger. If the network sees only cases in a certain range, its predictions outside this range may be completely invalid.

Second, neural networks do not have a built-in variable selection mechanism. This means that there is a need for careful consideration of predictors. A combination with classification and regression trees (see Chapter 9) and other dimension reduction techniques (e.g., principal components analysis in Chapter 4) is often used to identify key predictors.

Third, neural networks are extremely flexible, but they rely heavily on having sufficient data for training purposes. As our tiny example shows, a neural network performs poorly when the training set size is insufficient, even when the relationship between the response and predictors is very simple. A related issue is that in classification problems, the network requires sufficient records of the minority class in order to learn it. This is achieved by oversampling, as explained in Chapter 2.

Fourth, there is the risk of obtaining weights that lead to a local optimum rather than the global optimum, in the sense that the weights converge to values that do not provide the best fit to the training data. We described several parameters that are used to avoid this situation (e.g., controlling the learning rate and slowly reducing the momentum). However, there is no guarantee that the resulting weights are indeed the optimal ones.

Finally, a practical consideration that can determine the usefulness of a neural network is the timeliness of computation. Neural networks are relatively heavy on computation time, requiring a longer runtime than other classifiers. This runtime grows greatly when the number of predictors is increased (as there will be many more weights to compute). In applications where real-time or near-real-time prediction is required, runtime should be measured to make sure that it does not cause unacceptable delay in the decision making.

UNSUPERVISED FEATURE EXTRACTION AND DEEP LEARNING

Most of the data we have dealt with in this book has been either numeric or categorical, and the available predictor variables or features have been inherently informative (e.g., the size of a car's engine, or a car's gas mileage). It has been relatively straightforward to discover their impact on the target variable, and focus on the most meaningful ones. With some data, however, we begin with a huge mass of values that are not "predictors" in the same sense. With image data, one "variable" would be the color and intensity of the pixel in a particular position, and one 2-inch square image might have 40,000 pixels. We are now in the realm of "big data."

How can we derive meaningful higher level features such as edges, curves, shapes, and even higher level features such as faces? *Deep learning* has made big strides in this area. *Deep learning networks* (DLN) refer to neural nets with many hidden layers used to self-learn features from complex data. DLNs are especially effective at capturing local structure and dependencies in complex data, such as in images or audio. This is done by using neural nets in an unsupervised way, where the data are used as both the input (with some added noise) and the output. When the multiple hidden layers are used, the DLN's learning is hierarchical—typically, from a single pixel to edges, from edges to higher level features, and so on. The network then "pools" similar high-level features across images to build up clusters of similar features. If they appear often enough in large numbers of images, distinctive and regular structures like faces, vehicles and houses will emerge as highest level features. See Le at al. (2012) for a technical description of how this method yielded labels not just of faces but of different types of faces (e.g., cat faces, of which there are many on the Internet).

Other applications include speech and handwriting recognition. Facebook and Google are effectively using deep learning to identify faces from images, and high-end retail stores use facial recognition technology involving deep learning to identify VIP customers and give them special sales attention.[2]

While DLNs are not new, they have only gained momentum recently, thanks to dramatic improvements in computing power, allowing us to deal both with the huge amount of data and the extreme complexity of the networks. The abundance of unlabeled data such as images and audio has also helped tremendously. As with neural networks, DLNs suffer from overfitting and slow runtime. Solutions include using regularization (removing "useless" nodes by zeroing out their coefficients) and tricks to improve over the back-propagation algorithm.

---

[2] www.npr.org/sections/alltechconsidered/2013/07/21/203273764/high-end-stores-use-facial-recognition-tools-to-spot-vips, Accessed July 21, 2013.

PROBLEMS ██ ■

11.1   Credit Card Use. Consider the following hypothetical bank data on consumers' use of credit card credit facilities in Table 11.3. Create a small worksheet in Excel, like that used in Example 1, to illustrate one pass through a simple neural network.

TABLE 11.3	DATA FOR CREDIT CARD EXAMPLE AND VARIABLE DESCRIPTIONS[a]

Years	Salary	Used Credit
4	43	0
18	65	1
1	53	0
3	95	0
15	88	1
6	112	1

[a]Years = Number of years that a customer has been with the bank; Salary = customer's salary (in thousands of dollars); Used Credit 1 = customer has left an unpaid credit card balance at the end of at least one month in the prior year; and 0 = balance was paid off at the end of each month.

11.2   Neural Net Evolution. A neural net typically starts out with random coefficients; hence it produces essentially random predictions when presented with its first case. What is the key ingredient by which the net evolves to produce a more accurate prediction?

11.3   Car Sales. Consider again the data on used cars (ToyotaCorolla.xls) with 1436 records and details on 38 attributes, including Price, Age, KM, HP, and other specifications. The goal is to predict the price of a used Toyota Corolla based on its specifications.

   a. Use XLMiner's neural network routine to fit a model. Depending on XLMiner version, you can probably accept default values for the neural net parameters, except data should be normalized and error tolerance may need to be set to 0. Record the RMS error for the training data and the validation data. Repeat the process, changing the number of epochs (and only this) to 300, 3000, and 10,000.

      i. What happens to the RMS error for the training data as the number of epochs increases?

      ii. What happens to the RMS error for the validation data?

      iii. Comment on the appropriate number of epochs for the model.

11.4   Direct Mailing to Airline Customers. East-West Airlines has entered into a partnership with the wireless phone company Telcon to sell the latter's service via direct mail. The file EastWestAirlinesNN.xls contains a subset of a data sample of who has already received a test offer. About 13% accepted.

    You are asked to develop a model to classify East-West customers as to whether they purchased a wireless phone service contract (target variable Phone_Sale), a model that can be used to predict classifications for additional customers.

  a. Using XLMiner, run a neural net model on these data, using the option to normalize the data, setting the number of epochs at 3000, and requesting lift charts for both the training and validation data. Interpret the meaning (in business terms) of the leftmost bar of the validation lift chart (the bar chart).

  b. Comment on the difference between the training and validation lift charts.

  c. Run a second neural net model on the data, this time setting the number of epochs at 100. Comment now on the difference between this model and the model you ran earlier, and how overfitting might have affected results.

  d. What sort of information, if any, is provided about the effects of the various variables?

## CHAPTER 12

Discriminant Analysis

In this chapter we describe the method of discriminant analysis, which is a model-based approach to classification. We discuss the main principle, where classification is based on the distance of an observation from each of the class averages. We explain the underlying measure of "statistical distance," which takes into account the correlation between predictors. The output of a discriminant analysis procedure generates estimated "classification functions," which are then used to produce classification scores that can be translated into classifications or propensities (probabilities of class membership). One can also directly integrate misclassification costs into the discriminant analysis setup, and we explain how this is achieved. Finally, we discuss the underlying model assumptions, the practical robustness to some, and the advantages of discriminant analysis when the assumptions are reasonably met (e.g., the sufficiency of a small training sample).

12.1 INTRODUCTION

Discriminant analysis is a classification method. Like logistic regression, it is a classical statistical technique that can be used for classification and profiling. It uses sets of measurements on different classes of items to classify new items into one of those classes (*classification*). Common uses of the method have been in classifying organisms into species and subspecies; classifying applications for loans, credit cards, and insurance into low- and high-risk categories; classifying customers of new products into early adopters, early majority, late majority,

Data Mining for Business Analytics: Concepts, Techniques, and Applications with XLMiner®, Third Edition.
Galit Shmueli, Peter C. Bruce, and Nitin R. Patel.
© 2016 John Wiley & Sons Inc. Published 2016 by John Wiley & Sons, Inc.

and laggards; classifying bonds into bond-rating categories; classifying skulls of human fossils; research studies involving disputed authorship; decision on college admission; medical studies involving alcoholics and nonalcoholics; and methods to identify human fingerprints. Discriminant analysis can also be used to highlight the characteristics that distinguish the classes (*profiling*).

We return to two examples that were described in earlier chapters, the riding-mowers and personal loan acceptance examples. In each of these, the response has two classes. We close with a third example involving more than two classes.

Example 1: Riding Mowers

We return to the example from Chapter 7, where a riding-mower manufacturer would like to find a way of classifying families in a city into those likely to purchase a riding mower and those not likely to purchase one. A pilot random sample of 12 owners and 12 nonowners in the city is undertaken. The data are given in Chapter 7 (Table 7.1), and a scatter plot is shown in Figure 12.1. We can think of a linear classification rule as a line that separates the two-dimensional region into two parts, with most of the owners in one half-plane and most nonowners in the complementary half-plane. A good classification rule would separate the data so that the fewest points are misclassified: The line shown in Figure 12.1 seems to do a good job in discriminating between the two classes as it makes four misclassifications out of 24 points. Can we do better?

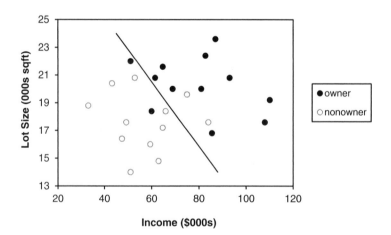

FIGURE 12.1   SCATTER PLOT OF LOT SIZE VS. INCOME FOR 24 OWNERS AND NONOWNERS OF RIDING MOWERS. THE (AD HOC) LINE TRIES TO SEPARATE OWNERS FROM NONOWNERS.

Example 2: Personal Loan Acceptance

The riding-mowers example is a classic example and is useful in describing the concept and goal of discriminant analysis. However, in today's business applications, the number of observations is much larger, and the separation of observations into classes is much less distinct. To illustrate this, we return to the Universal Bank example described in Chapter 9, where the bank's goal is to identify new customers most likely to accept a personal loan. For simplicity, we will consider only two predictor variables: the customer's annual income (Income, in $000s), and the average monthly credit card spending (CCAvg, in $000s). The first part of Figure 12.2 shows the acceptance of a personal loan by a subset of 200 customers from the bank's database as a function of Income and CCAvg. We use a logarithmic scale on both axes to enhance visibility because there are many points condensed in the low-income, low-CC spending area. Even for this small subset, the separation is not clear. The second figure shows all 5000 customers and the added complexity of dealing with large numbers of observations.

12.2 DISTANCE OF AN OBSERVATION FROM A CLASS

Finding the best separation between observations involves measuring their distance from their class. The general idea is to classify an observation to the class to which it is closest. Suppose that we are required to classify a new customer of Universal Bank as being an acceptor or a nonacceptor of their personal loan offer, based on an income of x. From the bank's database we find that the average income for loan acceptors was $144.75K and for nonacceptors $66.24K. We can use *Income* as a predictor of loan acceptance via a simple *Euclidean distance rule*: If x is closer to the average income of the acceptor class than to the average income of the nonacceptor class, classify the customer as an acceptor; otherwise, classify the customer as a nonacceptor. In other words, if $|x - 144.75| < |x - 66.24|$, then classification = acceptor; otherwise, nonacceptor. Moving from a single predictor variable (income) to two or more predictor variables, the equivalent of the mean of a class is the *centroid* of a class. This is simply the vector of means $\bar{\mathbf{x}} = [\bar{x}_1, \ldots, \bar{x}_p]$. The Euclidean distance between an observation with p measurements $\mathbf{x} = [x_1, \ldots, x_p]$ and the centroid $\bar{\mathbf{x}}$ is defined as the root of the sum of the squared differences between the individual values and the means:

$$D_{\text{Euclidean}}(\mathbf{x}, \bar{\mathbf{x}}) = \sqrt{(x_1 - \bar{x}_1)^2 + \cdots + (x_p - \bar{x}_p)^2}. \qquad (12.1)$$

Using the Euclidean distance has three drawbacks. First, the distance depends on the units we choose to measure the predictor variables. We will get different

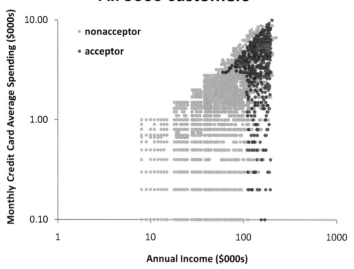

FIGURE 12.2 PERSONAL LOAN ACCEPTANCE AS A FUNCTION OF INCOME AND CREDIT CARD SPENDING FOR 5000 CUSTOMERS OF THE UNIVERSAL BANK (IN LOG SCALE).

answers if we decide to measure income in dollars, for instance, rather than in thousands of dollars.

Second, Euclidean distance does not take into account the variability of the variables. For example, if we compare the variability in income in the two classes, we find that for acceptors the standard deviation is lower than for nonacceptors ($31.6K vs. $40.6K). The income of a new customer might be closer to the acceptors' average income in dollars, but because of the large variability in income for nonacceptors, this customer is just as likely to be a nonacceptor. We therefore want the distance measure to take into account the variance of the different variables and measure a distance in standard deviations rather than in the original units. This is equivalent to z-scores.

Third, Euclidean distance ignores the correlation between the variables. This is often a very important consideration, especially when we are using many predictor variables to separate classes. In this case there will often be variables that alone are useful discriminators between classes but in the presence of other predictor variables are practically redundant, as they capture the same effects as the other variables.

A solution to these drawbacks is to use a measure called *statistical distance* (or *Mahalanobis distance*). Let us denote by S the covariance matrix between the p variables. The definition of a statistical distance is

$$D_{\text{Statistical}}(\mathbf{x}, \overline{\mathbf{x}}) = [\mathbf{x} - \overline{\mathbf{x}}]' S^{-1} [\mathbf{x} - \overline{\mathbf{x}}]$$

$$= [(x_1 - \overline{x}_1), (x_2 - \overline{x}_2), \dots, (x_p - \overline{x}_p)] S^{-1} \begin{bmatrix} x_1 - \overline{x}_1 \\ x_2 - \overline{x}_2 \\ \vdots \\ x_p - \overline{x}_p \end{bmatrix}$$

$$(12.2)$$

(the notation $'$, which represents *transpose operation*, simply turns the column vector into a row vector). S^{-1} is the inverse matrix of S, which is the p-dimension extension to division. When there is a single predictor ($p = 1$), this reduces to a z-score, since we subtract the mean and divide by the standard deviation. The statistical distance takes into account not only the predictor averages but also the spread of the predictor values and the correlations between the different predictors. To compute a statistical distance between an observation and a class, we first compute the predictor averages (the centroid) and the covariances between each pair of predictors. These are then used to construct the distances. The method of discriminant analysis uses statistical distance as the basis for finding a separating line (or, if there are more than two variables, a separating hyperplane) that is equally distant from the different class means.[1] The method is based on measuring the statistical distances of an observation to each of the classes and allocating it to the closest class. The measuring is done through *classification functions*, which are explained next.

---

[1] An alternative approach finds a separating line or hyperplane that is "best" at separating the different clouds of points. In the case of two classes, the two methods coincide.

12.3 FISHER'S LINEAR CLASSIFICATION FUNCTIONS

Linear classification functions were proposed in 1936 by the noted statistician R. A. Fisher as the basis for improved separation of observations into classes. The idea is to find linear functions of the measurements that maximize the ratio of between-class variability to within-class variability. In other words, we would obtain classes that are very homogeneous and that differ the most from each other. For each observation, these functions are used to compute scores that measure the proximity of that observation to each of the classes. An observation is classified as belonging to the class for which it has the highest classification score (equivalent to the smallest statistical distance).

USING CLASSIFICATION FUNCTION SCORES TO CLASSIFY

For each observation, we calculate the value of the classification function (one for each class); whichever class's function has the highest value (= score) is the class assigned to that observation.

The classification functions are estimated using software. For example, Figure 12.3 shows the output for the riding-mowers data in the case of two predictors. Note that the number of classification functions is equal to the number of classes (here two: owner/nonowner).

To classify a family into the class of owners or nonowners, we use the functions above to compute the family's classification scores: a family is classified into the class of *owners* if the owner function score is higher than the nonowner function score, and into *nonowners* if the reverse is the case. These functions are specified in a way that can be generalized easily to more than two classes. The values given for the functions are simply the weights to be associated with each variable in the linear function in a manner analogous to multiple linear regression. For instance, the first household has an income of $60K and a lot size of

Variables	Classification Function	
	Owner	Nonowner
Constant	−73.16020203	−51.42144394
Income	0.42958561	0.32935533
Lot Size	5.46674967	4.68156528

FIGURE 12.3 DISCRIMINANT ANALYSIS OUTPUT FOR RIDING-MOWER DATA, DISPLAYING THE ESTIMATED CLASSIFICATION FUNCTIONS (CALLED "LINEAR DISCRIMINANT FUNCTIONS" IN XLMINER.)

18.4K ft^2. Their *owner* score is therefore $-73.16 + (0.43)(60) + (5.47)(18.4)$ $= 53.2$, and their *nonowner* score is $-51.42 + (0.33)(60) + (4.68)(18.4) = 54.48$. Since the second score is higher, the household is (mis)classified by the model as a nonowner. The scores for all 24 households are given in Figure 12.4.

An alternative way for classifying an observation into one of the classes is to compute the probability of belonging to each of the classes and assigning the observation to the most likely class. If we have two classes, we need only compute a single probability for each observation (e.g., of belonging to *owners*). Using a cutoff of 0.5 is equivalent to assigning the observation to the class with the highest classification score. The advantage of this approach is that we obtain propensities, which can be used for goals such as ranking: we sort the observations in order of descending probabilities and generate lift curves.

Let us assume that there are m classes. To compute the probability of belonging to a certain class k, for a certain observation i, we need to compute all the classification scores $c_1(i), c_2(i), \ldots, c_m(i)$ and combine them using the

Classification Scores

Row ID	Predicted Class	Actual Class	Owners	Nonowners
1	nonowner	owner	53.2031285	54.48067701
2	owner	owner	55.4107621	55.38873348
3	owner	owner	72.7587384	71.04259149
4	owner	owner	66.9677061	66.21046668
5	owner	owner	93.2290383	87.71741038
6	owner	owner	79.0987673	74.72663127
7	owner	owner	69.449838	66.54448063
8	owner	owner	84.8646791	80.71623966
9	owner	owner	65.8161985	64.93537943
10	owner	owner	80.4996528	76.58515957
11	owner	owner	69.0171568	68.37011405
12	owner	owner	70.9712258	68.88764339
13	owner	nonowner	66.2070123	65.0388853
14	nonowner	nonowner	63.2303113	63.34507531
15	nonowner	nonowner	48.7050398	50.44370426
16	nonowner	nonowner	56.9195896	58.31063803
17	owner	nonowner	59.1397834	58.63995271
18	nonowner	nonowner	44.1902042	47.17838722
19	nonowner	nonowner	39.8251779	43.04730714
20	nonowner	nonowner	55.7806422	56.45680899
21	nonowner	nonowner	36.8568505	40.96766929
22	nonowner	nonowner	43.7910169	47.46070921
23	nonowner	nonowner	25.2831595	30.91759181
24	nonowner	nonowner	34.8115865	38.61510799

FIGURE 12.4   CLASSIFICATION SCORES FOR RIDING-MOWER DATA. THESE SCORES WERE COMPUTED MANUALLY.

following formula:

P[observation i(with measurements $x_1, x_2, ..., x_p$) belongs to class k]

$$= \frac{e^{c_k(i)}}{e^{c_1(i)} + e^{c_2(i)} + \cdots + e^{c_m(i)}}.$$

In XLMiner these probabilities are computed automatically, as can be seen in Figure 12.5.

We now have three misclassifications, compared to four in our original (ad hoc) classification. This can be seen in Figure 12.6, which includes the line resulting from the discriminant model.[2]

Cutoff Prob.Val. for Success (Updatable)	**0.5**

Row ID	Predicted Class	Actual Class	Prob. for Owner (success)	Income	Lot Size
1	nonowner	owner	0.21796781	60	18.4
2	owner	owner	0.505506928	85.5	16.8
3	owner	owner	0.847631864	64.8	21.6
4	owner	owner	0.680754087	61.5	20.8
5	owner	owner	0.995976726	87	23.6
6	owner	owner	0.987533139	110.1	19.2
7	owner	owner	0.948110638	108	17.6
8	owner	owner	0.984456382	82.8	22.4
9	owner	owner	0.706991915	69	20
10	owner	owner	0.980439587	93	20.8
11	owner	owner	0.656343749	51	22
12	owner	owner	0.889297203	81	20
13	owner	nonowner	0.762806287	75	19.6
14	nonowner	nonowner	0.47134045	52.8	20.8
15	nonowner	nonowner	0.149482655	64.8	17.2
16	nonowner	nonowner	0.199240432	43.2	20.4
17	owner	nonowner	0.622419543	84	17.6
18	nonowner	nonowner	0.047962588	49.2	17.6
19	nonowner	nonowner	0.038341401	59.4	16
20	nonowner	nonowner	0.337117362	66	18.4
21	nonowner	nonowner	0.016129906	47.4	16.4
22	nonowner	nonowner	0.024850999	33	18.8
23	nonowner	nonowner	0.003559986	51	14
24	nonowner	nonowner	0.021806029	63	14.8

FIGURE 12.5　DISCRIMINANT ANALYSIS OUTPUT FOR RIDING-MOWER DATA, DISPLAYING THE PROPENSITIES (ESTIMATED PROBABILITY) OF OWNERSHIP FOR EACH FAMILY

---

[2] The slope of the line is given by $-a_1/a_2$ and the intercept is $a_1/a_2\ \overline{x}_1 + \overline{x}_2$, where a_i is the difference between the ith classification function coefficients of owners and nonowners (e.g., here $a_{\text{income}} = 0.43 - 0.33$).

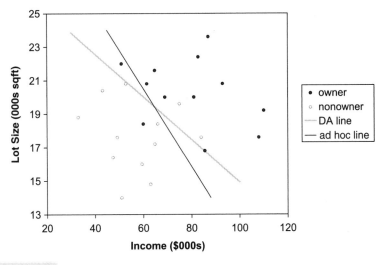

FIGURE 12.6   **CLASS SEPARATION OBTAINED FROM THE DISCRIMINANT MODEL (COMPARED TO AD HOC LINE FROM FIGURE 12.1)**

12.4 CLASSIFICATION PERFORMANCE OF DISCRIMINANT ANALYSIS

The discriminant analysis method relies on two main assumptions to arrive at classification scores. The first assumption is that the measurements in all classes come from a multivariate normal distribution. When this assumption is reasonably met, discriminant analysis has been proved to be a more powerful tool than other classification methods, such as logistic regression. In fact Efron (1975) showed that discriminant analysis is 30% more efficient than logistic regression if the data are multivariate normal, in the sense that we require 30% fewer observations to arrive at the same results. Moreover, in practice, it has been shown that this method is relatively robust to departures from normality in the sense that predictors can be nonnormal and even dummy variables. This is true as long as the smallest class is sufficiently large (approximately more than 20 observations). This method is also known to be sensitive to outliers in both the univariate space of single predictors and in the multivariate space. Exploratory analysis should therefore be used to locate extreme cases and determine whether they can be eliminated.

The second assumption behind discriminant analysis is that the correlation structure between the different predictors within a class is the same across classes. This can be roughly checked by computing the correlation matrix between the predictors for each class and comparing matrices. If the correlations differ substantially across classes, the classifier will tend to classify cases into the class with the largest variability. When the correlation structure differs significantly and the dataset is very large, an alternative is to use quadratic discriminant analysis.[3]

Notwithstanding the caveats embodied in these statistical assumptions, recall that in a predictive modeling environment, the ultimate test is whether the model works effectively. A reasonable approach is to conduct some exploratory analysis with respect to normality and correlation, train and evaluate a model, and then, depending on classification accuracy and what you learned from the initial exploration, circle back and explore further whether outliers should be examined or choice of variables revisited.

With respect to the evaluation of classification accuracy, we once again use the general measures of performance that were described in Chapter 5 (judging the performance of a classifier), with the principal ones based on the confusion matrix (accuracy alone or combined with costs) for classification and the lift chart for ranking. The same argument for using the validation set for evaluating performance still holds. For example, in the riding-mowers example, families 1, 13, and 17 are misclassified. This means that the model yields an error rate of 12.5% for these data. However, this rate is a biased estimate—it is overly optimistic, since we have used the same data for fitting the classification functions and for estimating the error. Therefore, as with all other models, we test performance on a validation set that includes data that were not involved in estimating the classification functions.

To obtain the confusion matrix from a discriminant analysis, we either use the classification scores directly or the propensities (probabilities of class membership) that are computed from the classification scores. In both cases we decide on the class assignment of each observation based on the highest score or probability. We then compare these classifications to the actual class memberships of these observations. This yields the confusion matrix.

12.5 PRIOR PROBABILITIES

So far we have assumed that our objective is to minimize the classification error. The method presented above assumes that the chances of encountering an observation from either class is the same. If the probability of encountering an observation for classification in the future is not equal for the different classes, we should modify our functions to reduce our expected (long-run average) error rate. The modification is done as follows: Let p_j denote the prior or

---

[3] In practice, quadratic discriminant analysis has not been found useful except when the difference in the correlation matrices is large and the number of observations available for training and testing is large. The reason is that the quadratic model requires estimating many more parameters that are all subject to error [for c classes and p variables, the total number of parameters to be estimated for all the different correlation matrices is $cp(p + 1)/2$].

future probability of membership in class j (in the two-class case we have p_1 and $p_2 = 1 - p_1$). We modify the classification function for each class by adding $\log(p_j)$.[4] To illustrate this, suppose that the percentage of riding-mower owners in the population is 15%, compared to 50% in the sample. This means that the model should classify fewer households as owners. To account for this distortion, adjust the constants in the classification functions from Figure 12.3 and obtain the adjusted constants $-73.16 + \log(0.15) = -75.06$ for owners and $-51.42 + \log(0.85) = -50.58$ for nonowners. To see how this can affect classifications, consider family 13, which was misclassified as an owner in the case involving equal probability of class membership. When we account for the lower probability of owning a mower in the population, family 13 is classified properly as a nonowner (its owner classification score is below the nonowner score).

12.6   UNEQUAL MISCLASSIFICATION COSTS

A second practical modification is needed when misclassification costs are not symmetrical. If the cost of misclassifying a class 1 item is very different from the cost of misclassifying a class 2 item, we may want to minimize the expected cost of misclassification rather than the simple error rate (which does not account for unequal misclassification costs). In the two-class case, it is easy to manipulate the classification functions to account for differing misclassification costs (in addition to prior probabilities). We denote by q_1 the cost of misclassifying a class 1 member (into class 2). Similarly, q_2 denotes the cost of misclassifying a class 2 member (into class 1). These costs are integrated into the constants of the classification functions by adding $\log(q_1)$ to the constant for class 1 and $\log(q_2)$ to the constant of class 2. To incorporate both prior probabilities and misclassification costs, add $\log(p_1 q_1)$ to the constant of class 1 and $\log(p_2 q_2)$ to that of class 2.

In practice, it is not always simple to come up with misclassification costs q_1 and q_2 for each class. It is usually much easier to estimate the ratio of costs q_2/q_1 (e.g., the cost of misclassifying a credit defaulter is 10 times more expensive than that of misclassifying a nondefaulter). Luckily, the relationship between the classification functions depends only on this ratio. Therefore we can set $q_1 = 1$ and $q_2 = ratio$ and simply add $\log(q_2/q_1)$ to the constant for class 2.

---

[4] XLMiner by default sets the prior class probabilities as the ratios in the dataset. This is based on the assumption that a random sample will yield a reasonable estimate of membership probabilities. However, for other prior probabilities the classification functions can be modified manually.

12.7 CLASSIFYING MORE THAN TWO CLASSES

Example 3: Medical Dispatch to Accident Scenes

Ideally, every automobile accident call to the emergency number 911 results in the immediate dispatch of an ambulance to the accident scene. However, in some cases the dispatch might be delayed (e.g., at peak accident hours or in some resource-strapped towns or shifts). In such cases, the 911 dispatchers must make decisions about which units to send based on sketchy information. It is useful to augment the limited information provided in the initial call with additional information in order to classify the accident as minor injury, serious injury, or death. For this purpose we can use data that were collected on automobile accidents in the United States in 2001 that involved some type of injury. For each accident, additional information is recorded, such as day of week, weather conditions, and road type. Figure 12.7 shows a small sample of observations with 11 measurements of interest.

The goal is to see how well the predictors can be used to classify injury type correctly. To evaluate this, a sample of 1000 observations was drawn and partitioned into training and validation sets, and a discriminant analysis was performed on the training data. The output structure is very similar to that for

Accident #	RushHour	WRK_ZONE	WKDY	INT_HWY	LGTCON	LEVEL	SPD_LIM	SUR_COND	TRAF_WAY	WEATHER	MAX_SEV
1	1	0	1	1	dark_light	1	70	ice	one_way	adverse	no-injury
2	1	0	1	0	dark_light	0	70	ice	divided	adverse	no-injury
3	1	0	1	0	dark_light	0	65	ice	divided	adverse	nonfatal
4	1	0	1	0	dark_light	0	55	ice	two_way	not_adverse	nonfatal
5	1	0	0	0	dark_light	0	35	snow	one_way	adverse	no-injury
6	1	0	1	0	dark_light	1	35	wet	divided	adverse	no-injury
7	0	0	1	1	dark_light	1	70	wet	divided	adverse	nonfatal
8	0	0	1	0	dark_light	1	35	wet	two_way	adverse	no-injury
9	1	0	1	0	dark_light	0	25	wet	one_way	adverse	nonfatal
10	1	0	1	0	dark_light	0	35	wet	divided	adverse	nonfatal
11	1	0	1	0	dark_light	0	30	wet	divided	adverse	nonfatal
12	1	0	1	0	dark_light	0	60	wet	divided	not_adverse	no-injury
13	1	0	1	0	dark_light	0	40	wet	two_way	not_adverse	no-injury
14	0	0	1	0	day	1	65	dry	two_way	not_adverse	fatal
15	1	0	0	0	day	0	55	dry	two_way	not_adverse	fatal
16	1	0	1	0	day	0	55	dry	two_way	not_adverse	nonfatal
17	1	0	0	0	day	0	55	dry	two_way	not_adverse	nonfatal
18	0	0	1	0	dark	0	55	ice	two_way	not_adverse	no-injury
19	0	0	0	0	dark	0	50	ice	two_way	adverse	no-injury
20	0	0	0	0	dark	1	55	snow	divided	adverse	no-injury

FIGURE 12.7   SAMPLE OF 20 AUTOMOBILE ACCIDENTS FROM THE 2001 DEPARTMENT OF TRANSPORTATION DATABASE. EACH ACCIDENT IS CLASSIFIED AS ONE OF THREE INJURY TYPES (NO-INJURY, NONFATAL, OR FATAL), AND HAS 10 MEASUREMENTS (EXTRACTED FROM A LARGER SET OF MEASUREMENTS).

the two-class case. The only difference is that each observation now has three classification functions (one for each injury type), and the confusion and error matrices are 3×3 to account for all the combinations of correct and incorrect classifications (see Figure 12.8). The rule for classification is still to classify an observation to the class that has the highest corresponding classification score. The classification scores are computed, as before, using the classification function coefficients. This can be seen in Figure 12.7. For instance, the *nonfatal* classification score for the first accident in the training set is $-24.51 + (1.95)(1) + (1.19)(0) + \cdots + (16.36)(1) = 30.93$. The *no-injury* score is similarly computed as 31.42 and the *fatal* score as 25.94. Since the *no-injury* score is highest, this accident is (correctly) classified as having no injuries.

(a) Classification Function

Variables	Classification Function		
	fatal	no-injury	nonfatal
Constant	−25.59584999	−24.51432228	−24.2336216
RushHour	0.92256236	1.95240343	1.9031992
WRK_ZONE	0.51786095	1.19506037	0.77056831
WKDY	4.78014898	6.41763353	6.11652184
INT_HWY	−1.84187829	−2.67303801	−2.53662229
LGTCON_day	3.70701218	3.66607523	3.7276206
LEVEL	2.62689376	1.56755066	1.71386576
SPD_LIM	0.50513172	0.46147966	0.45208475
SUR_COND_dry	9.99886131	15.8337965	16.25656509
TRAF_WAY_two_way	7.10797691	6.34214783	6.35494375
WEATHER_adverse	9.68802357	16.36388016	16.31727791

(b) Training Data Scoring—Summary Report

Classification Confusion Matrix			
	Predicted Class		
Actual Class	fatal	no-injury	nonfatal
fatal	1	1	3
no-injury	6	114	172
nonfatal	6	95	202

Error Report			
Class	# Cases	# Errors	% Error
fatal	5	4	80.00
no-injury	292	178	60.96
nonfatal	303	101	33.33
Overall	600	283	47.17

FIGURE 12.8   XLMINER'S DISCRIMINANT ANALYSIS OUTPUT FOR THE THREE-CLASS INJURY EXAMPLE: (a) CLASSIFICATION FUNCTIONS AND (b) CONFUSION MATRIX FOR TRAINING SET

We can also compute for each accident the propensities (estimated probabilities) of belonging to each of the three classes using the same relationship between classification scores and probabilities as in the two-class case. For instance, the probability of the accident above involving nonfatal injuries is estimated by the model as

$$\frac{e^{31.42}}{e^{31.42} + e^{30.93} + e^{25.94}} = 0.38. \tag{12.3}$$

The probabilities of an accident involving no injuries or fatal injuries are computed in a similar manner. For the first accident in the training set, the highest probability is that of involving no injuries, and therefore it is classified as a *no-injury* accident (see Figure 12.9). In general, membership probabilities can be obtained directly from XLMiner for the training set, the validation set, and new observations.

12.8   ADVANTAGES AND WEAKNESSES

Discriminant analysis is typically considered more of a statistical classification method than a data mining method. This is reflected in its absence or short mention in many data mining resources. However, it is very popular in social sciences and has shown good performance. The use and performance of discriminant analysis are similar to those of multiple linear regression. The two methods therefore share several advantages and weaknesses.

Like linear regression, discriminant analysis searches for the optimal weighting of predictors. In linear regression, weighting is with relation to the numerical response, whereas in discriminant analysis it is with relation to separating the classes. Both use least squares for estimation and the resulting estimates are robust to local optima.

In both methods, an underlying assumption is normality. In discriminant analysis we assume that the predictors are approximately from a multivariate normal distribution. Although this assumption is violated in many practical situations (e.g., with commonly used binary predictors), the method is surprisingly robust. According to Hastie et al. (2001), the reason may be that data can usually support only simple separation boundaries, such as linear boundaries. However, for continuous variables that are found to be very skewed (as can be seen through a histogram), transformations such as the log transform can improve performance. In addition, the method's sensitivity to outliers commands exploring the data for extreme values and removing those observations from the analysis.

Row Id.	Predicted Class	Actual Class	Score for fatal	Score for no-injury	Score for nonfatal	Prob. for class fatal	Prob. for class no-injury	Prob. for class nonfatal	RushHour	WRK_ZONE	WKDY	INT_HWY	LGTCON_day	LEVEL	SPD_LIM	SUR_COND_dry	TRAF_two_way	WEATHER_adverse
2	no-injury	no-injury	25.94	31.42	30.93	0.002583566	0.618769909	0.378646525	1	0	1	1	0	1	70	0	0	1
56	no-injury	nonfatal	15.00	15.58	15.01	0.263257586	0.471205318	0.265537095	1	0	1	0	0	0	55	0	1	0
79	no-injury	no-injury	2.69	9.95	9.81	0.000376892	0.535717942	0.463905165	1	0	0	0	0	0	35	0	0	1
141	no-injury	no-injury	10.10	17.94	17.64	0.000226522	0.574000564	0.425772914	1	0	1	0	0	1	35	0	0	1
203	no-injury	nonfatal	2.42	11.76	11.41	5.18896E-05	0.586851481	0.413096629	1	0	1	0	0	0	25	0	0	1
215	no-injury	nonfatal	7.47	16.37	15.93	8.33756E-05	0.609408333	0.390508291	1	0	1	0	0	0	35	0	0	1
281	no-injury	no-injury	10.41	11.54	10.91	0.174287408	0.539388146	0.286324446	1	0	1	0	0	0	60	0	0	0
660	nonfatal	nonfatal	22.57	28.37	28.46	0.001452278	0.476947897	0.521599825	1	0	1	1	0	1	45	0	1	0
838	nonfatal	nonfatal	15.12	20.45	20.47	0.002408934	0.493295725	0.504295341	0	0	0	1	0	0	55	1	0	0
878	no-injury	nonfatal	23.62	29.32	29.15	0.00181542	0.540936208	0.457248372	1	0	1	1	0	0	70	1	0	0
882	no-injury	no-injury	20.17	25.06	24.99	0.003898532	0.515946916	0.480154552	0	0	1	1	0	0	65	1	0	0
907	no-injury	nonfatal	10.31	18.15	18.13	0.00019948	0.505559651	0.494240868	1	0	0	1	0	0	40	1	0	0
1072	no-injury	nonfatal	12.84	20.46	20.39	0.000253881	0.517266017	0.482480102	1	0	1	0	0	0	45	1	0	0
1162	nonfatal	nonfatal	15.16	20.38	20.62	0.002378591	0.43818467	0.559436739	1	0	0	0	0	0	45	1	0	0
1245	no-injury	no-injury	19.94	26.80	26.74	0.000542513	0.513922638	0.485534849	1	1	1	1	0	0	45	1	1	0
1326	nonfatal	nonfatal	12.37	19.88	19.96	0.000262636	0.478861093	0.520876272	1	1	0	0	0	0	30	1	1	0
1468	no-injury	nonfatal	25.00	31.41	31.26	0.000877376	0.537149759	0.461972865	1	0	1	1	0	0	55	0	1	0

FIGURE 12.9 CLASSIFICATION SCORES, MEMBERSHIP PROBABILITIES, AND CLASSIFICATIONS FOR THE THREE-CLASS INJURY TRAINING DATASET

An advantage of discriminant analysis as a classifier is that it provides estimates of single-predictor contributions (it is like logistic regression in this respect).[5] This is useful for obtaining a ranking of predictor importance, and for variable selection.

Finally, the method is computationally simple, parsimonious, and especially useful for small datasets. With its parametric form, discriminant analysis makes the most out of the data and is therefore especially useful with small samples (as explained in Section 12.4).

---

[5] Comparing predictor contribution requires normalizing all the predictors before running discriminant analysis. Then each coefficient is compared across the two classification functions: coefficients with large differences indicate a predictor with high separation power.

PROBLEMS

12.1 **Personal Loan Acceptance.** Universal Bank is a relatively young bank growing rapidly in terms of overall customer acquisition. The majority of these customers are liability customers with varying sizes of relationship with the bank. The customer base of asset customers is quite small, and the bank is interested in expanding this base rapidly to bring in more loan business. In particular, it wants to explore ways of converting its liability customers to personal loan customers.

A campaign the bank ran for liability customers last year showed a healthy conversion rate of over 9% successes. This has encouraged the retail marketing department to devise smarter campaigns with better target marketing. The goal of our analysis is to model the previous campaign's customer behavior to analyze what combination of factors make a customer more likely to accept a personal loan. This will serve as the basis for the design of a new campaign.

The file `UniversalBank.xls` contains data on 5000 customers. The data include customer demographic information (e.g., age, income), the customer's relationship with the bank (e.g., mortgage, securities account), and the customer response to the last personal loan campaign (Personal Loan). Among these 5000 customers, only 480 (= 9.6%) accepted the personal loan that was offered to them in the previous campaign.

Partition the data (60% training and 40% validation) and then perform a discriminant analysis that models Personal Loan as a function of the remaining predictors (excluding zip code). Remember to turn categorical predictors with more than two categories into dummy variables first. Specify the success class as 1 (loan acceptance), and use the default cutoff value of 0.5.

  a. Compute summary statistics for the predictors separately for loan acceptors and nonacceptors. For continuous predictors, compute the mean and standard deviation. For categorical predictors, compute the percentages. Are there predictors where the two classes differ substantially?

  b. Examine the model performance on the validation set.

   i. What is the misclassification rate?

   ii. Is one type of misclassification more likely than the other?

   iii. Select three customers who were misclassified as acceptors and three who were misclassified as nonacceptors. The goal is to determine why they are misclassified. First, examine their probability of being classified as acceptors: is it close to the threshold of 0.5? If not, compare their predictor values to the summary statistics of the two classes to determine why they were misclassified.

  c. As in many marketing campaigns, it is more important to identify customers who will accept the offer rather than customers who will not accept it. Therefore a good model should be especially accurate at detecting acceptors. Examine the lift chart and decile chart for the validation set and interpret them in light of this ranking goal.

  d. Compare the results from the discriminant analysis with those from a logistic regression (both with cutoff 0.5 and the same predictors). Examine the confusion matrices, the lift charts, and the decile charts. Which method performs better on your validation set in detecting the acceptors?

e. The bank is planning to continue its campaign by sending its offer to 1000 additional customers. Suppose that the cost of sending the offer is $1 and the profit from an accepted offer is $50. What is the expected profitability of this campaign?

f. The cost of misclassifying a loan acceptor customer as a nonacceptor is much higher than the opposite misclassification cost. To minimize the expected cost of misclassification, should the cutoff value for classification (which is currently at 0.5) be increased or decreased?

12.2 **Identifying Good System Administrators.** A management consultant is studying the roles played by experience and training in a system administrator's ability to complete a set of tasks in a specified amount of time. In particular, she is interested in discriminating between administrators who are able to complete given tasks within a specified time and those who are not. Data are collected on the performance of 75 randomly selected administrators. They are stored in the file `SystemAdministrators.xls`.

Using these data, the consultant performs a discriminant analysis. The variable Experience measures months of full-time system administrator experience, while Training measures number of relevant training credits. The dependent variable Completed is either *Yes* or *No*, according to whether or not the administrator completed the tasks.

a. Create a scatter plot of Experience vs. Training using color or symbol to differentiate administrators who completed the tasks from those who did not complete them. See if you can identify a line that separates the two classes with minimum misclassification.

b. Run a discriminant analysis with both predictors using the entire dataset as training data. Among those who completed the tasks, what is the percentage of administrators who are classified incorrectly as failing to complete the tasks?

c. Compute the two classification scores for an administrator with four months of experience and 6 credits of training. Based on these, how would you classify this administrator?

d. How much experience must be accumulated by an administrator with 4 training credits before his or her estimated probability of completing the tasks exceeds 0.5?

e. Compare the classification accuracy of this model to that resulting from a logistic regression with cutoff 0.5.

12.3 **Detecting Spam Email (from the UCI Machine Learning Repository).** A team at Hewlett-Packard collected data on a large number of email messages from their postmaster and personal email for the purpose of finding a classifier that can separate email messages that are spam vs. nonspam (a.k.a. "ham"). The spam concept is diverse: it includes advertisements for products or websites, "make money fast" schemes, chain letters, pornography, and so on. The definition used here is "unsolicited commercial email." The file `Spambase.xls` contains information on 4601 email messages, among which 1813 are tagged "spam." The predictors include 57 attributes, most of them are the average number of times a certain word (e.g., mail, George) or symbol (e.g., #, !) appears in the email. A few predictors are related to the number and length of capitalized words.

a. To reduce the number of predictors to a manageable size, examine how each predictor differs between the spam and nonspam emails by comparing the spam-class average and nonspam-class average. Which are the 11 predictors that appear to vary the most between spam and nonspam emails? From these 11, which words or signs occur more often in spam?

b. Partition the data into training and validation sets; then perform a discriminant analysis on the training data using only the 11 predictors.

c. If we are interested mainly in detecting spam messages, is this model useful? Use the confusion matrix, lift chart, and decile chart for the validation set for the evaluation.

d. In the sample, almost 40% of the email messages were tagged as spam. However, suppose that the actual proportion of spam messages in these email accounts is 10%. Compute the constants of the classification functions to account for this information.

e. A spam filter that is based on your model is used, so that only messages that are classified as nonspam are delivered, while messages that are classified as spam are quarantined. In this case, misclassifying a nonspam email (as spam) has much heftier results. Suppose that the cost of quarantining a nonspam email is 20 times that of not detecting a spam message. Compute the constants of the classification functions to account for these costs (assume that the proportion of spam is reflected correctly by the sample proportion).

## CHAPTER 13

Combining Methods: Ensembles and Uplift Modeling

In this chapter we look at two useful approaches that combine methods for improving predictive power: *ensembles* and *uplift modeling*. An ensemble combines multiple supervised models into a "supermodel." The previous chapters in this part of the book introduced different supervised methods for prediction and classification. Earlier, in Chapter 5, we learned about evaluating predictive performance, which can be used to compare several models and choose the best one. An ensemble is based on the powerful notion of *combining models*. Instead of choosing a single predictive model, we can combine several models to achieve improved predictive accuracy. In this chapter, we explain the underlying logic and why ensembles can improve predictive accuracy and introduce popular approaches for combining models, including simple averaging, bagging, and boosting.

In *uplift modeling* we combine supervised modeling with A-B testing, which is a simple type of randomized experiment. We describe the basics of A-B testing and how it is used along with predictive models in persuasion messaging not to predict outcomes but to predict who should receive which message treatment.

Data Mining for Business Analytics: Concepts, Techniques, and Applications with XLMiner®, Third Edition.
Galit Shmueli, Peter C. Bruce, and Nitin R. Patel.
© 2016 John Wiley & Sons Inc. Published 2016 by John Wiley & Sons, Inc.

13.1 ENSEMBLES[1]

Ensembles played a major role in the million-dollar Netflix Prize contest that started in 2006. At the time, Netflix, the largest DVD rental service in the United States, wanted to improve their movie recommendation system (from www.netflixprize.com):

> Netflix is all about connecting people to the movies they love. To help customers find those movies, we've developed our world-class movie recommendation system: Cinematch^{SM} ... And while Cinematch is doing pretty well, it can always be made better.

In a bold move, the company decided to share a large amount of data on movie ratings by their users, and set up a contest, open to the public, aimed at improving their recommendation system:

> We provide you with a lot of anonymous rating data, and a prediction accuracy bar that is 10% better than what Cinematch can do on the same training data set.

During the contest, an active leader-board showed the results of the competing teams. An interesting behavior started appearing: Different teams joined forces to create combined, or *ensemble*, predictions, which proved more accurate than the individual predictions. The winning team, called "BellKor's Pragmatic Chaos" combined results from the "BellKor" and "Big Chaos" teams alongside additional members. In a 2010 article in *Chance* magazine, the Netflix Prize winners described the power of their ensemble approach:

> An early lesson of the competition was the value of combining sets of predictions from multiple models or algorithms. If two prediction sets achieved similar RMSEs, it was quicker and more effective to simply average the two sets than to try to develop a new model that incorporated the best of each method. Even if the RMSE for one set was much worse than the other, there was almost certainly a linear combination that improved on the better set.

Why Ensembles Can Improve Predictive Power

The principle of combining methods is popular for reducing risk. For example, in finance, portfolios are created for reducing investment risk. The return from

---

[1]This and subsequent sections in this chapter © 2016 Statistics.com and Galit Shmueli. Used by permission.

a portfolio is typically less risky because the variation is smaller than each of the individual components.

In predictive modeling, "risk" is equivalent to variation in prediction error. The more our prediction errors vary, the more volatile is our predictive model. Consider predictions from two different models for a set of n observations. $e_{1,i}$ is the prediction error for the ith observation by method 1 and $e_{2,i}$ is the prediction for the same observation by method 2.

Suppose that each model produces prediction errors that are, on average, zero (for some observations the model overpredicts and for some it underpredicts, but on average the error is zero):

$$E(e_{1,i}) = E(e_{2,i}) = 0.$$

If, for each observation, we take an average of the two predictions: $\bar{y}_i = (\hat{y}_{1,i} + \hat{y}_{2,i})/2$, then the expected average error will also be zero:

$$E\left(y_i - \overline{y}_i\right) = E\left(y_i - \frac{\hat{y}_{1,i} + \hat{y}_{2,i}}{2}\right) \tag{13.1}$$

$$= E\left(\frac{y_i - \hat{y}_{1,i}}{2} + \frac{y_i - \hat{y}_{2,i}}{2}\right) = E\left(\frac{e_{1,i} + e_{2,i}}{2}\right) = 0.$$

This means that the ensemble has the same average error as the individual models. Now let us examine the variance of the ensemble's prediction errors:

$$\text{Var}\left(\frac{e_{1,i} + e_{2,i}}{2}\right) = \frac{1}{4}\left(\text{Var}(e_{1,i}) + \text{Var}(e_{2,i})\right) + 2 \times \frac{1}{4}\text{Cov}\left(e_{1,i}, e_{2,i}\right). \tag{13.2}$$

This variance can be lower than each of the individual variances $\text{Var}(e_{1,i})$ and $\text{Var}(e_{2,i})$ under some circumstances. A key component is the covariance (or equivalently, correlation) between the two prediction errors. The case of no correlation leaves us with a quantity that can be smaller than each of the individual variances. The variance of the average prediction error will be even smaller when the two prediction errors are negatively correlated.

In summary, using an average of two predictions can potentially lead to smaller error variance, and therefore better predictive power. These results generalize to more than two methods; you can combine results from multiple prediction methods or classifiers.

THE WISDOM OF CROWDS

In his book *The Wisdom of Crowds*, James Surowiecki recounts how Francis Galton, a prominent statistician from the 19th century, watched a contest at a county fair in England. The contest's objective was to guess the weight of an ox. Individual contest entries were highly variable, but the mean of all the estimates was surprisingly accurate—within 1% of the true weight of the ox. On balance, the errors from multiple guesses tended to cancel one another out. You can think of the output of a predictive model as a more informed version of these guesses. Averaging together multiple guesses will yield a more accurate answer than the vast majority of the individual guesses. Note that in Galton's story there were a few (lucky) individuals who scored better than the average. An ensemble estimate will not always be more accurate than all the individual estimates in all cases, but it will be more accurate most of the time.

Simple Averaging

The simplest approach for creating an ensemble is to combine the predictions, classifications, or propensities from multiple models. For example, we might have a linear regression model, a regression tree, and a k-NN algorithm. We use each of the three methods to score, say, a test set. We then combine the three sets of results.

The three models can also be variations that use the same algorithm. For example, we might have three linear regression models, each using a different set of predictors.

Combining Predictions   In prediction tasks, where the outcome variable is numerical, we can combine the predictions from the different methods simply by taking an average. In the example above, for each observation in the test set we have three predictions (one from each model). The ensemble prediction is then the average of the three values.

One alternative to a simple average is taking the median prediction, which would be less affected by extreme predictions. Another possibility is computing a weighted average, where weights are proportional to a quantity of interest. For instance, weights can be proportional to the accuracy of the model, or if different data sources are used, the weights can be proportional to the quality of the data.

Ensembles for prediction are useful not only in cross-sectional prediction, but also in time series forecasting (see Chapters 16–18). In forecasting, the same approach of combining future forecasts from multiple methods can lead to more precise predictions. One example is the weather forecasting application Forecast.io (www.forecast.io), which describes their algorithm as follows:

Forecast.io is backed by a wide range of data sources, which are aggregated together statistically to provide the most accurate forecast possible for a given location.

Combining Classifications   In the case of classification, combining the results from multiple classifiers can be done using "voting": For each record, we have multiple classifications. A simple rule would be to choose the most popular class among these classifications. For example, we might use a classification tree, a naive Bayes classifier, and discriminant analysis for classifying a binary outcome. For each observation we then generate three predicted classes. Simple voting would choose the most common class among the three.

As in prediction, we can assign heavier weights to scores from some models, based on considerations such as model accuracy or data quality. This would be done by setting a "majority rule" that is different from 50%.

Combining Propensities   Similar to predictions, propensities can be combined by taking a simple (or weighted) average. Recall that some algorithms, such as naive Bayes (see Chapter 8), produce biased propensities and should therefore not be simply averaged with propensities from other methods.

Bagging

Another form of ensembles is based on averaging across multiple random data samples. *Bagging*, short for "bootstrap aggregating," comprises two steps:

1. Generate multiple random samples (by sampling with replacement from the original data)—this method is called "bootstrap sampling".
2. Running an algorithm on each sample and producing scores.

Bagging improves the performance stability of a model and helps avoid overfitting by separately modeling different data samples and then combining the results. It is therefore especially useful for algorithms such as trees and neural networks. In Chapter 9 we described random forests, an ensemble based on bagged trees. XLMiner offers bagged versions for a few algorithms (version V15 offers bagged trees, random forests, and bagged neural networks).

Boosting

Boosting is a slightly different approach to creating ensembles. Here the goal is to directly improve areas in the data where our model makes errors, by forcing the model to pay more attention to those records. The steps in boosting are:

1. Fit a model to the data.

2. Draw a sample from the data so that misclassified observations (or observations with large prediction errors) have higher probabilities of selection.

3. Fit the model to the new sample.

4. Repeat steps 2–3 multiple times.

XLMiner offers boosting versions for a few algorithms (version V15 offers boosted trees and boosted neural networks.)

Advantages and Weaknesses of Ensembles

Combining scores from multiple models is aimed at generating more precise predictions (lowering the prediction error variance). The ensemble approach is most useful when the combined models generate prediction errors that are negatively associated, but it can also be useful when the correlation is low. Ensembles can use simple averaging, weighted averaging, voting, medians, and the like. Models can be based on the same algorithm or on different algorithms, using the same sample or different samples. Ensembles have become a major strategy for participants in data mining contests, where the goal is to optimize some predictive measure. In that sense, ensembles also provide an operational way to obtain solutions with high predictive power in a fast way, by engaging multiple teams of "data crunchers" working in parallel and combining their results.

Ensembles that are based on different data samples help avoid overfitting. However, remember that you can also overfit the data with an ensemble if you tweak it (e.g., choosing the "best" weights when using a weighted average). The major disadvantage of an ensemble is the resources that it requires: computationally as well as in terms of software availability, and the analyst's skill and time investment. Ensembles that combine results from different algorithms require developing each of the models and evaluating them. Boosting-type ensembles and bagging-type ensembles do not require such effort, but they do have a computational cost (although boosting can easily be parallelized). And finally, ensembles are "black-box" methods, in that the relationship between the predictors and the output variable usually becomes nontransparent.

13.2 UPLIFT (PERSUASION) MODELING

Long before the advent of the Internet, sending messages directly to individuals (i.e., direct mail) held a big share of the advertising market. Direct marketing affords the marketer the ability to invite and monitor direct responses from consumers. This, in turn, allows the marketer to learn whether the messaging is paying off. A message can be tested with a small section of a large list, and if it pays off, the message can be rolled out to the entire list. With predictive modeling, we have seen that the rollout can be targeted to that portion of the list that is most likely to respond or behave in a certain way. None of this was possible with traditional media advertising (television, radio, newspaper, magazine).

Direct response also made it possible to test one message against another and find out which does better.

A-B Testing

A-B testing is the marketing industry's term for a standard scientific experiment in which results can be tracked for each individual. The idea is to test one treatment against another, or a treatment against a control. "Treatment" is simply the term for the intervention you are testing: in a medical trial it is typically a drug, device, or other therapy; in marketing it is typically an offering to a consumer such as an email or a web page shown to a consumer. A general display ad in a magazine would not generally qualify, unless it has a specific call to action that allows the marketer to trace the action (e.g., purchase) to a given ad, plus the ability to split the magazine distribution randomly and provide a different offer to each segment.

An important element of A-B testing is randomization—the treatments are assigned or delivered to individuals randomly. That way, any difference between treatment A and treatment B can be attributed to the treatment (unless it is due to chance).

Uplift

An A-B test tells you which treatment does better on average, but says nothing about which treatment does better for which individual. A classic example is in political campaigns. Consider the following scenario: The campaign director for Smith, a Democratic Congressional candidate, would like to know which voters should be called to encourage to support Smith. Voters that tend to vote Democratic but are not activists might be more inclined to vote for Smith if they got a call. Active Democrats are probably already supportive of him, and therefore a call to them would be wasted. Calls to Republicans are not only wasteful, but they could be harmful.

Campaigns now maintain extensive data on voters to help guide decisions about outreach to individual voters. Prior to the 2008 Obama campaign, the practice was to make rule-based decisions in accord with expert political judgment. Since 2008, it has increasingly been recognized that, rather than relying on judgment or supposition to determine whether an individual should be called, it is best to use the data to develop a model that can predict whether a voter will respond positively to outreach.

Gathering the Data

US states maintain publicly available files of voters, as part of the transparent oversight process for elections. The voter file contains data such as name, address, and date of birth. Political parties have "poll-watchers" at elections to record who votes, so they have additional data on which elections voters voted in. Census data for neighborhoods can be appended, based on voter address. Finally, commercial demographic data can be purchased and matched to the voter data. Table 13.1 shows a small extract of data derived from the voter file for the US state of Delaware.[2] The actual data used in this problem is in the file `Voter-Persuasion.xlsx`, and it contains 10,000 records and many additional variables beyond those shown in Table 13.1.

First, the campaign director conducts a survey of 10,000 voters to determine their inclination to vote Democratic. Then she conducts an experiment, randomly splitting the sample of 10,000 voters in half and mailing a flyer promoting Smith to half the list (treatment A), and nothing to the other half (treatment B). The control group that gets no flyer is essential, since other campaigns or

TABLE 13.1   **DATA ON VOTERS (SMALL SUBSET OF VARIABLES AND RECORDS) AND DATA DICTIONARY**

Voter	Age	NH_White	Comm_PT	H_F1	Reg_Days	PR_Pelig	E_Elig	Political_C
1	28	61	0	0	3997	0	20	1
2	23	67	3	0	300	0	0	1
3	57	64	4	0	2967	0	0	0
4	61	53	2	1	16620	100	90	1
5	37	76	2	0	3786	0	20	0

Data Dictionary

Age	voter age in years
NH_White	neighborhood average of % nonhispanic white in household
Comm_PT	neighborhood % of workers who take public transit
H_F1	single female household (1 = yes)
Reg_Days	days since voter registered at current address
PR_Pelig	voted in what % of nonpresidential primaries
E_Pelig	voted in what % of any primaries
Political_C	is there a political contributor in the home (1 = yes)

---

[2] Thanks to Ken Strasma, founder of the microtargeting firm HaystaqDNA and director of targeting for the 2004 Kerry campaign and the 2008 Obama campaign, for these data.

news events might cause a shift in opinion. The goal is to measure the change in opinion after the flyer goes out, relative to the no-flyer control group.

The next step is conducting a post-flyer survey of the same sample of 10,000 voters, to measure whether each voter's opinion of Smith has shifted in a positive direction. A binary variable, Moved_AD, will be added to the collected data, indicating whether opinion has moved in a Democratic direction (1) or not (0).

Table 13.2 summarizes the results of the survey, by comparing the movement in a Democratic direction for each of the treatments. Overall, the flyer (*flyer = 1*) is modestly effective.

TABLE 13.2   RESULTS OF SENDING A PRO-DEMOCRATIC FLYER TO VOTERS

	# sent	# Moved Dem.	% Moved
Message_A	5000	2012	40.2%
No message	5000	1722	34.4%

Movement in a Democratic direction among those who got no message is 34.4%. This probably reflects the approach of the election, the heightening campaign activity, and the reduction in the "no-opinion" category. It also illustrates the need for a control group. Among those who did get the flyer, the movement in a Democratic direction is 40.2%. So, overall, the lift from the flyer is 4.8%.

We can now append two variables to the voter data shown earlier in Table 13.1: *Flyer* (whether they received the flyer (1) or not (0)) and *Moved_AD* (whether they moved in a Democratic direction (1) or not (0)). The augmented data are shown in Table 13.3.

TABLE 13.3   TARGET VARIABLE (MOVED_AD) AND TREATMENT VARIABLE (FLYER) ADDED TO VOTER DATA

Voter	Age	NH_White	Comm_PT	H_F1	Reg_Days	PR_Pelig	E_Elig	Political_C	Flyer	Moved_AD
1	28	61	0	0	3997	0	20	1	1	0
2	23	67	3	0	300	0	0	1	1	1
3	57	64	4	0	2967	0	0	0	0	0
4	61	53	2	1	16620	100	90	1	0	0
5	37	76	2	0	3786	0	20	0	1	0

A Simple Model

We can develop a predictive model with *Moved_AD* as the target (output variable), and various predictor variables, including *Flyer*. Any classification method can be used; Table 13.4 shows the first few lines from the output of a logistic regression model that was used to predict MOVED_AD.

TABLE 13.4   CLASSIFICATIONS AND PROPENSITIES FROM PREDICTIVE MODEL (SMALL EXTRACT)

Voter	Message	Actual Moved	Predicted	Predicted Prob.
1	0	1	1	0.5975
2	1	1	1	0.5005
3	0	0	0	0.2235
4	0	0	0	0.3052
5	1	0	0	0.4140

However, our interest is not just how the flyer did overall, nor is it whether we can predict the probability that a voter's opinion will move in a favorable direction. Rather, our goal is to predict how much (positive) impact the flyer will have on a specific voter. That way the campaign can direct its limited resources toward the voters who are the most persuadable—those for whom mailing the flyer will have the greatest positive effect.

Modeling Individual Uplift

To answer the question about the flyer's impact on each voter, we need to model the effect of the flyer at the individual voter level. For each voter, uplift is defined as follows:

Uplift = increase in propensity of favorable opinion after getting flyer

To build an uplift model, we follow the following steps to estimate the change in probability of "success" (propensity) that comes from receiving the treatment (the flyer):

1. Randomly split a data sample into treatment and control groups, conduct an A-B test, and record the outcome (in our example: *Moved_AD*).

2. Recombining the data sample, partition it into training and validation sets; build a predictive model with this outcome as the target variable and include a predictor variable that denotes treatment status (in our example: *Flyer*).

3. Score this predictive model to a partition of the data; you can use the validation partition. This will yield for each validation record its propensity of success given its treatment.

4. Reverse the value of the treatment variable and re-score the same model to that partition. This will yield for each validation record its propensity of success had it received the other treatment.

5. Uplift is estimated for each individual by $P(\text{Success} \mid \text{Treatment} = 1)$ $- P(\text{Success} \mid \text{Treatment} = 0)$.

6. For new data where no experiment has been performed, simply add a synthetic predictor variable for treatment and assign first a "1," score the model, then a "0," and score the model again. Estimate uplift for the new record(s) as above.

Continuing with the small voter example, the results from step 3 were as shown in Table 13.4—the right column shows the propensities from the model. Next, we re-estimate the logistic model, but with the values of the treatment variable *Flyer* reversed for each row. Table 13.5 shows the propensities with variable *Flyer* reversed (you can see the reversed values in column *Flyer*). Finally, in step 5, we calculate the uplift for each voter.

TABLE 13.5 CLASSIFICATIONS AND PROPENSITIES FROM PREDICTIVE MODEL (SMALL EXTRACT) WITH FLYER VALUES REVERSED

Voter	Message	Actual Moved	Predicted	Predicted Prob.
1	1	1	1	0.6908
2	0	1	1	0.3996
3	1	0	0	0.3022
4	1	0	0	0.3980
5	0	0	0	0.3194

Table 13.6 shows the uplift for each voter – the success (Moved_AD = 1) propensity given *Flyer* = 1 minus the success propensity given *Flyer* = 0.

TABLE 13.6 UPLIFT: CHANGE IN PROPENSITIES FROM SENDING MESSAGE VS. NOT SENDING MESSAGE

Voter	Prob. if Msg.	Prob. if no msg	Uplift
1	0.6908	0.5975	0.0933
2	0.5005	0.3996	0.1009
3	0.3022	0.2235	0.0787
4	0.3980	0.3052	0.0928
5	0.4140	0.3194	0.0946

Using the Results of an Uplift Model

Once we have estimated the uplift for each individual, the results can be ordered by uplift. The flyer could then be sent to all those voters with a positive uplift, or, if resources are limited, only to a subset—those with the greatest uplift.

Uplift modeling is used mainly in marketing and, more recently, in political campaigns. It has two main purposes:

- To determine whether to send someone a persuasion message, or just leave them alone.
- When a message is definitely going to be sent, to determine which message, among several possibilities, to send.

Technically this amounts to the same thing—"send no message" is simply another category of treatment, and an experiment can be constructed with multiple treatments, such as no message, message A, and message B. However, practitioners tend to think of the two purposes as distinct, and tend to focus on the first. Marketers want to avoid sending discount offers to customers who would make a purchase anyway, or renew a subscription anyway. Political campaigns likewise want to avoid calling voters who would vote for their candidate in any case. And both parties especially want to avoid sending messages or offers where the effect might be antagonistic—where the uplift is negative.[3]

13.3 SUMMARY

In practice, the methods discussed in this book are often used not in isolation, but as building blocks in an analytic process whose goal is always to inform and provide insight.

In this chapter we looked at two ways that multiple models are deployed. In ensembles, multiple models are weighted and combined to produce improved predictions. In uplift modeling, the results of A-B testing are folded into the predictive modeling process as a predictor variable to guide choices, not just about whether to send an offer or persuasion message but also as to who to send it to.

---

[3] Note that the uplift model can be applied to validation data to assess the model. The validation process does not have individual record-level "ground truth" of the treatment's uplift, since no individual has both treatment and non-treatment, but it will provide an estimate of how well the uplift model does with new data as a whole, as a mitigation of possible overfitting to the training data.

PROBLEMS

13.1 **Acceptance of Consumer Loan.** Universal Bank has begun a program to encourage its existing customers to borrow via a consumer loan program. The bank has promoted the loan to 5000 customers, of whom 480 accepted the offer. The data are available in file `UniversalBank.xls`. The bank now wants to develop a model to predict which customers have the greatest probability of accepting the loan, to reduce promotion costs and send the offer only to a subset of its customers.

We will develop several models, then combine them in an ensemble. The models we will use are (1) logistic regression, (2) k-nearest-neighbors with $k = 3$, and (3) naive Bayes. Pre-process the data as follows:

- Bin the following variables so they can be used in naive Bayes: Age, Experience, Income, CC Average, and Mortgage. Set the # of bins to 20, use "equal count per bin," and show the bin mean in the resulting binned variable.
- Education and Family can be used as is, without binning.
- Zip code can be ignored.
- Partition the data: 60% training, 40% validation.

  a. Fit models to the data for (1) logistic regression, (2) k-nearest-neighbors with $k = 3$, and (3) Naive Bayes. Use Personal Loan as the target variable. Report the confusion matrix for each of the three models.

  b. In a new worksheet, copy from each of the three model outputs the columns for actual outcome, predicted outcome, and probability. Report the first 10 rows of these columns.

  c. Add two columns to this worksheet for (1) a majority vote of predicted outcomes, and (2) the average of the predicted probabilities. Using the classifications generated by these two methods derive a confusion matrix for each method and note the overall error rate.

  d. Compare the error rates for the three individual methods and the two ensemble methods.

13.2 **eBay Auctions—Boosting and Bagging.** Using the eBay auction data (file `eBayAuctions.xls`) with variable *Competitive* as the target, partition the data into training (50%), validation (30%) and test sets (20%).

  a. Run a classification tree, setting the minimum number of records in terminal nodes to 0. Looking at the test set, what is the overall accuracy? What is the lift on the first decile?

  b. Run the same tree with the boosting option selected. For the test set, what is the overall accuracy? What is the lift on the first decile?

  c. Now try the same tree with the bagging option selected. For the test set, what is the overall accuracy? What is the lift on the first decile?

13.3 **Hair Care Product—Uplift Modeling.** This problem uses the data set in `Hair-Care-Product.xlsx`, courtesy of SAS. In this hypothetical case, a promotion for a hair care product was sent to some members of a buyers club. Purchases were then recorded for both the members who got the promotion and those who did not.

a. What is the purchase propensity

    i. among those who received the promotion?

    ii. among those who did not receive the promotion?

b. Partition the data into training (60%) and validation (40%) and fit a model of your choice, with *Purchase* as the target. Report the predicted class and propensities for the first 10 records in the validation set.

c. Copy the validation data to a new worksheet and reverse the values of the *Promotion* variable (call it *Promotion-R* to avoid confusion). This means that for every record, if the original had *Promotion* = 1, the copy will have *Promotion-R* = 0, and vice versa.

    Score the model to the new copy of the validation data. In XLMiner you can do this either with the Score function, or by re-running the model and selecting the copy as new data in the Score New Data area. You will need to match *Promotion* to *Promotion-R*. Report the predicted class and propensities for the first 10 records in the validation set.

d. In a new worksheet, copy the purchase propensities from the original validation data alongside those from the reversed-promotion validation data. Subtract the purchase propensity when *Promotion* = 0 from the purchase propensity when *Promotion* = 1. This is the uplift. Report the uplift for the first 10 records in your validation set.

## PART V

Mining Relationships among Records

## CHAPTER 14

Association Rules and Collaborative Filtering

In this chapter we describe the unsupervised learning methods of association rules (also called "affinity analysis" and "market basket analysis") and collaborative filtering. Both methods are popular in marketing for cross-selling products associated with an item that a consumer is considering.

In association rules, the goal is to identify item clusters in transaction-type databases. Association rule discovery in marketing is termed "market basket analysis" and is aimed at discovering which groups of products tend to be purchased together. These items can then be displayed together, offered in post-transaction coupons, or recommended in online shopping. We describe the two-stage process of rule generation and then assessment of rule strength to choose a subset. We look at the popular rule-generating Apriori algorithm, and then criteria for judging the strength of rules.

In collaborative filtering, the goal is to provide personalized recommendations that leverage user-level information. User-based collaborative filtering starts with a user, then finds users who have purchased a similar set of items or ranked items in similar fashion, and makes a recommendation to the initial user based on what the similar users purchase or like. Item-based collaborative filtering starts with an item being considered by a user, then locates other items that tend to be co-purchased with that first item. We explain the technique and the requirements for applying it in practice.

Data Mining for Business Analytics: Concepts, Techniques, and Applications with XLMiner®, Third Edition.
Galit Shmueli, Peter C. Bruce, and Nitin R. Patel.
© 2016 John Wiley & Sons Inc. Published 2016 by John Wiley & Sons, Inc.

14.1 ASSOCIATION RULES

Put simply, association rules, or *affinity analysis*, constitute a study of "what goes with what." This method is also called *market basket analysis* because it originated with the study of customer transactions databases to determine dependencies between purchases of different items. Association rules are heavily used in retail for learning about items that are purchased together, but they are also useful in other fields. For example, a medical researcher might want to learn what symptoms go with what confirmed diagnoses. In law, word combinations that appear too often might indicate plagiarism.

Discovering Association Rules in Transaction Databases

The availability of detailed information on customer transactions has led to the development of techniques that automatically look for associations between items that are stored in the database. An example is data collected using bar-code scanners in supermarkets. Such *market basket databases* consist of a large number of transaction records. Each record lists all items bought by a customer on a single-purchase transaction. Managers are interested to know if certain groups of items are consistently purchased together. They could use such information for making decisions on store layouts and item placement, for cross-selling, for promotions, for catalog design, and for identifying customer segments based on buying patterns. Association rules provide information of this type in the form of "if–then" statements. These rules are computed from the data; unlike the if–then rules of logic, association rules are probabilistic in nature.

Association rules are commonly encountered in online *recommendation systems* (or *recommender systems*), where customers examining an item or items for possible purchase are shown other items that are often purchased in conjunction with the first item(s). The display from Amazon.com's online shopping system illustrates the application of rules like this under "Frequently bought together." In the example shown in Figure 14.1, a user browsing a Samsung Galaxy S5 cell phone is shown a case and a screen protector that are often purchased along with this phone.

We introduce a simple artificial example and use it throughout the chapter to demonstrate the concepts, computations, and steps of association rules. We end by applying association rules to a more realistic example of book purchases.

Example 1: Synthetic Data on Purchases of Phone Faceplates

A store that sells accessories for cellular phones runs a promotion on faceplates. Customers who purchase multiple faceplates from a choice of six different colors get a discount. The store managers, who would like to know what colors of

Cell Phones & Accessories › Cell Phones › Unlocked Cell Phones

Samsung Galaxy S5 SM-G900H 16GB Factory Unlocked International Version - WHITE
by Samsung

    • 926 customer reviews | 731 answered questions

List Price: ~~$599.99~~
Price: **$396.98**
You Save: $203.01 (34%)

In Stock.
Sold by DeltaMobiles and Fulfilled by Amazon.

This item does not ship to **Taipei City, Taiwan; Republic of China**. Please check other sellers who may ship internationally. Learn more
Color: **White**

| $409.99 | $399.75 | $396.98 |

- 5.1" Full HD Super AMOLED(TM) (1080 x 1920)
- Exyon Quad Core; 1.9GHz,1.3GHz
- 16 MP Camera with LED Flash
- Must be activated with an Americas-region SIM
- 16GB of Internal Memory
- Unlocked cell phones are compatible with GSM carriers but are not compatible with CDMA Carriers.

Roll over image to zoom in

14 new from $381.87   33 used from $315.00
21 refurbished from $345.99

Frequently Bought Together

 + +

Price for all three: $422.96

[Add all three to Cart]
[Add all three to Wish List]

Show availability and shipping details

☑ **This item:** Samsung Galaxy S5 SM-G900H 16GB Factory Unlocked International Version - WHITE $396.98
☑ Galaxy S5 Case, Spigen Slim Armor Case for Galaxy S5 - Shimmery White (SGP10755) $16.99
☑ Galaxy S5 Screen Protector, Spigen◆ [Full HD] Samsung Galaxy S5 Screen Protector [Crystal ... $8.99

Customers Who Bought This Item Also Bought   Page 1 of 25

[Galaxy S5 Screen Protector] Poweradd™ Tempered Glass HD Clear Screen Protector Guard...	Samsung Galaxy S5 G900H 16GB Unlocked GSM Octa-Core Android Smartphone - Black	Galaxy S5 Screen Protector, Spigen◆ [Full HD] Samsung Galaxy S5 Screen Protector...	MPERO Collection 5 Pack of Ultra Clear Screen Protectors for Samsung Galaxy S5 / GS5
☆☆☆☆☆ 515	☆☆☆☆☆ 411	☆☆☆☆☆ 428	☆☆☆☆☆ 1,489
$7.99 *Prime*	$387.00 *Prime*	$8.99 *Prime*	$0.99

FIGURE 14.1   RECOMMENDATIONS UNDER "FREQUENTLY BOUGHT TOGETHER" ARE BASED ON ASSOCIATION RULES

faceplates customers are likely to purchase together, collected the transaction database as shown in Table 14.1.

TABLE 14.1   TRANSACTIONS FOR PURCHASES OF DIFFERENT-COLORED CELLULAR PHONE FACEPLATES

Transaction	Faceplate Colors Purchased			
1	red	white	green	
2	white	orange		
3	white	blue		
4	red	white	orange	
5	red	blue		
6	white	blue		
7	white	orange		
8	red	white	blue	green
9	red	white	blue	
10	yellow			

Generating Candidate Rules

The idea behind association rules is to examine all possible rules between items in an if–then format, and select only those that are most likely to be indicators of true dependence. We use the term *antecedent* to describe the IF part, and *consequent* to describe the THEN part. In association analysis, the antecedent and consequent are sets of items (called *itemsets*) that are disjoint (do not have any items in common). Note that itemsets are not records of what people buy; they are simply possible combinations of items, including single items.

Returning to the phone faceplate purchase example, one example of a possible rule is "if red, then white," meaning that if a red faceplate is purchased, a white one is too. Here the antecedent is *red* and the consequent is *white*. The antecedent and consequent each contain a single item in this case. Another possible rule is "if red and white, then green." Here the antecedent includes the itemset {*red, white*} and the consequent is {*green*}.

The first step in association rules is to generate all the rules that would be candidates for indicating associations between items. Ideally, we might want to look at all possible combinations of items in a database with p distinct items (in the phone faceplate example, $p = 6$). This means finding all combinations of single items, pairs of items, triplets of items, and so on, in the transactions database. However, generating all these combinations requires a long computation time that grows exponentially[1] in p. A practical solution is to consider only

---

[1] The number of rules that one can generate for p items is $3^p - 2^{p+1} + 1$. Computation time therefore grows by a factor for each additional item. For 6 items we have 602 rules, and for 7 items the number of rules grows to 1932.

combinations that occur with higher frequency in the database. These are called *frequent itemsets.*

Determining what qualifies as a frequent itemset is related to the concept of *support.* The support of a rule is simply the number of transactions that include both the antecedent and consequent itemsets. It is called a support because it measures the degree to which the data "support" the validity of the rule. The support is sometimes expressed as a percentage of the total number of records in the database. For example, the support for the itemset {red,white} in the phone faceplate example is 4 ($100 \times \frac{4}{10} = 40\%$).

What constitutes a frequent itemset is therefore defined as an itemset that has a support that exceeds a selected minimum support, determined by the user.

The Apriori Algorithm

Several algorithms have been proposed for generating frequent itemsets, but the classic algorithm is the *Apriori algorithm* of Agrawal et al. (1993). The key idea of the algorithm is to begin by generating frequent itemsets with just one item (one-item sets) and to recursively generate frequent itemsets with two items, then with three items, and so on, until we have generated frequent itemsets of all sizes.

It is easy to generate frequent one-itemsets. All we need to do is to count, for each item, how many transactions in the database include the item. These transaction counts are the supports for the one-itemsets. We drop one-itemsets that have support below the desired minimum support to create a list of the frequent one-itemsets.

To generate frequent two-itemsets, we use the frequent one-itemsets. The reasoning is that if a certain one-itemset did not exceed the minimum support, any larger size itemset that includes it will not exceed the minimum support. In general, generating k-itemsets uses the frequent $(k-1)$-itemsets that were generated in the preceding step. Each step requires a single run through the database, and therefore the Apriori algorithm is very fast even for a large number of unique items in a database.

Selecting Strong Rules

From the abundance of rules generated, the goal is to find only the rules that indicate a strong dependence between the antecedent and consequent itemsets. To measure the strength of association implied by a rule, we use the measures of *confidence* and *lift ratio*, as described below.

Support and Confidence In addition to support, which we described earlier, there is another measure that expresses the degree of uncertainty about the if–then rule. This is known as the *confidence*[2] of the rule. This measure compares the co-occurrence of the antecedent and consequent itemsets in the database to the occurrence of the antecedent itemsets. Confidence is defined as the ratio of the number of transactions that include all antecedent and consequent itemsets (i.e., the support) to the number of transactions that include all the antecedent itemsets:

$$\text{Confidence} = \frac{\text{Number of transactions with both antecedent and consequent itemsets}}{\text{Number of transactions with antecedent itemset}}.$$

For example, suppose that a supermarket database has 100,000 point-of-sale transactions. Of these transactions, 2000 include both orange juice and (over-the-counter) flu medication, and 800 of these include soup purchases. The association rule "IF orange juice and flu medication are purchased, THEN soup is purchased on the same trip" has a support of 800 transactions (alternatively, 0.8% = 800/100,000) and a confidence of 40% (= 800/2000).

To see the relationship between support and confidence, let us think about what each is measuring (estimating). One way to think of support is that it is the (estimated) probability that a transaction selected randomly from the database will contain all items in the antecedent and the consequent:

$$P(\text{antecedent AND consequent}).$$

In comparison, the confidence is the (estimated) *conditional probability* that a transaction selected randomly will include all the items in the consequent *given* that the transaction includes all the items in the antecedent:

$$\frac{P(\text{antecedent AND consequent})}{P(\text{antecedent})} = P(\text{consequent} \mid \text{antecedent}).$$

A high value of confidence suggests a strong association rule (in which we are highly confident). However, this can be deceptive because if the antecedent and/or the consequent has a high level of support, we can have a high value for confidence even when the antecedent and consequent are independent! For example, if nearly all customers buy bananas and nearly all customers buy ice cream, the confidence level will be high regardless of whether there is an association between the items.

Lift Ratio A better way to judge the strength of an association rule is to compare the confidence of the rule with a benchmark value, where we assume that the occurrence of the consequent itemset in a transaction is independent of the occurrence of the antecedent for each rule. In other words, if the antecedent and consequent itemsets are independent, what confidence values would we

---

[2] The concept of confidence is different from and unrelated to the ideas of confidence intervals and confidence levels used in statistical inference.

expect to see? Under independence, the support would be

$$P(\text{antecedent AND consequent}) = P(\text{antecedent}) \times P(\text{consequent}),$$

and the benchmark confidence would be

$$\frac{P(\text{antecedent}) \times P(\text{consequent})}{P(\text{antecedent})} = P(\text{consequent}).$$

The estimate of this benchmark from the data, called the *benchmark confidence value* for a rule, is computed by

$$\text{Benchmark confidence} = \frac{\text{Number of transactions with consequent itemset}}{\text{Number of transactions in database}}.$$

We compare the confidence to the benchmark confidence by looking at their ratio: this is called the *lift ratio* of a rule. The lift ratio is the confidence of the rule divided by the confidence, assuming independence of consequent from antecedent:

$$\text{Lift ratio} = \frac{\text{Confidence}}{\text{Benchmark confidence}}.$$

A lift ratio greater than 1.0 suggests that there is some usefulness to the rule. In other words, the level of association between the antecedent and consequent itemsets is higher than would be expected if they were independent. The larger the lift ratio, the greater is the strength of the association.

To illustrate the computation of support, confidence, and lift ratio for the cellular phone faceplate example, we introduce a presentation of the data better suited to this purpose.

Data Format

Transaction data are usually displayed in one of two formats: a list of items purchased (each row representing a transaction), or a binary matrix in which columns are items, rows again represent transactions, and each cell has either a 1 or a 0, indicating the presence or absence of an item in the transaction. For example, Table 14.1 displays the data for the cellular faceplate purchases in item list format. We translate these into a binary matrix format in Table 14.2.

Now suppose that we want association rules between items for this database that have a support count of at least 2 (equivalent to a percentage support of $2/10 = 20\%$): that is, rules based on items that were purchased together in at least 20% of the transactions. By enumeration, we can see that only the itemsets listed in Table 14.3 have a count of at least 2.

The first itemset {red} has a support of 6 because six of the transactions included a red faceplate. Similarly, the last itemset {red, white, green} has a support of 2 because only two transactions included red, white, and green faceplates.

TABLE 14.2 PHONE FACEPLATE DATA IN BINARY MATRIX FORMAT

Transaction	Red	White	Blue	Orange	Green	Yellow
1	1	1	0	0	1	0
2	0	1	0	1	0	0
3	0	1	1	0	0	0
4	1	1	0	1	0	0
5	1	0	1	0	0	0
6	0	1	1	0	0	0
7	1	0	1	0	0	0
8	1	1	1	0	1	0
9	1	1	1	0	0	0
10	0	0	0	0	0	1

In XLMiner the user can choose to input data using the *Associate > Association Rules* facility in either item-list format or binary matrix format.

TABLE 14.3 ITEMSETS WITH SUPPORT COUNT OF AT LEAST TWO

Itemset	Support (Count)
{red}	6
{white}	7
{blue}	6
{orange}	2
{green}	2
{red, white}	4
{red, blue}	4
{red, green}	2
{white, blue}	4
{white, orange}	2
{white, green}	2
{red, white, blue}	2
{red, white, green}	2

The Process of Rule Selection

The process of selecting strong rules is based on generating all association rules that meet stipulated support and confidence requirements. This is done in two stages. The first stage, described earlier, consists of finding all "frequent" itemsets, those itemsets that have a requisite support. In the second stage we generate, from the frequent itemsets, association rules that meet a confidence requirement. The first step is aimed at removing item combinations that are rare in the database. The second stage then filters the remaining rules and selects only those with high confidence. For most association analysis data, the computational challenge is the first stage, as described in the discussion of the Apriori algorithm.

The computation of confidence in the second stage is simple. Since any subset (e.g., {red} in the phone faceplate example) must occur at least as frequently as the set it belongs to (e.g., {red, white}), each subset will also be in the list. It is then straightforward to compute the confidence as the ratio of the support for the itemset to the support for each subset of the itemset. We retain the corresponding association rule only if it exceeds the desired cutoff value for confidence. For example, from the itemset {red, white, green} in the phone faceplate purchases, we get the following association rules and confidence values:

Rule		Confidence
Rule 1:	{red, white} $\Rightarrow$ {green}	$\dfrac{\text{support of \{red, white, green\}}}{\text{support of \{red, white\}}} = 2/4 = 50\%$
Rule 2:	{red, green} $\Rightarrow$ {white}	$\dfrac{\text{support of \{red, white, green\}}}{\text{support of \{red, green\}}} = 2/2 = 100\%$
Rule 3:	{white, green} $\Rightarrow$ {red}	$\dfrac{\text{support of \{red, white, green\}}}{\text{support of \{white, green\}}} = 2/2 = 100\%$
Rule 4:	{red} $\Rightarrow$ {white, green}	$\dfrac{\text{support of \{red, white, green\}}}{\text{support of \{red\}}} = 2/6 = 33\%$
Rule 5:	{white} $\Rightarrow$ {red, green}	$\dfrac{\text{support of \{red, white, green\}}}{\text{support of \{white\}}} = 2/7 = 29\%$
Rule 6:	{green} $\Rightarrow$ {red, white}	$\dfrac{\text{support of \{red, white, green\}}}{\text{support of \{green\}}} = 2/2 = 100\%$

If the desired minimum confidence is 70%, we would report only the second, third, and last rules.

We can generate association rules in XLMiner by specifying the minimum support count (2) and minimum confidence level percentage (70%). Figure 14.2 shows the output. Note that here we consider all possible itemsets, not just {red, white, green} as above.

The output includes information on the support of the antecedent, the support of the consequent, and the support of the combined set. It also gives the confidence of the rule (in %) and the lift ratio.

Inputs

Data	
# Transactions in Input Data	10
# Columns in Input Data	6
# Items in Input Data	6
# Association Rules	6
Minimum Support	2
Minimum Confidence	70.00%

List of Rules

Rule: If all Antecedent items are purchased, then with Confidence percentage Consequent items will also be purchased.

Row ID	Confidence %	Antecedent (A)	Consequent (C)	Support for A	Support for C	Support for A & C	Lift Ratio
1	100	orange	white	2	7	2	1.428571429
2	100	green	red	2	6	2	1.666666667
3	100	green	white	2	7	2	1.428571429
4	100	white & green	red	2	6	2	1.666666667
5	100	red & green	white	2	7	2	1.428571429
6	100	green	red & white	2	4	2	2.5

FIGURE 14.2 ASSOCIATION RULES FOR PHONE FACEPLATE TRANSACTIONS: XLMINER OUTPUT

Interpreting the Results

We can translate each of the rules from Figure 14.2 into an understandable sentence that provides information about performance. For example, we can read the first rule as follows:

> If orange is purchased, then with confidence 100% white will also be purchased. This rule has a lift ratio of 1.43.

In interpreting results, it is useful to look at the various measures. The support for the rule indicates its impact in terms of overall size: How many transactions are affected? If only a small number of transactions are affected, the rule may be of little use (unless the consequent is very valuable and/or the rule is very efficient in finding it).

The lift ratio indicates how efficient the rule is in finding consequents, compared to random selection. A very efficient rule is preferred to an inefficient rule, but we must still consider support: a very efficient rule that has very low support may not be as desirable as a less efficient rule with much greater support.

The confidence tells us at what rate consequents will be found, and it is useful in determining the business or operational usefulness of a rule. A rule with low confidence may find consequents at too low a rate to be worth the cost of (say) promoting the consequent in all the transactions that involve the antecedent.

TABLE 14.4				FIFTY TRANSACTIONS OF RANDOMLY ASSIGNED ITEMS								
Transaction		Items				Transaction		Items				

Transaction	Items					Transaction	Items					Transaction	Items			
1	8					18	8					35	3	4	6	8
2	3	4	8			19						36	1	4	8	
3	8					20	9					37	4	7	8	
4	3	9				21	2	5	6	8		38	8	9		
5	9					22	4	6	9			39	4	5	7	9
6	1	8				23	4	9				40	2	8	9	
7	6	9				24	8	9				41	2	5	9	
8	3	5	7	9		25	6	8				42	1	2	7	9
9	8					26	1	6	8			43	5	8		
10						27	5	8				44	1	7	8	
11	1	7	9			28	4	8	9			45	8			
12	1	4	5	8	9	29	9					46	2	7	9	
13	5	7	9			30	8					47	4	6	9	
14	6	7	8			31	1	5	8			48	9			
15	3	7	9			32	3	6	9			49	9			
16	1	4	9			33	7	9				50	6	7	8	
17	6	7	8			34	7	8	9							

Rules and Chance

What about confidence in the nontechnical sense? How sure can we be that the rules we develop are meaningful? Considering the matter from a statistical perspective, we can ask: are we finding associations that are really just chance occurrences?

Let us examine the output from an application of this algorithm to a small database of 50 transactions, where each of the nine items is assigned randomly to each transaction. The data are shown in Table 14.4, and the association rules generated are shown in Table 14.1. In looking at these tables, remember that "a" and "c" refer to itemsets, not records. So (*a&c*) means the union of itemsets "a" and "c," not the union of records with those items.

In this example, the lift ratios highlight rule 6 as most interesting, as it suggests that purchase of item 4 is almost five times as likely when items 3 and 8 are purchased than if item 4 was not associated with the itemset {3, 8}. Yet we know there is no fundamental association underlying these data—they were generated randomly.

Two principles can guide us in assessing rules for possible spuriousness due to chance effects:

1. The more records the rule is based on, the more solid is the conclusion. The key evaluative statistics are based on ratios and proportions, and we can look to statistical confidence intervals on proportions, such as political polls, for a rough preliminary idea of how variable rules might

TABLE 14.5   ASSOCIATION RULES OUTPUT FOR RANDOM DATA

Input Data: $A$5:$E$54
Min. Support: 2 = 4%
Min. Conf. % : 70

Rule	Confidence (%)	Antecedent a		Consequent c	Support (a)	Support (c)	Support (a&c)	Benchmark Confidence (%)	Lift Ratio (Conf./ Benchmark)
1	80	2	→	9	5	27	4	54	1.5
2	100	5, 7	→	9	3	27	3	54	1.9
3	100	6, 7	→	8	3	29	3	58	1.7
4	100	1, 5	→	8	2	29	2	58	1.7
5	100	2, 7	→	9	2	27	2	54	1.9
6	100	3, 8	→	4	2	11	2	22	4.5
7	100	3, 4	→	8	2	29	2	58	1.7
8	100	3, 7	→	9	2	27	2	547	1.9
9	100	4, 5	→	9	2	27	2	54	1.9

be owing to chance sampling variation. Polls based on 1500 respondents, for example, yield margins of error in the range of ±1.5%.

2. The more distinct are the rules we consider seriously (perhaps consolidating multiple rules that deal with the same items), the more likely it is that at least some will be based on chance sampling results. For one person to toss a coin 10 times and get 10 heads would be quite surprising. If 1000 people toss a coin 10 times each, it would not be nearly so surprising to have one get 10 heads. Formal adjustment of "statistical significance" when multiple comparisons are made is a complex subject in its own right, and beyond the scope of this book. A reasonable approach is to consider rules from the top down in terms of business or operational applicability, and not consider more than can reasonably be incorporated in a human decision-making process. This will impose a rough constraint on the dangers that arise from an automated review of hundreds or thousands of rules in search of "something interesting."

We now consider a more realistic example, using a larger database and real transactional data.

Example 2: Rules for Similar Book Purchases

The following example (drawn from the Charles Book Club case) examines associations among transactions involving various types of books. The database includes 2000 transactions, and there are 11 different types of books. The data, in binary matrix form, are shown in Table 14.6.

For instance, the first transaction included *YouthBks* (youth books) *DoItYBks* (do-it-yourself books), and *GeogBks* (geography books). Figure 14.3 shows (part of) the rules generated by XLMiner's *Association Rules* on these data. We spec-

TABLE 14.6 SUBSET OF BOOK PURCHASE TRANSACTIONS IN BINARY MATRIX FORMAT

ChildBks	YouthBks	CookBks	DoItYBks	cefBks	ArtBks	GeogBks	ItalCook	ItalAtlas	ItalArt	Florence
0	1	0	1	0	0	1	0	0	0	0
1	0	0	0	0	0	0	0	0	0	0
0	0	0	0	0	0	0	0	0	0	0
1	1	1	0	1	0	1	0	0	0	0
0	0	1	0	0	0	1	0	0	0	0
1	0	0	0	0	1	0	0	0	0	1
0	1	0	0	0	0	0	0	0	0	0
0	1	0	0	1	0	0	0	0	0	0
1	0	0	1	0	0	0	0	0	0	0
1	1	1	0	0	0	1	0	0	0	0
0	0	0	0	0	0	0	0	0	0	0

Inputs

Data	
# Transactions in Input Data	2000
# Columns in Input Data	11
# Items in Input Data	11
# Association Rules	49
Minimum Support	200
Minimum Confidence	50.00%

List of Rules

Rule: If all Antecedent items are purchased, then with Confidence percentage Consequent items will also be purchased.

Row ID	Confidence %	Antecedent (A)	Consequent (C)	Support for A	Support for C	Support for A & C	Lift Ratio
49	100.00	ItalCook	CookBks	227	862	227	2.320185615
45	62.77	ChildBks & ArtBks	GeogBks	325	552	204	2.274247492
28	54.13	CookBks & DoItYBks	ArtBks	375	482	203	2.246196404
48	61.98	CookBks & ArtBks	GeogBks	334	552	207	2.245508982
47	53.77	CookBks & GeogBks	ArtBks	385	482	207	2.230964057
19	57.11	RefBks	ChildBks & CookBks	429	512	245	2.230842075
44	52.31	ChildBks & GeogBks	ArtBks	390	482	204	2.170443664
27	60.78	CookBks & ArtBks	DoItYBks	334	564	203	2.15526394
35	58.40	ChildBks & CookBks	GeogBks	512	552	299	2.115885417
36	54.17	GeogBks	ChildBks & CookBks	552	512	299	2.115885417
42	57.87	CookBks & DoItYBks	GeogBks	375	552	217	2.096618357
39	56.79	ChildBks & DoItYBks	GeogBks	368	552	209	2.05773472
25	52.49	ArtBks	ChildBks & CookBks	482	512	253	2.050376037
6	50.39	ChildBks & CookBks	YouthBks	512	495	258	2.035984848
8	52.12	YouthBks	ChildBks & CookBks	495	512	258	2.035984848
13	57.03	ChildBks & CookBks	DoItYBks	512	564	292	2.022384752
14	51.77	DoItYBks	ChildBks & CookBks	564	512	292	2.022384752
41	56.36	CookBks & GeogBks	DoItYBks	385	564	217	1.998710509
32	52.90	ArtBks	GeogBks	482	552	255	1.91683204
26	82.19	DoItYBks & ArtBks	CookBks	247	862	203	1.906873198
38	53.59	ChildBks & GeogBks	DoItYBks	390	564	209	1.900345517
40	81.89	DoItYBks & GeogBks	CookBks	265	862	217	1.899925579
17	80.33	CookBks & RefBks	ChildBks	305	846	245	1.899003992
43	80.00	ArtBks & GeogBks	ChildBks	255	846	204	1.891252955
46	81.18	ArtBks & GeogBks	CookBks	255	862	207	1.883444793
5	79.63	YouthBks & CookBks	ChildBks	324	846	258	1.882497154

FIGURE 14.3   ASSOCIATION RULES FOR BOOK PURCHASE TRANSACTIONS: XLMINER OUTPUT

ified a minimal support of 200 transactions and a minimal confidence of 50%. This resulted in 49 rules (the 26 rules with the highest lift ratio are shown in Figure 14.3).

In reviewing these rules, we see that the the information can be compressed. First, rule 49 (top), which appears from the confidence level to be a very promising rule, is probably meaningless. It says: "If Italian cooking books have been purchased, then cookbooks are purchased." It seems likely that Italian cooking books are simply a subset of cookbooks. Rules 45 and 44 involve the same trio of books with different antecedents and consequents. The same is true of rules 8 and also 13 and 35 and 36. (Pairs and groups like this are easy to track down by looking for rows that share the same support.) This does not mean that the rules are not useful. On the contrary, it can reduce the number of itemsets to be considered for possible action from a business perspective.

14.2 COLLABORATIVE FILTERING[3]

Recommendation systems are a critically important part of websites that offer a large variety of products or services. Examples include Amazon.com, which offers millions of different products; Netflix has thousands of movies for rental; Google searches over huge numbers of web pages; Internet radio websites such as Spotify and Pandora include a large variety of music albums by various artists; travel websites offer many destinations and hotels; social network websites have many groups. The recommender engine provides personalized recommendations to a user based on the user's information as well as on similar users' information. Information means behaviors indicative of preference, such as purchase, ratings, and clicking.

The value that recommendation systems provide to users helps online companies convert browsers into buyers, increases cross-selling, and builds loyalty.

Collaborative filtering is a popular technique used by such recommendation systems. The term *collaborative filtering* is based on the notions of identifying relevant items for a specific user from the very large set of items ("filtering") by considering preferences of many users ("collaboration").

The Fortune.com article "Amazon's Recommendation Secret" (June 30, 2012) describes the company's use of collaborative filtering not only for providing personalized product recommendations, but also for customizing the entire website interface for each user:

> At root, the retail giant's recommendation system is based on a number of simple elements: what a user has bought in the past, which items they have in their virtual shopping cart, items they've rated and liked, and what other customers have viewed and purchased. Amazon calls this homegrown math "item-to-item collaborative filtering," and it's used this algorithm to heavily customize the browsing experience for returning customers.

Data Type and Format

Collaborative filtering requires availability of all item–user information. Specifically, for each item–user combination, we should have some measure of the user's preference for that item. Preference can be a numerical rating or a binary behavior such as a purchase, a "like," or a click.

---

[3]This section copyright © 2016 Statistics.com, Galit Shmueli, and Peter Bruce.

User ID	Item ID			
	I_1	I_2	$\cdots$	I_p
U_1	$r_{1,1}$	$r_{1,2}$	$\cdots$	$r_{1,p}$
U_2	$r_{2,1}$	$r_{2,2}$	$\cdots$	$r_{2,p}$
$\vdots$				
U_n	$r_{n,1}$	$r_{n,2}$	$\cdots$	$r_{n,p}$

TABLE 14.7   SCHEMATIC OF MATRIX FORMAT WITH RATINGS DATA

For n users $(u_1, u_2, \ldots, u_n)$ and p items $(i_1, i_2, \ldots, i_p)$, we can think of the data as an $n \times p$ matrix of n rows (users) by p columns (items). Each cell includes the rating or the binary event corresponding to the user's preference of the item (see the schematic in Table 14.7). Typically, not every user purchases or rates every item, and therefore a purchase matrix will have many zeros (it is sparse), and a rating matrix will have many missing values. Such missing values sometimes convey "uninterested" (as opposed to non-missing values that convey interest).

When both n and p are large, it is not practical to store the preferences data $(r_{u,i})$ in an $n \times p$ table. Instead, the data can be stored in many rows of triplets of the form $(U_u, I_i, r_{u,i})$, where each triplet contains the user ID, the item ID, and the preference information.

Example 3: Netflix Prize Contest

We have been considering both association rules and collaborative filtering as unsupervised techniques, but it is possible to judge how well they do by looking at holdout data to see what users purchase and how they rate items. The famous Netflix contest, mentioned in Chapter 13, did just this and provides a useful example to illustrate collaborative filtering, though the extension into training and validation is beyond the scope of this book.

In 2006, Netflix, the largest movie rental service in North America, announced a one million USD contest (www.netflixprize.com) for the purpose of improving its recommendation system called *Cinematch*. Participants were provided with a number of datasets, one for each movie. Each dataset included all the customer ratings for that movie (and the timestamp). We can think of one large combined dataset of the form [customer ID, movie ID, rating, date] where each record includes the rating given by a certain customer to a certain movie on a certain date. Ratings were on a 1–5 star scale. Contestants were asked to develop a recommendation algorithm that would improve over the existing Netflix system. Table 14.8 shows a small sample from the contest data, organized in matrix format. Rows indicate customers and columns are different movies.

It is interesting to note that the winning team was able to improve their system by considering not just the ratings for a movie, but whether a movie

					Movie ID				
Customer ID	**1**	**5**	**8**	**17**	**18**	**28**	**30**	**44**	**48**
30878	4	1			3	3	4	5	
124105	4								
822109	5								
823519	3		1	4		4	5		
885013	4	5							
893988	3						4	4	
1248029	3					2	4		3
1503895	4								
1842128	4						3		
2238063	3								

TABLE 14.8   SAMPLE OF RECORDS FROM THE NETFLIX PRIZE CONTEST, FOR A SUBSET OF 10 CUSTOMERS AND 9 MOVIES

was rated by a particular customer or not. In other words, the information on which movies a customer decided to rate turned out to be critically informative of customers' preferences, more than simply considering the 1–5 rating information:[4]

> Collaborative filtering methods address the sparse set of rating values. However, much accuracy is obtained by also looking at other features of the data. First is the information on which movies each user chose to rate, regardless of specific rating value ("the binary view"). This played a decisive role in our 2007 solution, and reflects the fact that the movies to be rated are selected deliberately by the user, and are not a random sample.

This is an example where converting the rating information into a binary matrix of rated/unrated proved to be useful.

User-Based Collaborative Filtering: "People Like You"

One approach to generating personalized recommendations for a user using collaborative filtering is based on finding users with similar preferences, and recommending items that they liked but the user hasn't purchased. The algorithm has two steps:

1. Find users who are most similar to the user of interest (neighbors). This is done by comparing the preference of our user to the preferences of other users.

2. Considering only the items that the user has *not* yet purchased, recommend the ones that are most preferred by the user's neighbors.

---

[4] "The BellKor 2008 Solution to the Netflix Prize," Bell, R. M., Koren, Y., and Volinsky, C., www.netflixprize.com/assets/ProgressPrize2008_BellKor.pdf.

This is the approach behind Amazon's "Customers Who Bought This Item Also Bought ..." (see Figure 14.1). It is also used in a Google search for generating the "Similar pages" link shown near each search result.

Step 1 requires choosing a distance (or proximity) metric to measure the distance between our user and the other users. Once the distances are computed, we can use a threshold on the distance or on the number of required neighbors to determine the nearest neighbors to be used in step 2. This approach is called "user-based top-N recommendation."

A nearest-neighbors approach measures the distance of our user to each of the other users in the database, similar to the k-nearest-neighbors algorithm (see Chapter 7). The Euclidean distance measure we discussed in that chapter does not perform as well for collaborative filtering as some other measures. A popular proximity measure between two users is the Pearson correlation between their ratings. We denote the ratings of items $I_1, \ldots, I_p$ by user U_1 as $r_{1,1}, r_{1,2}, \ldots, r_{1,p}$ and their average by $\bar{r}_1$. Similarly, the ratings by user U_2 are $r_{2,1}, r_{2,2}, \ldots, r_{2,p}$, with average $\bar{r}_2$. The correlation proximity between the two users is defined by

$$\text{Corr}(U_1, U_2) = \frac{\sum(r_{1,i} - \bar{r}_1)(r_{2,i} - \bar{r}_2)}{\sqrt{\sum(r_{1,i} - \bar{r}_1)^2}\sqrt{\sum(r_{2,i} - \bar{r}_2)^2}}, \tag{14.1}$$

where the summations are only over the items co-rated by both users.

To illustrate this, let us compute the correlation between customer 30878 and customer 823519 in the small Netflix sample in Table 14.8. We'll assume that the data shown in the table is the entire information. First, we compute the average rating by each of these users:

$$\bar{r}_{30878} = (4 + 1 + 3 + 3 + 4 + 5)/6 = 3.333.$$

$$\bar{r}_{823519} = (3 + 1 + 4 + 4 + 5)/5 = 3.4.$$

Note that the average is computed over a different number of movies for each of these customers, because they each rated a different set of movies. The average for a customer is computed over *all* the movies that a customer rated. The calculations for the correlation involve the departures from the average, but *only for the items that they co-rated*. In this case the co-rated movie IDs are 1, 28, and 30:

$$\text{Corr}(U_{30878}, U_{823519})$$

$$= \frac{(4 - 3.333)(3 - 3.4) + (3 - 3.333)(4 - 3.4) + (4 - 3.333)(5 - 3.4)}{\sqrt{(4 - 3.333)^2 + (3 - 3.333)^2 + (4 - 3.333)^2}\sqrt{(3 - 3.4)^2 + (4 - 3.4)^2 + (5 - 3.4)^2}}$$

$$= 0.6/1.75 = 0.34.$$

The same approach can be used when the data are in the form of a binary matrix (e.g., purchased or didn't purchase.)

Another popular measure is a variant of the Pearson correlation called *cosine similarity*. It differs from the correlation formula by not subtracting the means. Subtracting the mean in the correlation formula adjusts for users' different overall approaches to rating–for example, a customer who always rates highly versus one who tends to give low ratings.[5]

For example, the cosine similarity between the two Netflix customers is

$$\text{Cos Sim}(U_{30878}, U_{823519}) = \frac{4 \times 3 + 3 \times 4 + 4 \times 5}{\sqrt{4^2 + 3^2 + 4^2}\sqrt{3^2 + 4^2 + 5^2}}$$
$$= 44/45.277 = 0.972.$$

Note that when the data are in the form of a binary matrix, say, for purchase or no-purchase, the cosine similarity must be calculated over all items that either user has purchased; it cannot be limited to just the items that were co-purchased.

> Collaborative filtering suffers from what is called a *cold start*: it cannot be used as-is to create recommendations for new users or new items. For a user who rated a single item, the correlation coefficient between this and other users (in user-generated collaborative filtering) will have a denominator of zero and the cosine proximity will be 1 regardless of the rating. In a similar vein, users with just one item, and items with just one user, do not qualify as candidates for nearby neighbors.

For a user of interest, we compute his/her similarity to each of the users in our database using a correlation, cosine similarity, or another measure. Then, in step 2, we look only at the k-nearest users, and among all the other items that they rated/purchased, we choose the best one and recommend it to our user. What is the best one? For binary purchase data, it is the item most purchased. For rating data, it could be the highest rated, most rated, or a weighting of the two.

The nearest-neighbors approach can be computationally expensive when we have a large database of users. One solution is to use clustering methods (see Chapter 15) to group users into homogeneous clusters in terms of their preferences, and then to measure the distance of our user to each of the clusters. This approach places the computational load on the clustering step that can take place earlier and offline; it is then cheaper (and faster) to compare our user to each of the clusters in realtime. The price of clustering is less accurate

---

[5] Correlation and cosine similarity are popular in collaborative filtering because they are computationally fast for high-dimensional sparse data, and they account both for the rating values and the number of rated items.

recommendations, because not all the members of the closest cluster are the most similar to our user.

> Collaborative filtering is not implemented in XLMiner as well as in many other data mining software packages because the algorithm typically operates directly on a large user database and provides real-time recommendations. Such an implementation is beyond the scope of data mining software designed for model building and evaluation.

Item-Based Collaborative Filtering

When the number of users is much larger than the number of items, it is computationally cheaper (and faster) to find similar items rather than similar users. Specifically, when a user expresses interest in a particular item, the item-based collaborative filtering algorithm has two steps:

1. Find the items that were co-rated, or co-purchased, (by any user) with the item of interest.

2. Recommend the most popular or correlated item(s) among the similar items.

Similarity is now computed between items, instead of users. For example, in our small Netflix sample (Table 14.8), the correlation between movie 1 (with average $\bar{r}_1 = 3.7$) and movie 5 (with average $\bar{r}_5 = 3$) is

$$\text{Corr}(I_1, I_5) = \frac{(4 - 3.7)(1 - 3) + (4 - 3.7)(5 - 3)}{\sqrt{(4 - 3.7)^2 + (4 - 3.7)^2}\sqrt{(1 - 3)^2 + (5 - 3)^2}} = 0. \quad (14.2)$$

The zero correlation is due to the two opposite ratings of movie 5 by the users who also rated 1. One user rated it 5 stars and the other gave it a 1 star.

In like fashion, we can compute similarity between all the movies. This can be done offline. In realtime, for a user who rates a certain movie highly, we can look up the movie correlation table and recommend the movie with the highest positive correlation to the user's newly rated movie.

According to an industry report[6] by researchers who developed the Amazon item-to-item recommendation system,

> [The item-based] algorithm produces recommendations in real time, scales to massive data sets, and generates high quality recommendations.

---

[6] "Amazon.com Recommendations: Item-to-Item Collaborative Filtering" by Greg Linden, Brent Smith, and Jeremy York, *IEEE Internet Computing*, 2003, vol. 7, no. 1, pp. 76–80.

The disadvantage of item-based recommendations is that there is less diversity between items (compared to users' taste), and therefore the recommendations are often obvious.

Advantages and Weaknesses of Collaborative Filtering

Collaborative filtering relies on the availability of subjective information regarding users' preferences. It provides useful recommendations, even for "long tail" items, if our database contains sufficient similar users (not necessarily many, but at least a few per user), so that each user can find others users with similar tastes. Similarly, the data should include sufficient per-item ratings or purchases. One limitation of collaborative filtering is therefore that it cannot generate recommendations for new users, nor for new items. There are various approaches for tackling this challenge.

User-based collaborative filtering looks for similarity in terms of highly rated or preferred items. However, it is blind to data on low-rated or unwanted items. We can therefore not expect to use it as-is for detecting unwanted items.

User-based collaborative filtering helps leverage similarities between people's tastes for providing personalized recommendations. However, when the number of users becomes very large, collaborative filtering becomes computationally difficult. Solutions include item-based algorithms, clustering of users, and dimension reduction. The most popular dimension reduction method used in such cases is *singular value decomposition*, a computationally superior form of principal components analysis (see Chapter 4).

Although the term "prediction" is often used to describe the output of collaborative filtering, this method is unsupervised by nature. It can be used to generate predicted ratings or purchase indication for a user, but usually we do not have the true outcome value in practice. One important way to improve recommendations generated by collaborative filtering is by getting user feedback. Once a recommendation is generated, the user can indicate whether the recommendation was adequate or not. For this reason, many recommender systems entice users to provide feedback on their recommendations.

Collaborative Filtering vs. Association Rules

While collaborative filtering and association rules are both unsupervised methods used for generating recommendations, they differ in several ways:

Frequent itemsets vs. personalized recommendations Association rules look for frequent item combinations and will provide recommendations only for those items. In contrast, collaborative filtering provides personalized recommendations for every item, thereby catering to users with unusual taste. In this sense, collaborative filtering is useful for capturing the "long tail" of user preferences, while association rules look for the "head." This difference has implications for the data needed: association rules require data on a very large number of "baskets" (transactions) in order to find a sufficient number of baskets that contain certain combinations of items. In contrast, collaborative filtering does not require many "baskets," but it does require data on as many items as possible for many users. Also association rules operate at the basket level (our database can include multiple transactions for each user), while collaborative filtering operates at the user level.

Because association rules produce generic, impersonal rules (association-based recommendations such as Amazon's "Frequently Bought Together" display the same recommendations to all users searching for a specific item), they can be used for setting common strategies such as product placement in a store or sequencing of diagnostic tests in hospitals. In contrast, collaborative filtering generates user-specific recommendations (e.g., Amazon's "Customers Who Bought This Item Also Bought . . . ") and is therefore a tool designed for personalization.

Transactional data vs. user data Association rules provide recommendations of items based on their co-purchase with other items in *many transactions/baskets*. In contrast, collaborative filtering provides recommendations of items based on their co-purchase or co-rating by even a small number of other *users*. Considering distinct baskets is useful when the same items are purchased over and over again (e.g., in grocery shopping). Considering distinct users is useful when each item is typically purchased/rated once (e.g., purchases of books, music, and movies).

Binary data and ratings data Association rules treat items as binary data (1 = purchase, 0 = nonpurchase), whereas collaborative filtering can operate on either binary data or on numerical ratings.

Two or more items In association rules, the antecedent and consequent can each include one or more items (e.g., IF milk THEN cookies and cornflakes). Hence a recommendation might be a bundle of the item of interest with multiple items ("buy milk, cookies, and cornflakes and receive 10% discount"). In contrast, in collaborative filtering similarity is measured between *pairs* of items or pairs of users. A recommendation will therefore be either for a single item (the most popular item purchased by people like you, which you haven't purchased), or for multiple single items that do not necessarily relate to each other (the top two most popular items purchased by people like you, which you haven't purchased).

These distinctions are sharper for purchases and recommendations of nonpopular items, especially when comparing association rules to user-based collaborative filtering. When considering what to recommend to a user who purchased a popular item, then association rules and item-based collaborative filtering might yield the same recommendation for a single item. But a user-based recommendation will likely differ. Consider a customer who purchases milk every week as well as gluten-free products (which are rarely purchased by other customers). Suppose that using association rules on the transaction database we identify the rule "IF milk THEN cookies." Then the next time our customer purchases milk, s/he will receive a recommendation (e.g., a coupon) to purchase cookies, whether or not s/he purchased cookies, and regardless of his/her gluten-free item purchases. In item-based collaborative filtering, we would look at all items co-purchased with milk across all users and recommend the most popular item among them (which was not purchased by our customer). This might also lead to a recommendation of cookies, because this item was not purchased by our customer.[7] Now consider user-based collaborative filtering. User-based collaborative filtering searches for similar customers—those who purchased the same set of items—and then recommend the item most commonly purchased by these neighbors, which *was not purchased by our customer*. The user-based recommendation is therefore unlikely to recommend cookies and more likely to recommend popular gluten-free items that the customer has not purchased.

14.3 SUMMARY

Association rules (also called market basket analysis) and collaborative filtering are unsupervised methods for deducing associations between purchased items from databases of transactions. Association rules search for generic rules about items that are purchased together. The main advantage of this method is that it generates clear, simple rules of the form "IF X is purchased, THEN Y is also likely to be purchased." The method is very transparent and easy to understand.

The process of creating association rules is two-staged. First, a set of candidate rules based on frequent itemsets is generated (the Apriori algorithm being the most popular rule generating algorithm). Then from these candidate rules, the rules that indicate the strongest association between items are selected. We use the measures of support and confidence to evaluate the uncertainty in a rule. The user also specifies minimal support and confidence values to be used in the rule generation and selection process. A third measure, the lift ratio, compares the efficiency of the rule to detect a real association compared to a random combination.

---

[7] If the rule is "IF milk, THEN cookies and cornflakes," then the association rules would recommend cookies and cornflakes to a milk purchaser, while item-based collaborative filtering would recommend the most popular single item purchased with milk.

One shortcoming of association rules is the profusion of rules that are generated. There is therefore a need for ways to reduce these to a small set of useful and strong rules. An important nonautomated method to condense the information involves examining the rules for uninformative and trivial rules as well as for rules that share the same support. Another issue that needs to be kept in mind is that rare combinations tend to be ignored because they do not meet the minimum support requirement. For this reason it is better to have items that are approximately equally frequent in the data. This can be achieved by using higher-level hierarchies as the items. An example is to use types of books rather than titles of individual books in deriving association rules from a database of bookstore transactions.

Collaborative filtering is a popular technique used in online recommendation systems. It is based on the relationship between items formed by users who acted similarly on an item, such as purchasing or rating an item highly. User-based collaborative filtering operates on data on item-user combinations, calculates the similarities between users, and provides personalized recommendations to users. An important component for the success of collaborative filtering is that users provide feedback about the recommendations provided and have sufficient information on each item. One disadvantage of collaborative filtering methods is that they cannot generate recommendations for new users or new items. Also, with a huge number of users, user-based collaborative filtering becomes computationally challenging, and alternatives such as item-based methods or dimension reduction are popularly used.

PROBLEMS

14.1 **Satellite Radio Customers.** An analyst at a subscription-based satellite radio company has been given a sample of data from their customer database, with the goal of finding groups of customers who are associated with one another. The data consist of company data, together with purchased demographic data that are mapped to the company data (see Table 14.9). The analyst decides to apply association rules to learn more about the associations between customers. Comment on this approach.

TABLE 14.9 **SAMPLE OF DATA ON SATELLITE RADIO CUSTOMERS**

Row ID	zipconvert_2	zipconvert_3	zipconvert_4	zipconvert_5	homeowner dummy	NUMCHLD	INCOME	gender dummy	WEALTH
17	0	1	0	0	1	1	5	1	9
25	1	0	0	0	1	1	1	0	7
29	0	0	0	1	0	2	5	1	8
38	0	0	0	1	1	1	3	0	4
40	0	1	0	0	1	1	4	0	8
53	0	1	0	0	1	1	4	1	8
58	0	0	0	1	1	1	4	1	8
61	1	0	0	0	1	1	1	0	7
71	0	0	1	0	1	1	4	0	5
87	1	0	0	0	1	1	4	1	8
100	0	0	0	1	1	1	4	1	8
104	1	0	0	0	1	1	1	1	5
121	0	0	1	0	1	1	4	1	5
142	1	0	0	0	0	1	5	0	8

14.2 **Identifying Course Combinations.** The Institute for Statistics Education at Statistics.com offers online courses in statistics and analytics, and it is seeking information that will help in packaging and sequencing courses. Consider the data in the file Course-Topics.xls, the first few rows of which are shown in Table 14.10. These data are

TABLE 14.10 **DATA ON PURCHASES OF ONLINE STATISTICS COURSES**

Intro	DataMining	Survey	CatData	Regression	Forecast	DOE	SW
1	1	0	0	0	0	0	0
0	0	1	0	0	0	0	0
0	1	0	1	1	0	0	1
1	0	0	0	0	0	0	0
1	1	0	0	0	0	0	0
0	1	0	0	0	0	0	0
1	0	0	0	0	0	0	0
0	0	0	1	0	1	1	1
1	0	0	0	0	0	0	0
0	0	0	1	0	0	0	0
1	0	0	0	0	0	0	0

for purchases of online statistics courses at Statistics.com. Each row represents the courses attended by a single customer. The firm wishes to assess alternative sequencings and bundling of courses. Use association rules to analyze these data, and interpret several of the resulting rules.

14.3   Cosmetics Purchases. The data shown in Table 14.11 and the output in Figure 14.4 are based on a subset of a dataset on cosmetic purchases (`Cosmetics.xls`) at a large chain drugstore. The store wants to analyze associations among purchases of these items for purposes of point-of-sale display, guidance to sales personnel in promoting cross sales, and guidance for piloting an eventual time-of-purchase electronic recommender system to boost cross sales. Consider first only the data shown in Table 14.11, given in binary matrix form.

    a. Select several values in the matrix and explain their meaning.

    b. Consider the results of the association rules analysis shown in Figure 14.4.

        i. For the first row, explain the "Conf. %" output and how it is calculated.

        ii. For the first row, explain the "Support for A," "Support for C," and "Support for A & C" output and how it is calculated.

        iii. For the first row, explain the "Lift Ratio" and how it is calculated.

        iv. For the first row, explain the rule that is represented there in words.
        Now use the complete dataset on the cosmetics purchases (in the file `Cosmetics.xls`).

        v. Using XLMiner, apply association rules to these data (use the default parameters).

        vi. Interpret the first three rules in the output in words.

        vii. Reviewing the first couple of dozen rules, comment on their redundancy and how you would assess their utility.

TABLE 14.11   **EXCERPT FROM DATA ON COSMETICS PURCHASES IN BINARY MATRIX FORM**

Trans. #	Bag	Blush	Nail Polish	Brushes	Concealer	Eyebrow Pencils	Bronzer
1	0	1	1	1	1	0	1
2	0	0	1	0	1	0	1
3	0	1	0	0	1	1	1
4	0	0	1	1	1	0	1
5	0	1	0	0	1	0	1
6	0	0	0	0	1	0	0
7	0	1	1	1	1	0	1
8	0	0	1	1	0	0	1
9	0	0	0	0	1	0	0
10	1	1	1	1	0	0	0
11	0	0	1	0	0	0	1
12	0	0	1	1	1	0	1

Row ID	Confidence %	Antecedent (A)	Consequent (C)	Support for A	Support for C	Support for A & C	Lift Ratio
50	80.52	Brushes & Concealer	Nail Polish & Bronzer	77	103	62	3.908712647
51	60.19	Nail Polish & Bronzer	Brushes & Concealer	103	77	62	3.908712647
46	81.58	Nail Polish & Concealer & Bronzer	Brushes	76	110	62	3.708133971
54	56.36	Brushes	Nail Polish & Concealer & Bronzer	110	76	62	3.708133971
29	76.36	Brushes	Nail Polish & Bronzer	110	103	84	3.706972639
27	81.55	Nail Polish & Bronzer	Brushes	103	110	84	3.706972639
49	73.81	Brushes & Bronzer	Nail Polish & Concealer	84	109	62	3.385757973
52	56.88	Nail Polish & Concealer	Brushes & Bronzer	109	84	62	3.385757973
15	70.64	Nail Polish & Concealer	Brushes	109	110	77	3.211009174
17	70.00	Brushes	Nail Polish & Concealer	110	109	77	3.211009174
6	67.07	Blush & Nail Polish	Brushes	82	110	55	3.048780488
7	50.00	Brushes	Blush & Nail Polish	110	82	55	3.048780488

FIGURE 14.4   ASSOCIATION RULES FOR COSMETICS PURCHASES DATA

14.4 **Course Ratings.** The Institute for Statistics Education at Statistics.com asks students to rate a variety of aspects of a course as soon as the student completes it. The Institute is contemplating instituting a recommendation system that would provide students with recommendations for additional courses as soon as they submit their rating for a completed course. Consider the excerpt from student ratings of online statistics courses shown in Table 14.12, and the problem of what to recommend to student E.N.

TABLE 14.12 **RATINGS OF ONLINE STATISTICS COURSES: 4 = BEST, 1 = WORST, BLANK = NOT TAKEN**

	SQL	Spatial	PA 1	DM in R	Python	Forecast	R Prog	Hadoop	Regression
L N	4				3	2	4		2
M H	3	4			4				
J H	2	2							
E N	4			4			4		3
D U	4	4							
F L		4							
G L		4							
A H		3							
S A			4						
R W			2					4	
B A			4						
M G			4			4			
A F			4						
K G			3						
D S	4			2			4		

a. First consider a user-based collaborative filter. This requires computing correlations between all student pairs. For which students is it possible to compute correlations with E.N.? Compute them.

b. Based on the single nearest student to E.N., which single course should we recommend to E.N.? Explain why.

c. Replace the ratings with a binary matrix indicating whether or not a student took the course. Compute the cosine similarity of E.N. from each of the other students for which such a calculation is feasible.

d. Based on the cosine similarities of the nearest students to E.N., which course should be recommended to E.N.?

e. What is the conceptual difference between using the actual ratings as opposed to a binary matrix showing whether a student took each course or not? (Hint: How are the missing values in the matrix handled in each case?)

f. With large datasets, it is computationally difficult to compute user-based recommendations in real time, and an item-based approach is used instead. Returning to the rating data (not the binary matrix), let's now take that approach.

  i. If the goal is still to find a recommendation for E.N., for which course pairs is it possible and useful to calculate correlations?

  ii. Just looking at the data, and without yet calculating course pair correlations, which course would you recommend to E.N., relying on item-based filtering? Calculate two course pair correlations involving your guess and report the results.

## CHAPTER 15

Cluster Analysis

This chapter is about the popular unsupervised learning task of clustering, where the goal is to segment the data into a set of homogeneous clusters of observations for the purpose of generating insight. Separating a dataset into clusters of homogeneous records is also useful for improving performance of supervised methods, by modeling each cluster separately rather than the entire, heterogeneous dataset. Clustering is used in a vast variety of business applications, from customized marketing to industry analysis. We describe two popular clustering approaches: *hierarchical clustering* and *k-means clustering*. In hierarchical clustering, observations are sequentially grouped to create clusters, based on distances between observations and distances between clusters. We describe how the algorithm works in terms of the clustering process and mention several common distance metrics used. Hierarchical clustering also produces a useful graphical display of the clustering process and results, called a dendrogram. We present dendrograms and illustrate their usefulness. *k*-means clustering is widely used in large dataset applications. In *k*-means clustering, observations are allocated to one of a pre-specified set of clusters, according to their distance from each cluster. We describe the *k*-means clustering algorithm and its computational advantages. Finally, we present techniques that assist in generating insight from clustering results.

Data Mining for Business Analytics: Concepts, Techniques, and Applications with XLMiner®, Third Edition.
Galit Shmueli, Peter C. Bruce, and Nitin R. Patel.

15.1   INTRODUCTION

Cluster analysis is used to form groups or clusters of similar observations based on several measurements made on these observations. The key idea is to characterize the clusters in ways that would be useful for the aims of the analysis. This idea has been applied in many areas, including astronomy, archaeology, medicine, chemistry, education, psychology, linguistics, and sociology. Biologists, for example, have made extensive use of classes and subclasses to organize species. A spectacular success of the clustering idea in chemistry was Mendeleev's periodic table of the elements.

One popular use of cluster analysis in marketing is for *market segmentation:* customers are segmented based on demographic and transaction history information, and a marketing strategy is tailored for each segment. In countries such as India, where customer diversity is extremely location-sensitive, chain stores often perform market segmentation at the store level, rather than chain-wide (called "micro segmentation"). Another use is for *market structure analysis:* identifying groups of similar products according to competitive measures of similarity. In marketing and political forecasting, clustering of neighborhoods using US postal zip codes has been used successfully to group neighborhoods by lifestyles. Claritas, a company that pioneered this approach, grouped neighborhoods into 40 clusters using various measures of consumer expenditure and demographics. Examining the clusters enabled Claritas to come up with evocative names, such as "Bohemian Mix," "Furs and Station Wagons," and "Money and Brains," for the groups that captured the dominant lifestyles. Knowledge of lifestyles can be used to estimate the potential demand for products (e.g., sports utility vehicles) and services (e.g., pleasure cruises). Similarly, sales organizations will derive customer segments and give them names—"personas"—to focus sales efforts.

In finance, cluster analysis can be used for creating *balanced portfolios*: given data on a variety of investment opportunities (e.g., stocks), one may find clusters based on financial performance variables such as return (daily, weekly, or monthly), volatility, beta, and other characteristics, such as industry and market capitalization. Selecting securities from different clusters can help create a balanced portfolio. Another application of cluster analysis in finance is for *industry analysis*: for a given industry, we are interested in finding groups of similar firms based on measures such as growth rate, profitability, market size, product range, and presence in various international markets. These groups can then be analyzed in order to understand industry structure and to determine, for instance, who is a competitor.

An interesting and unusual application of cluster analysis, described in Berry and Linoff (1997), is the design of a new set of sizes for army uniforms for women in the US Army. The study came up with a new clothing size system with only 20 sizes, where different sizes fit different body types. The 20 sizes are combinations of five measurements: chest, neck, and shoulder circumference, sleeve outseam, and neck-to-buttock length (for further details, see McCullugh et al., 1998). This example is important because it shows how a completely new insightful view can be gained by examining clusters of records.

Cluster analysis can be applied to huge amounts of data. For instance, Internet search engines use clustering techniques to cluster queries that users submit. These can then be used for improving search algorithms. The objective of this chapter is to describe the key ideas underlying the most commonly used techniques for cluster analysis and to lay out their strengths and weaknesses.

Typically, the basic data used to form clusters are a table of measurements on several variables, where each column represents a variable and a row represents an observation. Our goal is to form groups of records so that similar observations are in the same group. The number of clusters may be prespecified or determined from the data.

Example: Public Utilities

Table 15.1 gives corporate data on 22 public utilities in the United States (the variable definitions are given in the table footnote). We are interested in forming groups of similar utilities. The observations to be clustered are the utilities, and the clustering will be based on the eight measurements on each utility. An example where clustering would be useful is a study to predict the cost impact of deregulation. To do the requisite analysis, economists would need to build a detailed cost model of the various utilities. It would save a considerable amount of time and effort if we could cluster similar types of utilities and build detailed cost models for just one "typical" utility in each cluster and then scale up from these models to estimate results for all utilities.

For simplicity, let us consider only two of the measurements: *Sales* and *Fuel Cost*. Figure 15.1 shows a scatterplot of these two variables, with labels marking each company. At first glance, there appear to be two or three clusters of utilities: one with utilities that have high fuel costs, a second with utilities that have lower fuel costs and relatively low sales, and a third with utilities with low fuel costs but high sales. We can therefore think of cluster analysis as a more formal algorithm that measures the distance between records and, according to these distances (here, two-dimensional distances), forms clusters.

Two general types of clustering algorithms for a dataset of n observations are hierarchical and non-hierarchical clustering:

Hierarchical methods can be either *agglomerative* or *divisive*. Agglomerative methods begin with n clusters and sequentially merge similar clusters until a single cluster is obtained. Divisive methods work in the opposite

TABLE 15.1	DATA ON 22 PUBLIC UTILITIES[a]

Company	Fixed	RoR	Cost	Load	Demand	Sales	Nuclear	Fuel Cost
Arizona Public Service	1.06	9.2	151	54.4	1.6	9,077	0.0	0.628
Boston Edison Co.	0.89	10.3	202	57.9	2.2	5,088	25.3	1.555
Central Louisiana Co.	1.43	15.4	113	53.0	3.4	9,212	0.0	1.058
Commonwealth Edison Co.	1.02	11.2	168	56.0	0.3	6,423	34.3	0.700
Consolidated Edison Co. (NY)	1.49	8.8	192	51.2	1.0	3,300	15.6	2.044
Florida Power & Light Co.	1.32	13.5	111	60.0	−2.2	11,127	22.5	1.241
Hawaiian Electric Co.	1.22	12.2	175	67.6	2.2	7,642	0.0	1.652
daho Power Co.	1.10	9.2	245	57.0	3.3	13,082	0.0	0.309
Kentucky Utilities Co.	1.34	13.0	168	60.4	7.2	8,406	0.0	0.862
Madison Gas & Electric Co.	1.12	12.4	197	53.0	2.7	6,455	39.2	0.623
Nevada Power Co.	0.75	7.5	173	51.5	6.5	17,441	0.0	0.768
New England Electric Co.	1.13	10.9	178	62.0	3.7	6,154	0.0	1.897
Northern States Power Co.	1.15	12.7	199	53.7	6.4	7,179	50.2	0.527
Oklahoma Gas & Electric Co.	1.09	12.0	96	49.8	1.4	9,673	0.0	0.588
Pacific Gas & Electric Co.	0.96	7.6	164	62.2	−0.1	6,468	0.9	1.400
Puget Sound Power & Light Co.	1.16	9.9	252	56.0	9.2	15,991	0.0	0.620
San Diego Gas & Electric Co.	0.76	6.4	136	61.9	9.0	5,714	8.3	1.920
The Southern Co.	1.05	12.6	150	56.7	2.7	10,140	0.0	1.108
Texas Utilities Co.	1.16	11.7	104	54.0	−2.1	13,507	0.0	0.636
Wisconsin Electric Power Co.	1.20	11.8	148	59.9	3.5	7,287	41.1	0.702
United Illuminating Co.	1.04	8.6	204	61.0	3.5	6,650	0.0	2.116
Virginia Electric & Power Co.	1.07	9.3	174	54.3	5.9	10,093	26.6	1.306

[a]Fixed = fixed-charge covering ratio (income/debt); RoR = rate of return on capital; Cost = cost per kilowatt capacity in place; Load = annual load factor; Demand = peak kilowatthour demand growth from 1974 to 1975; Sales = sales (kilowatthour use per year); Nuclear = percent nuclear; Fuel Cost = total fuel costs (cents per kilowatthour).

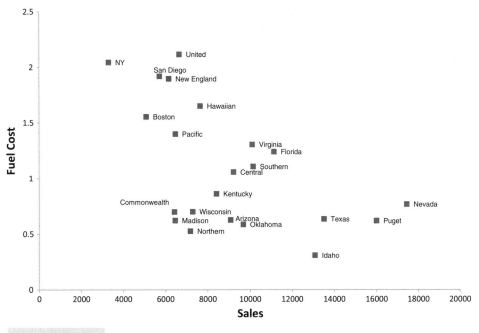

FIGURE 15.1	SCATTER PLOT OF FUEL COST VS. SALES FOR THE 22 UTILITIES.

direction, starting with one cluster that includes all observations. Hierarchical methods are especially useful when the goal is to arrange the clusters into a natural hierarchy.

Non-hierarchical methods, such as k-means. Using a prespecified number of clusters, the method assigns observations to each cluster. These methods are generally less computationally intensive and are therefore preferred with very large datasets.

We concentrate here on the two most popular methods: hierarchical agglomerative clustering and k-means clustering. In both cases we need to define two types of distances: distance between two observations and distance between two clusters. In both cases there is a variety of metrics that can be used.

15.2 MEASURING DISTANCE BETWEEN TWO OBSERVATIONS

We denote by d_{ij} a *distance metric*, or *dissimilarity measure*, between observations i and j. For observation i we have the vector of p measurements $(x_{i1}, x_{i2}, \ldots, x_{ip})$, while for observation j we have the vector of measurements $(x_{j1}, x_{j2}, \ldots, x_{jp})$. For example, we can write the measurement vector for Arizona Public Service as $[1.06, 9.2, 151, 54.4, 1.6, 9077, 0, 0.628]$.

Distances can be defined in multiple ways, but in general, the following properties are required:

Nonnegative: $d_{ij} \geq 0$

Self-proximity: $d_{ii} = 0$ (the distance from an observation to itself is zero)

Symmetry: $d_{ij} = d_{ji}$

Triangle inequality: $d_{ij} \leq d_{ik} + d_{kj}$ (the distance between any pair cannot exceed the sum of distances between the other two pairs)

Euclidean Distance

The most popular distance measure is the *Euclidean distance*, d_{ij}, which between two observations, i and j, is defined by

$$d_{ij} = \sqrt{(x_{i1} - x_{j1})^2 + (x_{i2} - x_{j2})^2 + \cdots + (x_{ip} - x_{jp})^2}.$$

For instance, the Euclidean distance between Arizona Public Service and Boston Edison Co. can be computed from the raw data by

$$d_{12} = \sqrt{(1.06 - 0.89)^2 + (9.2 - 10.3)^2 + (151 - 202)^2 + \cdots + (0.628 - 1.555)^2}$$

$$= 3989.408.$$

Normalizing Numerical Measurements

The measure computed above is highly influenced by the scale of each variable, so that variables with larger scales (e.g., Sales) have a much greater influence over the total distance. It is therefore customary to *normalize* continuous measurements before computing the Euclidean distance. This converts all measurements to the same scale. Normalizing a measurement means subtracting the average and dividing by the standard deviation (normalized values are also called *z-scores*). For instance, the average sales amount across the 22 utilities is 8914.045 and the standard deviation is 3549.984. The normalized sales for Arizona Public Service is therefore $(9077 - 8914.045)/3549.984 = 0.046$.

Returning to the simplified utilities data with only two measurements (Sales and Fuel Cost), we first normalize the measurements (see Table 15.2), and then compute the Euclidean distance between each pair. Table 15.3 gives these pairwise distances for the first five utilities. A similar table can be constructed for all 22 utilities.

Other Distance Measures for Numerical Data

It is important to note that the choice of the distance measure plays a major role in cluster analysis. The main guideline is domain dependent: What exactly is being measured? How are the different measurements related? What scale should each measurement be treated as (numerical, ordinal, or nominal)? Are there outliers? Finally, depending on the goal of the analysis, should the clusters be distinguished mostly by a small set of measurements, or should they be separated by multiple measurements that weight moderately?

Although Euclidean distance is the most widely used distance, it has three main features that need to be kept in mind. First, as mentioned above, it is highly scale dependent. Changing the units of one variable (e.g., from cents to dollars) can have a huge influence on the results. Normalizing is therefore a common solution. But unequal weighting should be considered if we want the clusters to depend more on certain measurements and less on others. The second feature of Euclidean distance is that it completely ignores the relationship between the measurements. Thus, if the measurements are in fact strongly correlated, a different distance (e.g., the statistical distance, described below) is likely to be a better choice. Third, Euclidean distance is sensitive to outliers. If the data are believed to contain outliers and careful removal is not a choice, the use of more robust distances (e.g., the Manhattan distance described below) is preferred.

TABLE 15.2	ORIGINAL AND NORMALIZED MEASUREMENTS FOR SALES AND FUEL COST

Company	Sales	Fuel Cost	NormSales	NormFuel
Arizona Public Service	9,077	0.628	0.0459	−0.8537
Boston Edison Co.	5,088	1.555	−1.0778	0.8133
Central Louisiana Co.	9,212	1.058	0.0839	−0.0804
Commonwealth Edison Co.	6,423	0.7	−0.7017	−0.7242
Consolidated Edison Co. (NY)	3,300	2.044	−1.5814	1.6926
Florida Power & Light Co.	11,127	1.241	0.6234	0.2486
Hawaiian Electric Co.	7,642	1.652	−0.3583	0.9877
Idaho Power Co.	13,082	0.309	1.1741	−1.4273
Kentucky Utilities Co.	8,406	0.862	−0.1431	−0.4329
Madison Gas & Electric Co.	6,455	0.623	−0.6927	−0.8627
Nevada Power Co.	17,441	0.768	2.4020	−0.6019
New England Electric Co.	6,154	1.897	−0.7775	1.4283
Northern States Power Co.	7,179	0.527	−0.4887	−1.0353
Oklahoma Gas & Electric Co.	9,673	0.588	0.2138	−0.9256
Pacific Gas & Electric Co.	6,468	1.4	−0.6890	0.5346
Puget Sound Power & Light Co.	15,991	0.62	1.9935	−0.8681
San Diego Gas & Electric Co.	5,714	1.92	−0.9014	1.4697
The Southern Co.	10,140	1.108	0.3453	0.0095
Texas Utilities Co.	13,507	0.636	1.2938	−0.8393
Wisconsin Electric Power Co.	7,287	0.702	−0.4583	−0.7206
United Illuminating Co.	6,650	2.116	−0.6378	1.8221
Virginia Electric & Power Co.	10,093	1.306	0.3321	0.3655
Mean	8,914.05	1.10	0.00	0.00
Standard deviation	3,549.98	0.56	1.00	1.00

TABLE 15.3	DISTANCE MATRIX BETWEEN PAIRS OF THE FIRST FIVE UTILITIES, USING EUCLIDEAN DISTANCE AND NORMALIZED MEASUREMENTS

	Arizona	Boston	Central	Commonwealth	Consolidated
Arizona	0				
Boston	2.01	0			
Central	0.77	1.47	0		
Commonwealth	0.76	1.58	1.02	0	
Consolidated	3.02	1.01	2.43	2.57	0

Additional popular distance metrics are often used (for reasons such as the ones above):

Correlation-based similarity Sometimes it is more natural or convenient to work with a similarity measure between observations rather than distance, which measures dissimilarity. A popular similarity measure is the square of the Pearson correlation coefficient, r_{ij}^2, where the correlation coefficient is defined by

$$r_{ij} = \frac{\sum_{m=1}^{p}(x_{im} - \overline{x}_m)(x_{jm} - \overline{x}_m)}{\sqrt{\sum_{m=1}^{p}(x_{im} - \overline{x}_m)^2 \sum_{m=1}^{p}(x_{jm} - \overline{x}_m)^2}}. \tag{15.1}$$

Such measures can always be converted to distance measures. In the example above we could define a distance measure $d_{ij} = 1 - r_{ij}^2$.

Statistical distance (also called *Mahalanobis distance*) This metric has an advantage over the other metrics mentioned in that it takes into account the correlation between measurements. With this metric, measurements that are highly correlated with other measurements do not contribute as much as those that are uncorrelated or mildly correlated. The statistical distance between observations i and j is defined as

$$d_{i,j} = \sqrt{(\mathbf{x}_i - \mathbf{x}_j)'S^{-1}(\mathbf{x}_i - \mathbf{x}_j)},$$

where $\mathbf{x}_i$ and $\mathbf{x}_j$ are p-dimensional vectors of the measurements values for observations i and j, respectively, and S is the covariance matrix for these vectors. ($'$, a transpose operation, simply turns a column vector into a row vector). S^{-1} is the inverse matrix of S, which is the p-dimension extension to division. For further information on statistical distance, see Chapter 12.

Manhattan distance ("city block") This distance looks at the absolute differences rather than squared differences, and is defined by

$$d_{ij} = \sum_{m=1}^{p} | x_{im} - x_{jm} | .$$

Maximum coordinate distance This distance looks only at the measurement on which observations i and j deviate most. It is defined by

$$d_{ij} = \max_{m=1,2,\dots,p} | x_{im} - x_{jm} | .$$

Distance Measures for Categorical Data

In the case of measurements with binary values, it is more intuitively appealing to use similarity measures than distance measures. Suppose that we have binary values for all the x_{ij}'s, and for observations i and j we have the following 2×2 table:

Observation j

		0	1	
Observation i	0	a	b	$a + b$
	1	c	d	$c + d$
		$a + c$	$b + d$	n

(15.2)

where a denotes the number of variables for which observations i and j do not have that attribute (they each have value 0 on that attribute), d is the number of variables for which the two observations have the attribute present, and so on. The most useful similarity measures in this situation are:

Matching coefficient: $(a + d)/n$.

Jaccard's coefficient: $d/(b + c + d)$. This coefficient ignores zero matches. This is desirable when we do not want to consider two people to be similar simply because a large number of characteristics are absent in both. For example, if *owns a Corvette* is one of the variables, a matching "yes" would be evidence of similarity, but a matching "no" tells us little about whether the two people are similar.

Distance Measures for Mixed Data

When the measurements are mixed (some continuous and some binary), a similarity coefficient suggested by Gower is very useful. *Gower's similarity measure* is a weighted average of the distances computed for each variable, after scaling each variable to a [0,1] scale. It is defined as

$$s_{ij} = \frac{\sum_{m=1}^{p} w_{ijm} s_{ijm}}{\sum_{m=1}^{p} w_{ijm}},$$

where s_{ijm} is the similarity between observations i and j on measurement m, and w_{ijm} is a binary weight given to the corresponding distance.

The similarity measures s_{ijm} and weights w_{ijm} are computed as follows:

1. For continuous measurements, $s_{ijm} = 1 - \frac{|x_{im} - x_{jm}|}{\max(x_m) - \min(x_m)}$ and $w_{ijm} = 1$ unless the value for measurement m is unknown for one or both of the observations, in which case $w_{ijm} = 0$.

2. For binary measurements, $s_{ijm} = 1$ if $x_{im} = x_{jm} = 1$ and 0 otherwise. $w_{ijm} = 1$ unless $x_{im} = x_{jm} = 0$.

3. For nonbinary categorical measurements, $s_{ijm} = 1$ if both observations are in the same category, and otherwise $s_{ijm} = 0$. As in continuous measurements, $w_{ijm} = 1$ unless the category for measurement m is unknown for one or both of the observations, in which case $w_{ijm} = 0$.

15.3 MEASURING DISTANCE BETWEEN TWO CLUSTERS

We define a cluster as a set of one or more observations. How do we measure distance between clusters? The idea is to extend measures of *distance between observations* into *distances between clusters*. Consider cluster A, which includes the m observations $A_1, A_2, \ldots, A_m$, and cluster B, which includes n observations $B_1, B_2, \ldots, B_n$. There are four widely used measures of distance between clusters: minimum distance, maximum distance, average distance, and centroid distance.

Minimum Distance

The distance between the pair of observations A_i and B_j that are closest:

$$\min(\text{distance}(A_i, B_j)), \quad i = 1, 2, \ldots, m; \ j = 1, 2, \ldots, n.$$

Maximum Distance

The distance between the pair of observations A_i and B_j that are farthest:

$$\max(\text{distance}(A_i, B_j)), \quad i = 1, 2, \ldots, m; \ j = 1, 2, \ldots, n.$$

Average Distance

The average distance of all possible distances between observations in one cluster and observations in the other cluster:

$$\text{Average}(\text{distance}(A_i, B_j)), \quad i = 1, 2, \ldots, m; \ j = 1, 2, \ldots, n.$$

Centroid Distance

The distance between the two cluster centroids. A *cluster centroid* is the vector of measurement averages across all the observations in that cluster. For cluster A, this is the vector $\overline{x}_A = \left[(1/m \sum_{i=1}^{m} x_{1i}, \ldots, 1/m \sum_{i=1}^{m} x_{pi}) \right]$. The centroid distance between clusters A and B is

$$\text{distance}(\overline{x}_A, \overline{x}_B).$$

Minimum distance, maximum distance, and centroid distance are illustrated visually for two dimensions using a map of Portugal and France in Figure 15.2.

FIGURE 15.2 TWO-DIMENSIONAL REPRESENTATION OF SEVERAL DIFFERENT DISTANCE MEASURES BETWEEN PORTUGAL AND FRANCE.

As a numerical example, consider again the first two utilities (Arizona, Boston) as cluster A, and the next three utilities (Central, Commonwealth, Consolidated) as cluster B. Using the normalized scores in Table 15.2 and the distance matrix in Table 15.3, we can compute each of the distances described above. Using Euclidean distance for each distance calculation, we get:

- The closest pair is Arizona and Commonwealth, and therefore the minimum distance between clusters A and B is 0.76.

- The farthest pair is Arizona and Consolidated, and therefore the maximum distance between clusters A and B is 3.02.

- The average distance is $(0.77 + 0.76 + 3.02 + 1.47 + 1.58 + 1.01)/6 = 1.44$.

- The centroid of cluster A is

$$\left[\frac{0.0459 - 1.0778}{2}, \frac{-0.8537 + 0.8133}{2} \right] = [-0.516, -0.020],$$

and the centroid of cluster B is

$$\left[\frac{0.0839 - 0.7017 - 1.5814}{3}, \frac{-0.0804 - 0.7242 + 1.6926}{3} \right]$$
$$= [-0.733, 0.296].$$

The distance between the two centroids is then

$$\sqrt{(-0.516 + 0.733)^2 + (-0.020 - 0.296)^2} = 0.38.$$

In deciding among clustering methods, domain knowledge is key. If you have good reason to believe that the clusters might be chain- or sausage-like, minimum distance would be a good choice. This method does not require that cluster members all be close to one another, only that the new members being added be close to one of the existing members. An example of an application where this might be the case would be characteristics of crops planted in long rows, or disease outbreaks along navigable waterways that are the main areas of settlement in a region. Another example is laying and finding mines (land or marine). Minimum distance is also fairly robust to small deviations in the distances. However, adding or removing data can influence it greatly.

Maximum and average distance are better choices if you know that the clusters are more likely to be spherical (e.g., customers clustered on the basis of numerous attributes). If you do not know the probable nature of the cluster, these are good default choices, since most clusters tend to be spherical in nature.

We now move to a more detailed description of the two major types of clustering algorithms: hierarchical (agglomerative) and non-hierarchical.

15.4 HIERARCHICAL (AGGLOMERATIVE) CLUSTERING

The idea behind hierarchical agglomerative clustering is to start with each cluster comprising exactly one observation and then progressively agglomerating (combining) the two nearest clusters until there is just one cluster left at the end, which consists of all the observations.

Returning to the small example of five utilities and two measures (Sales and Fuel Cost) and using the distance matrix (Table 15.3), the first step in the hierarchical clustering would join Arizona and Commonwealth, which are the closest (using normalized measurements and Euclidean distance). Next we would recalculate a 4×4 distance matrix that would have the distances between these four clusters: {Arizona, Commonwealth}, {Boston}, {Central}, and {Consolidated}. At this point we use a measure of distance between clusters, such as the ones described in Section 15.3. Each of these distances (minimum, maximum, average, and centroid distance) can be implemented in the hierarchical scheme as described below.

HIERARCHICAL AGGLOMERATIVE CLUSTERING ALGORITHM:

1. Start with *n* clusters (each observation = cluster).
2. The two closest observations are merged into one cluster.
3. At every step, the two clusters with the smallest distance are merged. This means that either single observations are added to existing clusters or two existing clusters are combined.

Single Linkage

In *single linkage clustering*, the distance measure that we use is the minimum distance (the distance between the nearest pair of observations in the two clusters, one observation in each cluster). In our utilities example, we would compute the distances between each of {Boston}, {Central}, and {Consolidated} with {Arizona, Commonwealth} to create the 4 × 4 distance matrix shown in Table 15.4.

The next step would consolidate {Central} with {Arizona, Commonwealth} because these two clusters are closest. The distance matrix will again be recomputed (this time it will be 3 × 3), and so on.

This method has a tendency to cluster together at an early stage observations that are distant from each other because of a chain of intermediate observations in the same cluster. Such clusters have elongated sausage-like shapes when visualized as objects in space.

TABLE 15.4   **DISTANCE MATRIX AFTER ARIZONA AND COMMONWEALTH CONSOLIDATION CLUSTER TOGETHER, USING SINGLE LINKAGE**

	Arizona–Commonwealth	Boston	Central	Consolidated
Arizona–Commonwealth	0			
Boston	min(2.01,1.58)	0		
Central	min(0.77,1.02)	1.47	0	
Consolidated	min(3.02,2.57)	1.01	2.43	0

Complete Linkage

In *complete linkage clustering*, the distance between two clusters is the maximum distance (between the farthest pair of observations). If we used complete linkage with the five-utilities example, the recomputed distance matrix would be equivalent to Table 15.4, except that the "min" function would be replaced with a "max."

This method tends to produce clusters at the early stages with observations that are within a narrow range of distances from each other. If we visualize them as objects in space, the observations in such clusters would have roughly spherical shapes.

Average Linkage (in XLMiner: "Group Average Linkage")

Average linkage clustering is based on the average distance between clusters (between all possible pairs of observations). If we used average linkage with the five-utilities example, the recomputed distance matrix would be equivalent to Table 15.4, except that the "min" function would be replaced with "average." This method is also called *unweighted pair-group method using averages* (UPGMA).

Note that unlike average linkage, the results of the single and complete linkage methods depend only on the ordering of the inter-observation distances. Linear transformations of the distances (and other transformations that do not change the ordering) do not affect the results.

Centroid Linkage

Centroid linkage clustering is based on centroid distance, where clusters are represented by their mean values for each variable, which forms a vector of means. The distance between two clusters is the distance between these two vectors. In average linkage, each pairwise distance is calculated, and the average of all such distances is calculated. In contrast, in centroid distance clustering, just one distance is calculated: the distance between group means. This method is also called *unweighted pair-group method using centroids* (UPGMC).

Ward's Method

Ward's method is also agglomerative, in that it joins observations and clusters together progressively to produce larger and larger clusters, but operates slightly differently from the general approach described above. Ward's method considers the "loss of information" that occurs when observations are clustered together. When each cluster has one observation, there is no loss of information and all individual values remain available. When observations are joined together and represented in clusters, information about an individual observation is replaced by the information for the cluster to which it belongs. To measure loss of information, Ward's method employs a measure "error sum of squares" (ESS) that measures the difference between individual observations and a group mean.

This is easiest to see in univariate data. For example, consider the values (2, 6, 5, 6, 2, 2, 2, 2, 0, 0, 0) with a mean of 2.5. Their ESS is equal to

$$(2 - 2.5)^2 + (6 - 2.5)^2 + (5 - 2.5)^2 + \ldots + (0 - 2.5)^2 = 50.5.$$

The loss of information associated with grouping the values into a single group is therefore 50.5. Now group the observations into four groups: (0, 0, 0), (2, 2, 2, 2), (5), (6, 6). The loss of information is the sum of the ESS's for each group, which is 0 (each observation in each group is equal to the mean for that group, so the ESS for each group is 0). Thus clustering the 10 observations into 4 clusters results in no loss of information, and this would be the first step in Ward's method. In moving to a smaller number of clusters, Ward's method would choose the configuration that results in the smallest incremental loss of information.

Ward's method tends to result in convex clusters that are of roughly equal size, which can be an important consideration in some applications (e.g., in establishing meaningful customer segments).

Dendrograms: Displaying Clustering Process and Results

A *dendrogram* is a treelike diagram that summarizes the process of clustering. At the bottom are the observations. Similar observations are joined by lines whose vertical length reflects the distance between the observations. Figure 15.3 shows the dendrogram that results from clustering all 22 utilities using the eight normalized measurements, Euclidean distance, and single linkage. By choosing a cutoff distance on the y-axis, a set of clusters is created. Visually, this means drawing a horizontal line on a dendrogram. Observations with connections below the horizontal line (i.e., their distance is smaller than the cutoff distance) belong to the same cluster. For example, setting the cutoff distance to 2.7 in Figure 15.3 results in six clusters. The six clusters are (from left to right on the dendrogram):

{12,21,7,15,2,4,10,13,20,22,1,18,14,19,9,6}
{3} = {Central}
{8, 16} = {Idaho, Puget}
{17} = {San Diego}
{11} = { Nevada }
{5} = {NY}

Note that if we wanted five clusters, they would be identical to the six above, with the exception that the first two clusters would be merged into one cluster. In general, all hierarchical methods have clusters that are nested within each other as we decrease the number of clusters. This is a valuable property for interpreting clusters and is essential in certain applications, such as taxonomy of varieties of living organisms.

The average linkage dendrogram is shown in Figure 15.4.

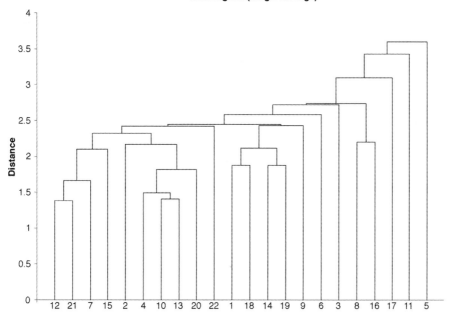

FIGURE 15.3 DENDROGRAM: SINGLE LINKAGE FOR ALL 22 UTILITIES, USING ALL 8 MEASUREMENTS.

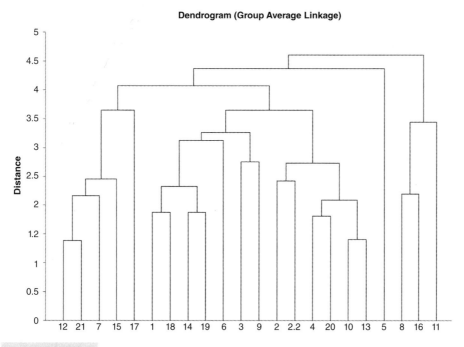

FIGURE 15.4 DENDROGRAM: AVERAGE LINKAGE FOR ALL 22 UTILITIES, USING ALL 8 MEASUREMENTS.

If we want six clusters using average linkage, we can choose a cutoff distance of 3.5. The resulting six clusters would be (from left to right in the dendrogram): {12,21,7,15}; {17}; {1,18,14,19,6,3,9}; {2,22,4,20,10,13}; {5}; {8,16,11}.

Validating Clusters

One important goal of cluster analysis is to come up with *meaningful clusters*. Since there are many variations that can be chosen, it is important to make sure that the resulting clusters are valid, in the sense that they really generate some insight. To see whether the cluster analysis is useful, consider each of the following aspects:

1. *Cluster interpretability*. Is the interpretation of the resulting clusters reasonable? To interpret the clusters, explore the characteristics of each cluster by
   a. obtaining summary statistics (e.g., average, min, max) from each cluster on each measurement that was used in the cluster analysis,
   b. examining the clusters for separation along some common feature (variable) that was not used in the cluster analysis, and
   c. labeling the clusters: based on the interpretation, trying to assign a name or label to each cluster.

2. *Cluster stability*. Do cluster assignments change significantly if some of the inputs are altered slightly? Another way to check stability is to partition the data and see how well clusters formed based on one part apply to the other part. To do this:
   a. Cluster partition A.
   b. Use the cluster centroids from A to assign each observation in partition B (each observation is assigned to the cluster with the closest centroid).
   c. Assess how consistent the cluster assignments are compared to the assignments based on all the data.

3. *Cluster separation*. Examine the ratio of between–cluster variation to within–cluster variation to see whether the separation is reasonable. There exist statistical tests for this task (an F-ratio), but their usefulness is somewhat controversial.

4. *Number of clusters*. The number of resulting clusters must be useful, given the purpose of the analysis. For example, suppose the goal of the clustering is to identify categories of customers and assign labels to them for market segmentation purposes. If the marketing department can only

Company	Cluster ID	Sub-Cluster	Fixed	RoR	Cost	Load	Demand	Sales	Nuclear	Fuel Cost
Arizona	1	1	1.06	9.2	151	54.4	1.6	9077	0	0.628
Central	1	3	1.43	15.4	113	53	3.4	9212	0	1.058
Florida	1	6	1.32	13.5	111	60	-2.2	11127	22.5	1.241
Kentucky	1	9	1.34	13	168	60.4	7.2	8406	0	0.862
Oklahoma	1	14	1.09	12	96	49.8	1.4	9673	0	0.588
Southern	1	18	1.05	12.6	150	56.7	2.7	10140	0	1.108
Texas	1	19	1.16	11.7	104	54	-2.1	13507	0	0.636
Boston	2	2	0.89	10.3	202	57.9	2.2	5088	25.3	1.555
Commonwealth	2	4	1.02	11.2	168	56	0.3	6423	34.3	0.7
Madison	2	10	1.12	12.4	197	53	2.7	6455	39.2	0.623
Northern	2	13	1.15	12.7	199	53.7	6.4	7179	50.2	0.527
Wisconsin	2	20	1.2	11.8	148	59.9	3.5	7287	41.1	0.702
Virginia	2	22	1.07	9.3	174	54.3	5.9	10093	26.6	1.306
NY	3	5	1.49	8.8	192	51.2	1	3300	15.6	2.044
Hawaiian	4	7	1.22	12.2	175	67.6	2.2	7642	0	1.652
New England	4	12	1.13	10.9	178	62	3.7	6154	0	1.897
Pacific	4	15	0.96	7.6	164	62.2	-0.1	6468	0.9	1.4
United	4	21	1.04	8.6	204	61	3.5	6650	0	2.116
Idaho	5	8	1.1	9.2	245	57	3.3	13082	0	0.309
Nevada	5	11	0.75	7.5	173	51.5	6.5	17441	0	0.768
Puget	5	16	1.16	9.9	252	56	9.2	15991	0	0.62
San Diego	6	17	0.76	6.4	136	61.9	9	5714	8.3	1.92

FIGURE 15.5   HEATMAP FOR THE 22 UTILITIES (IN ROWS). ROWS ARE SORTED BY THE SIX CLUSTERS FROM AVERAGE LINKAGE CLUSTERING. DARKER DENOTES HIGHER VALUES WITHIN A COLUMN.

manage to sustain three different marketing presentations, it would probably not make sense to identify more than three clusters.

Returning to the utilities example, we notice that both methods (single and average linkage) identify {5} (NY) and {17} (San Diego) as singleton clusters. Also both dendrograms imply that a reasonable number of clusters in this dataset is four. One insight that can be derived from the average linkage clustering is that clusters tend to group geographically. The four non–singleton clusters form (approximately) a southern group, a northern group, an east/west seaboard group, and a west group.

We can further characterize each of the clusters by examining the summary statistics of their measurements, or visually looking at a heatmap of their individual measurements. Figure 15.5 shows a heatmap of the four clusters and two singletons, highlighting the different profile that each cluster has in terms of the eight measurements. We see, for instance, that cluster 2 is characterized by utilities with a high percent of nuclear power; cluster 1 is characterized by high fixed cost and RoR; cluster 4 has high fuel costs.

Limitations of Hierarchical Clustering

Hierarchical clustering is very appealing in that it does not require specification of the number of clusters, and in that sense is purely data driven. The ability to represent the clustering process and results through dendrograms is also an advantage of this method, as it is easier to understand and interpret. There are, however, a few limitations to consider:

1. Hierarchical clustering requires the computation and storage of a $n \times n$ distance matrix. For very large datasets, this can be expensive and slow.

2. The hierarchical algorithm makes only one pass through the data. This means that observations that are allocated incorrectly early in the process cannot be reallocated subsequently.

3. Hierarchical clustering also tends to have low stability. Reordering data or dropping a few observations can lead to a different solution.

4. With respect to the choice of distance between clusters, single and complete linkage are robust to changes in the distance metric (e.g., Euclidean, statistical distance) as long as the relative ordering is kept. In contrast, average linkage is more influenced by the choice of distance metric, and might lead to completely different clusters when the metric is changed.

5. Hierarchical clustering is sensitive to outliers.

15.5 NON-HIERARCHICAL CLUSTERING: THE *k*-MEANS ALGORITHM

A non-hierarchical approach to forming good clusters is to pre-specify a desired number of clusters, k, and assign each case to one of the k clusters so as to minimize a measure of dispersion within the clusters. In other words, the goal is to divide the sample into a predetermined number k of non-overlapping clusters so that clusters are as homogeneous as possible with respect to the measurements used.

A common measure of within-cluster dispersion is the sum of distances (or sum of squared Euclidean distances) of observations from their cluster centroid. The problem can be set up as an optimization problem involving integer programming, but because solving integer programs with a large number of variables is time-consuming, clusters are often computed using a fast, heuristic method that produces good (although not necessarily optimal) solutions. The k-means algorithm is one such method.

The k-means algorithm starts with an initial partition of the observations into k clusters. Subsequent steps modify the partition to reduce the sum of the distances of each observation from its cluster centroid. The modification consists of allocating each observation to the nearest of the k centroids of the previous partition. This leads to a new partition for which the sum of distances is smaller than before. The means of the new clusters are computed and the improvement step is repeated until the improvement is very small.

K -MEANS CLUSTERING ALGORITHM:

1. Start with k initial clusters (user chooses k).
2. At every step, each observation is reassigned to the cluster with the "closest" centroid.
3. Recompute the centroids of clusters that lost or gained an observation, and repeat step 2.
4. Stop when moving any more observations between clusters increases cluster dispersion.

Returning to the example with the five utilities and two measurements, let us assume that $k = 2$ and that the initial clusters are A = {Arizona, Boston} and B = {Central, Commonwealth, Consolidated}. The cluster centroids were computed in Section 15.4:

$$\bar{x}_A = [-0.516, -0.020] \text{ and } \bar{x}_B = [-0.733, 0.296].$$

The distance of each observation from each of these two centroids is shown in Table 15.5.

TABLE 15.5   DISTANCE OF EACH OBSERVATION FROM EACH CENTROID

	Distance from Centroid A	Distance from Centroid B
Arizona	1.0052	1.3887
Boston	1.0052	0.6216
Central	0.6029	0.8995
Commonwealth	0.7281	1.0207
Consolidated	2.0172	1.6341

We see that Boston is closer to cluster B, and that Central and Commonwealth are each closer to cluster A. We therefore move each of these observations to the other cluster and obtain A = {Arizona, Central, Commonwealth} and B = {Consolidated, Boston}. Recalculating the centroids gives

$$\bar{x}_A = [-0.191, -0.553] \text{ and } \bar{x}_B = [-1.33, 1.253].$$

The distance of each observation from each of the newly calculated centroids is given in Table 15.6. At this point we stop because each observation is allocated to its closest cluster.

TABLE 15.6	DISTANCE OF EACH OBSERVATION FROM EACH NEWLY CALCULATED CENTROID

	Distance from Centroid A	Distance from Centroid B
Arizona	0.3827	2.5159
Boston	1.6289	0.5067
Central	0.5463	1.9432
Commonwealth	0.5391	2.0745
Consolidated	2.6412	0.5067

Initial Partition into k Clusters

The choice of the number of clusters can either be driven by external considerations (e.g., previous knowledge, practical constraints, etc.), or we can try a few different values for k and compare the resulting clusters. After choosing k, the n observations are partitioned into these initial clusters. If there is external reasoning that suggests a certain partitioning, this information should be used. Alternatively, if there exists external information on the centroids of the k-clusters, this can be used to initially allocate the observations.

In many cases, there is no information to be used for the initial partition. In these cases, the algorithm can be rerun with different randomly generated starting partitions to reduce chances of the heuristic producing a poor solution.

XLMiner : k-Means Clustering—Predicted Clusters

Record ID	Cluster ID	Dist. Clust-1	Dist. Clust-2	Dist. Clust-3	Dist. Clust-4	Dist. Clust-5	Dist. Clust-6
17	1	0.00	5165.86	1414.91	9791.06	358.84	2038.83
3	2	3498.11	1667.78	2084.47	6293.64	3144.30	1461.25
6	2	5413.09	247.95	3999.13	4379.18	5058.97	3375.26
14	2	3959.24	1206.81	2545.77	5833.06	3605.59	1922.48
19	2	7793.08	2627.26	6378.97	2001.25	7438.84	5755.17
5	3	2414.70	7580.25	3828.48	12204.72	2768.69	4452.41
7	3	1928.44	3238.52	513.68	7862.83	1573.45	115.55
9	3	2692.21	2474.55	1277.62	7098.89	2337.46	654.50
12	3	442.13	4726.31	974.42	9350.78	86.67	1598.67
18	3	4426.04	741.09	3011.69	5365.17	4071.54	2387.99
8	4	7368.82	2206.65	5954.04	2422.77	7013.56	5330.25
11	4	11727.07	6561.60	10312.60	1936.99	11372.35	9688.72
16	4	10277.66	5113.35	8862.96	487.19	9922.53	8239.11
2	5	629.76	5792.58	2040.73	10416.72	980.88	2664.50
15	5	754.61	4412.14	660.49	9036.87	400.27	1284.77
21	5	938.52	4230.90	479.45	8854.69	581.57	1103.25
1	6	3363.06	1803.33	1948.73	6428.08	3008.60	1325.23
4	6	710.29	4457.27	706.12	9081.90	355.97	1329.35
10	6	744.25	4425.81	674.84	9049.79	387.68	1297.58
13	6	1466.99	3702.19	74.30	8325.85	1111.18	574.23
20	6	1573.41	3593.18	164.93	8218.12	1219.49	466.11
22	6	4379.21	789.99	2964.70	5411.96	4024.41	2340.68

FIGURE 15.6	OUTPUT FOR k-MEANS CLUSTERING OF 22 UTILITIES INTO $k = 6$ CLUSTERS (SORTED BY CLUSTER ID).

The number of clusters in the data is generally not known, so it is a good idea to run the algorithm with different values for k that are near the number of clusters that one expects from the data to see how the sum of distances reduces with increasing values of k. Note that the clusters obtained using different values of k will not be nested (unlike those obtained by hierarchical methods).

The results of running the k-means algorithm for all 22 utilities and eight measurements with $k = 6$ are shown in Figure 15.6. As in the results from the hierarchical clustering, we see once again that $\{17\}$ (San Diego) is a singleton cluster. Two more clusters (clusters 4 and 6) are nearly identical to those that emerged in the hierarchical clustering with average linkage. Cluster 3 has the largest within-cluster variability

To characterize the resulting clusters, we examine the cluster centroids (numerically in Figure 15.7 or in the line chart ("profile plot") in Figure 15.8). We can see, for instance, that cluster 6 has the highest average Nuclear, a very high RoR, and a slow Demand Growth. In contrast, cluster 4 has the highest Sales, with no Nuclear, a high Demand Growth, and the highest average Cost.

We can also inspect the information on the within-cluster dispersion. From Figure 15.7 we see that cluster 3 has the largest average distance, and it includes five records. This is true for both normalized measurements (bottom right table) and original units (bottom left table). In comparison, cluster 6, with six records, has a smaller within-cluster average distance. This means that cluster 6 is more homogeneous than cluster 3.

When the number of clusters is not predetermined by domain requirements, we can use a graphical approach to evaluate different numbers of clusters. An "elbow chart" is a line chart depicting the decline in cluster heterogeneity as we add more clusters. Figure 15.9 shows the overall average within-cluster distance (normalized) for different choices of k. Moving from 3 to 4 tightens clusters considerably (reflected by the large reduction in within-cluster distance), and so does moving from 4 to 5. Adding more clusters beyond 5 brings less improvement to cluster homogeneity.

From the distances between clusters we can learn about the separation of the different clusters. For instance, we see that cluster 4 is very different from the other clusters except cluster 2. This might lead us to examine the possibility of merging the two. Cluster 1, which is a singleton cluster, appears to be very far from all the other clusters.

Finally, we can use the information on the distance between the final clusters to evaluate the cluster validity. The ratio of the sum of squared distances for a given k to the sum of squared distances to the mean of all the records ($k = 1$) is a useful measure for the usefulness of the clustering. If the ratio is near 1.0, the clustering has not been very effective, whereas if it is small, we have well-separated groups.

Cluster Centers

Cluster	Fixed	RoR	Cost	Load	Demand	Sales	Nuclear	Fuel Cost
Cluster-1	0.76	6.40	136.00	61.90	9.00	5714.00	8.30	1.92
Cluster-2	1.25	13.15	106.00	54.20	0.13	10879.75	5.63	0.88
Cluster-3	1.25	11.50	172.60	59.58	3.36	7128.40	3.12	1.51
Cluster-4	1.00	8.87	223.33	54.83	6.33	15504.67	0.00	0.57
Cluster-5	0.96	8.83	190.00	60.37	1.87	6068.67	8.73	1.69
Cluster-6	1.10	11.10	172.83	55.22	3.40	7752.33	31.90	0.75

Distance Between Centers	Cluster-1	Cluster-2	Cluster-3	Cluster-4	Cluster-5	Cluster-6
Cluster-1	0.00	5165.86	1414.91	9791.06	358.84	2038.83
Cluster-2	5165.86	0.00	3751.95	4626.41	4811.82	3128.24
Cluster-3	1414.91	3751.95	0.00	8376.42	1059.90	624.61
Cluster-4	9791.06	4626.41	8376.42	0.00	9436.07	7752.56
Cluster-5	358.84	4811.82	1059.90	9436.07	0.00	1683.92
Cluster-6	2038.83	3128.24	624.61	7752.56	1683.92	0.00

Data Summary

Original coordinates				Normalized coordinates		
Cluster	#Obs	Avg. Dist		Cluster	#Obs	Avg. Dist
Cluster-1	1	0		Cluster-1	1	0
Cluster-2	4	1437.451559		Cluster-2	4	1.733202
Cluster-3	5	1921.178091		Cluster-3	5	2.011386
Cluster-4	3	1615.648559		Cluster-4	3	1.757778
Cluster-5	3	654.2403723		Cluster-5	3	1.331169
Cluster-6	6	1222.19713		Cluster-6	6	1.608398
Overall	22	1340.843012		Overall	22	1.632135

FIGURE 15.7 CLUSTER CENTROIDS AND DISTANCES FOR k-MEANS WITH $k= 6$.

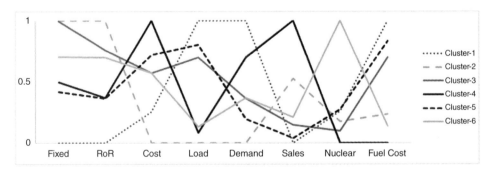

FIGURE 15.8 VISUAL PRESENTATION OF CLUSTER CENTROIDS (SCALED TO [0, 1] ON EACH MEASURE, FOR EASY COMPARISON). CREATED USING EXCEL'S *LINE CHART*.

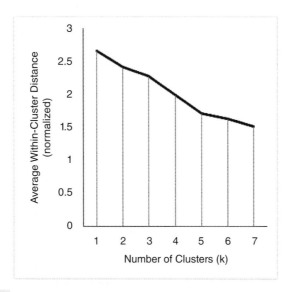

FIGURE 15.9   COMPARING DIFFERENT CHOICES OF k IN TERMS OF OVERALL AVERAGE WITHIN-CLUSTER DISTANCE. CREATED USING EXCEL'S *LINE CHART*.

PROBLEMS

15.1 **University Rankings.** The dataset on American College and University Rankings (available from www.dataminingbook.com) contains information on 1302 American colleges and universities offering an undergraduate program. For each university there are 17 measurements, including continuous measurements (e.g., tuition and graduation rate) and categorical measurements (e.g., location by state and whether it is a private or public school). 3 variables (Math, Verbal, ACT) from the original data were dropped because they had too many missing values.

Note that many observations are missing some measurements. Our first goal is to estimate these missing values from "similar" records. This will be done by clustering the complete records and then finding the closest cluster for each of the partial records. The missing values will be imputed from the information in that cluster.

a. Remove all observations with missing measurements from the dataset (by creating a new worksheet).

b. For all the continuous measurements, run hierarchical clustering using complete linkage and Euclidean distance. Make sure to normalize the measurements. Examine the dendrogram: How many clusters seem reasonable for describing these data?

c. Compare the summary statistics for each cluster and describe each cluster in this context (e.g., "Universities with high tuition, low acceptance rate..."). *Hint:* To obtain cluster statistics for hierarchical clustering, use Excel's Pivot Table on the "Predicted Clusters" table.

d. Use the categorical measurements that were not used in the analysis (State and Private/Public) to characterize the different clusters. Is there any relationship between the clusters and the categorical information?

e. Can you think of other external information that explains the contents of some or all of these clusters?

f. Consider Tufts University, which is missing some information. Compute the Euclidean distance of this observation from each of the clusters that you found above (using only the measurements that you have). Which cluster is it closest to? Impute the missing values for Tufts by taking the average of the cluster on those measurements.

15.2 **Pharmaceutical Industry.** An equities analyst is studying the pharmaceutical industry and would like your help in exploring and understanding the financial data collected by her firm. Her main objective is to understand the structure of the pharmaceutical industry using some basic financial measures.

Financial data gathered on 21 firms in the pharmaceutical industry are available in the file `Pharmaceuticals.xls`. For each firm, the following variables are recorded:

1. Market capitalization (in billions of dollars)
2. Beta
3. Price/earnings ratio
4. Return on equity
5. Return on assets
6. Asset turnover
7. Leverage
8. Estimated revenue growth
9. Net profit margin
10. Median recommendation (across major brokerages)

11. Location of firm's headquarters

12. Stock exchange on which the firm is listed

Use cluster analysis to explore and analyze the given dataset as follows:

a. Use only the quantitative variables (1 to 9) to cluster the 21 firms. Justify the various choices made in conducting the cluster analysis, such as weights accorded different variables, the specific clustering algorithm(s) used, the number of clusters formed, and so on.

b. Interpret the clusters with respect to the quantitative variables that were used in forming the clusters.

c. Is there a pattern in the clusters with respect to the qualitative variables (10 to 12) (those not used in forming the clusters)?

d. Provide an appropriate name for each cluster using any or all of the variables in the dataset.

15.3 Customer Rating of Breakfast Cereals. The dataset `Cereals.xls` includes nutritional information, store display, and consumer ratings for 77 breakfast cereals.

Data Preprocessing. Remove all cereals with missing values.

a. Apply hierarchical clustering to the data using Euclidean distance to the normalized measurements. Compare the dendrograms from single linkage and complete linkage, and look at cluster centroids. Comment on the structure of the clusters and on their stability. **Hints:** (1) To obtain cluster centroids for hierarchical clustering, apply Excel's Pivot Table to the "Predicted Clusters" table. (2) Running hierarchical clustering in XLMiner is an iterative process—run it once with a guess at the right number of clusters, then run it again after looking at the dendrogram, adjusting the number of clusters if needed.

b. Which method leads to the most insightful or meaningful clusters?

c. Choose one of the methods. How many clusters would you use? What distance is used for this cutoff? (Look at the dendrogram.)

d. The elementary public schools would like to choose a set of cereals to include in their daily cafeterias. Every day a different cereal is offered, but all cereals should support a healthy diet. For this goal you are requested to find a cluster of "healthy cereals." Should the data be normalized? If not, how should they be used in the cluster analysis?

15.4 Marketing to Frequent Fliers. The file `EastWestAirlinesCluster.xls` contains information on 3999 passengers who belong to an airline's frequent flier program. For each passenger the data include information on their mileage history and on different ways they accrued or spent miles in the last year. The goal is to try to identify clusters of passengers that have similar characteristics for the purpose of targeting different segments for different types of mileage offers.

a. Apply hierarchical clustering with Euclidean distance and Ward's method. Make sure to normalize the data first. How many clusters appear?

b. What would happen if the data were not normalized?

c. Compare the cluster centroid to characterize the different clusters, and try to give each cluster a label.

d. To check the stability of the clusters, remove a random 5% of the data (by taking a random sample of 95% of the records), and repeat the analysis. Does the same picture emerge?

e. Use k-means clustering with the number of clusters that you found above. Does the same picture emerge?

f. Which clusters would you target for offers, and what types of offers would you target to customers in that cluster?

## PART VI

Forecasting Time Series

## CHAPTER 16

Handling Time Series

In this chapter we describe the context of business time series forecasting and introduce the main approaches that are detailed in the next chapters, and in particular regression-based forecasting and smoothing-based methods. Our focus is on forecasting future values of a single time series. This chapter and the next two chapters are meant as an introduction to the general forecasting approach and methods.

In this chapter we start with a discussion of the difference between the predictive nature of time series forecasting vs. the descriptive or explanatory task of time series analysis. A general discussion of combining forecasting methods or results for added precision follows. Next we present a time series in terms of four components (level, trend, seasonality, and noise) and present methods for visualizing the different components and for exploring time series data. We close with a discussion of data partitioning (creating training and validation sets), which is performed differently from cross-sectional data partitioning.

16.1 INTRODUCTION[1]

Time series forecasting is performed in nearly every organization that works with quantifiable data. Retail stores use it to forecast sales. Energy companies use it to forecast reserves, production, demand, and prices. Educational institutions use it to forecast enrollment. Governments use it to forecast tax receipts

---

[1]This and subsequent sections in this chapter © 2016 Statistics.com and Galit Shmueli. Used by permission.

Data Mining for Business Analytics: Concepts, Techniques, and Applications with XLMiner®, Third Edition.
Galit Shmueli, Peter C. Bruce, and Nitin R. Patel.
© 2016 John Wiley & Sons Inc. Published 2016 by John Wiley & Sons, Inc.

and spending. International financial organizations like the World Bank and International Monetary Fund use it to forecast inflation and economic activity. Transportation companies use time series forecasting to forecast future travel. Banks and lending institutions use it (sometimes badly!) to forecast new home purchases. And venture capital firms use it to forecast market potential and to evaluate business plans.

Previous chapters in this book deal with classifying and predicting data where time is not a factor, in the sense that it is not treated differently from other variables, and where the sequence of measurements over time does not matter. These are typically called cross-sectional data. In contrast, this chapter deals with a different type of data: time series.

With today's technology, many time series are recorded on very frequent time scales. Stock data are available at ticker-level. Purchases at online and offline stores are recorded in real time. Although data might be available at a very frequent scale, for the purpose of forecasting it is not always preferable to use this time scale. In considering the choice of time scale, one must consider the scale of the required forecasts and the level of noise in the data. For example, if the goal is to forecast next-day sales at a grocery store, using minute-by-minute sales data is likely to be less useful for forecasting than using daily aggregates. The minute-by-minute series will contain many sources of noise (e.g., variation by peak and nonpeak shopping hours) that degrade its forecasting power, and these noise errors, when the data are aggregated to a cruder level, are likely to average out.

The focus in this part of the book is on forecasting a single time series. In some cases, multiple time series are to be forecasted (e.g., the monthly sales of multiple products). Even when multiple series are being forecasted, the most popular forecasting practice is to forecast each series individually. The advantage of single-series forecasting is its simplicity. The disadvantage is that it does not take into account possible relationships between series. The statistics literature contains models for multivariate time series that directly model the cross-correlations between series. Such methods tend to make restrictive assumptions about the data and the cross-series structure. They also require statistical expertise for estimation and maintenance. Econometric models often include information from one or more series as inputs into another series. However, such models are based on assumptions of causality that are based on theoretical models. An alternative approach is to capture the associations between the series of interest and external information more heuristically. An example is using the sales of lipstick to forecast some measure of the economy, based on the observation by Ronald Lauder, chairman of Estee Lauder, that lipstick sales tend to increase before tough economic times (a phenomenon called the "leading lipstick indicator").

16.2   Descriptive vs. Predictive Modeling

As with cross-sectional data, modeling time series data is done for either descriptive or predictive purposes. In descriptive modeling, or *time series analysis*, a time series is modeled to determine its components in terms of seasonal patterns, trends, relation to external factors, etc. These can then be used for decision making and policy formulation. In contrast, *time series forecasting* uses the information in a time series (and perhaps other information) to forecast future values of that series. The difference between the goals of time series analysis and time series forecasting leads to differences in the type of methods used and in the modeling process itself. For example, in selecting a method for describing a time series, priority is given to methods that produce understandable results (rather than "black-box" methods) and sometimes to models based on causal arguments (explanatory models). Furthermore describing can be done in retrospect, while forecasting is *prospective* in nature. This means that descriptive models might use "future" information (e.g., averaging the values of yesterday, today, and tomorrow to obtain a smooth representation of today's value) whereas forecasting models cannot.

In time series forecasting of this chapter, our goal is to predict future values of a time series. For information on time series analysis, see Chatfield (2003).

16.3   Popular Forecasting Methods in Business

Two types of forecasting methods are popular in business applications. Both are versatile and powerful, yet relatively simple to understand and deploy. One type of forecasting tool is linear regression, where the user specifies a certain model and then estimates it from the time series. The other is the more data-driven tool of smoothing, where the method learns patterns from the data. Each of the two types of tools has advantages and disadvantages, as detailed in Chapters 17 and 18. We also note that data mining methods such as neural networks and others that are intended for cross-sectional data and are sometimes used for time series forecasting, especially for incorporating external information into the forecasts (see Shmueli, 2011).

Combining Methods

A popular approach for improving predictive performance is in fact to combine forecasting methods. This is similar to the ensembles approach described in Chapter 13. Combining forecasting methods can be done via two-level (or multi-level) forecasters, where the first method uses the original time series to generate forecasts of future values, and the second method uses the residuals from the first model to generate forecasts of future forecast errors, thereby "correcting" the first level forecasts. Another combination approach is via "ensembles," where multiple methods are applied to the time series, and their resulting forecasts are averaged in some way to produce the final forecast. Combining methods can take advantage of the strengths of different forecasting methods to capture different aspects of the time series (also true in cross-sectional data). The averaging across multiple methods can lead to forecasts that are more robust and of higher precision.

16.4 TIME SERIES COMPONENTS

In both types of forecasting methods, regression models and smoothing, and in general, it is customary to dissect a time series into four components: *level*, *trend*, *seasonality*, and *noise*. The first three components are assumed to be invisible, as they characterize the underlying series, which we only observe with added noise. Level describes the average value of the series, trend is the change in the series from one period to the next, and seasonality describes a short-term cyclical behavior of the series that can be observed several times within the given series. Last, noise is the random variation that results from measurement error or other causes not accounted for. It is always present in a time series to some degree.

In order to identify the components of a time series, the first step is to examine a time plot. In its simplest form, a time plot is a line chart of the series values over time, with temporal labels (e.g., calendar date) on the horizontal axis. To illustrate this, consider the following example.

Example: Ridership on Amtrak Trains

Amtrak, a US railway company, routinely collects data on ridership. We will forecast future ridership using the series of monthly ridership between January 1991 and March 2004. These data are publicly available at www.forecastingprinciples.com[2].

A time plot for the monthly Amtrak ridership series is shown in Figure 16.1. Note that the values are in thousands of riders. Looking at the time plot reveals the nature of the series components: the overall level is around 1,800,000 passengers per month. A slight U-shaped trend is discernible during this period, with pronounced annual seasonality, with peak travel during summer (July and August).

---

[2] To get this series: click on Data. Under T-Competition Data click "time-series data" (the direct URL to the data file is www.forecastingprinciples.com/files/MHcomp1.xlsas of July 2015). This file contains many time series. In the Monthly worksheet, column AI contains series M034.

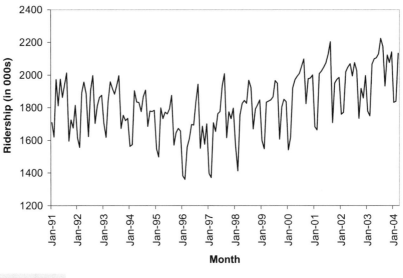

FIGURE 16.1   MONTHLY RIDERSHIP ON AMTRAK TRAINS (in thousands) FROM
JAN-1991 TO MAR-2004

A second step in visualizing a time series is to examine it more carefully. A
few tools are useful:

Zoom in: Zooming in to a shorter period within the series can reveal patterns that are hidden when viewing the entire series. This is especially important when the time series is long. Consider a series of the daily number of
vehicles passing through the Baregg tunnel in Switzerland (data are available
in the same location as the Amtrak Ridership data; series D028). The series
from Nov 1, 2003 to Nov 16, 2005 is shown in the top panel of Figure
16.2. Zooming in to a four-month period (bottom panel) reveals a strong
day-of-week pattern that is not visible in the time plot of the complete time
series.

Change scale of series: In order to better identify the shape of a trend, it is
useful to change the scale of the series. One simple option is to change the
vertical scale to an exponential scale (in Excel Ribbon select Chart Tools >
Axes > Primary Vertical Axis > More Primary Vertical Axis Options, and
click "logarithmic scale" in the Format Axis menu). If the trend on the new
scale appears more linear, then the trend in the original series is closer to an
exponential trend.

Add trend lines: Another possibility for better discerning the shape of the
trend is to add a trend line (Excel Ribbon: Chart Tools > Trendline).
By trying different trendlines, one can see what type of trend (e.g., linear,
exponential, quadratic) best approximates the data.

FIGURE 16.2   TIME PLOTS OF THE DAILY NUMBER OF VEHICLES PASSING THROUGH THE BAREGG TUNNEL, SWITZERLAND. THE BOTTOM PANEL ZOOMS IN TO A 4-MONTH PERIOD, REVEALING A DAY-OF-WEEK PATTERN.

Suppress seasonality: It is often easier to see trends in the data when seasonality is suppressed. Suppressing seasonality patterns can be done by plotting the series at a cruder time scale (e.g., aggregating monthly data into years) or creating separate lines or time plots for each season (e.g., separate lines for each day of week). Another popular option is to use moving average charts. We will discuss these in Chapter 18 (Section 18.2).

Continuing our example of Amtrak ridership, the charts in Figure 16.3 help make the series' components more visible.

Some forecasting methods directly model these components by making assumptions about their structure. For example, a popular assumption about trend is that it is linear or exponential over the given time period or part of it. Another common assumption is about the noise structure: many statistical

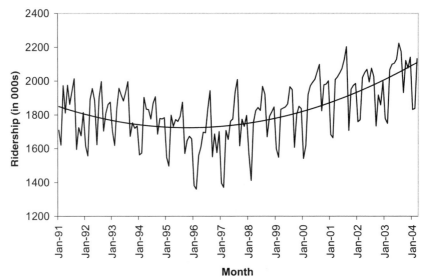

FIGURE 16.3 PLOTS THAT ENHANCE THE DIFFERENT COMPONENTS OF THE TIME SERIES. TOP: ZOOM-IN TO THREE YEARS OF DATA. BOTTOM: ORIGINAL SERIES WITH OVERLAID QUADRATIC TRENDLINE

methods assume that the noise follows a normal distribution. The advantage of methods that rely on such assumptions is that when the assumptions are reasonably met, the resulting forecasts will be more robust and the models more understandable. Other forecasting methods that are data adaptive make fewer assumptions about the structure of these components, and instead try to estimate them only from the data. Data-adaptive methods are advantageous when such assumptions are likely to be violated, or when the structure of the time series changes over time. Another advantage of many data-adaptive methods is their simplicity and computational efficiency.

A key criterion for deciding between model-driven and data-driven forecasting methods is the nature of the series in terms of global vs. local patterns. A global pattern is one that is relatively constant throughout the series. An example is a linear trend throughout the entire series. In contrast, a local pattern is one that occurs only in a short period of the data, and then changes, for example, a trend that is approximately linear within four neighboring time points, but whose trend size (slope) changes slowly over time.

Model-driven methods are generally preferable for forecasting series with global patterns because they use all the data to estimate the global pattern. For a local pattern, a model-driven model would require specifying how and when the patterns change, which is usually impractical and often unknown. Therefore data-driven methods are preferable for local patterns. Such methods "learn" patterns from the data and their memory length can be set to best adapt to the rate of change in the series. Patterns that change quickly warrant a "short memory," whereas patterns that change slowly warrant a "long memory." In short, the time plot should be used not only to identify the time series component but also the global/local nature of the trend and seasonality.

16.5 DATA PARTITIONING AND PERFORMANCE EVALUATION

As in the case of cross-sectional data, in order to avoid overfitting and to be able to assess the predictive performance of the model on new data, we first partition the data into a training set and a validation set (and perhaps an additional test set). However, there is one important difference between data partitioning in cross-sectional and time series data. In cross-sectional data the partitioning is usually done randomly, with a random set of observations designated as training data and the remainder as validation data. However, in time series a random partition would create two time series with "holes." Nearly all standard forecasting methods cannot handle time series with missing values. Therefore we partition a time series into training and validation sets differently. The series is trimmed into two periods: the earlier period is set as the training data and the later period as the validation data. Methods are then trained on the earlier training period, and their predictive performance assessed on the later validation period. Evaluation measures typically use the same metrics used in cross-sectional evaluation (see Chapter 5) with *MAPE* and *RMSE* being the most popular metrics in practice. In evaluating and comparing forecasting methods, another important tool is visualization: examining time plots of the actual and predicted series can shed light on performance and hint toward possible improvements.

Benchmark Performance: Naive Forecasts

While it is tempting to apply "sophisticated" forecasting methods, we must evaluate their value added compared to a very simple approach: the *naive forecast*. A naive forecast is simply the most recent value of the series. In other words, at time t, our forecast for any future period $t + k$ is simply the value of the series at time t. While simple, naive forecasts are sometimes surprisingly difficult to outperform with more sophisticated models. It is therefore important to benchmark against results from a naive forecasting approach.

Generating Future Forecasts

Another important difference between cross-sectional and time series partitioning occurs when creating the actual forecasts. Before attempting to forecast future values of the series, the training and validation sets are re-combined into one long series, and the chosen method/model is re-run on the complete data. This final model is then used to forecast future values. The three advantages in re-combining are (1) the validation set, which is the most recent period, usually contains the most valuable information in terms of being the closest in time to the forecast period; (2) with more data (the complete time series compared to only the training set), some models can be estimated more accurately; (3) if only the training set is used to generate forecasts, then it will require forecasting farther into the future (e.g., if the validation set contains 4 time points, forecasting the next observation will require a 5-step-ahead forecast from the training set).

In XLMiner, partitioning a time series is performed within the Time Series menu. Figure 16.4 shows a screenshot of the time series partitioning dialog box. After the final model is chosen, the same model should be re-run on the original, unpartitioned series in order to obtain forecasts. XLMiner (v. 2014) will only generate future forecasts if it is run on an unpartitioned series (see Figure 16.5).

FIGURE 16.4 PARTITIONING A TIME SERIES. THE DEFAULT IN XLMINER PARTITIONS THE DATA INTO 60% TRAINING (THE EARLIEST 60% TIME POINTS) AND 40% VALIDATION (THE LATER 40% TIME POINTS).

FIGURE 16.5 IN XLMINER'S TIME SERIES FUNCTIONS, FUTURE FORECASTS CAN ONLY BE GENERATED IF A FORECASTING METHOD IS APPLIED TO THE UNPARTITIONED SERIES (LEFT PANEL). IF APPLIED TO THE PARTITIONED SERIES, ONLY FORECASTS FOR THE VALIDATION SERIES WILL BE GENERATED (XLMINER VERSION 2014).

PROBLEMS ▬▬▬▬▬▬▬▬▬▬▬▬▬▬▬▬▬▬▬▬▬▬▬▬▬ ■

16.1 **Impact of September 11 on Air Travel in the United States.** The Research and Innovative Technology Administration's Bureau of Transportation Statistics conducted a study to evaluate the impact of the September 11, 2001 terrorist attack on US transportation. The 2006 study report and the data can be found at http://goo.gl/w2lJPV. The goal of the study was stated as follows:

> The purpose of this study is to provide a greater understanding of the passenger travel behavior patterns of persons making long distance trips before and after 9/11.

> The report analyzes monthly passenger movement data between January 1990 and May 2004. Data on three monthly time series are given in file Sept11-Travel.xls for this period:

1. Actual airline revenue passenger miles (Air),
2. Rail passenger miles (Rail), and
3. Vehicle miles traveled (Car).

In order to assess the impact of September 11, BTS took the following approach: using data before September 11, they forecasted future data (under the assumption of no terrorist attack). Then, they compared the forecasted series with the actual data to assess the impact of the event. Our first step therefore is to split each of the time series into two parts: pre– and post–September 11. We now concentrate only on the earlier time series.

a. Is the goal of this study descriptive or predictive?

b. Plot each of the three pre-event time series (air, rail, car).

    i. What time series components appear from the plot?

    ii. What type of trend appears? Change the scale of the series, add trendlines and suppress seasonality to better visualize the trend pattern.

16.2 **Performance on Training and Validation Data.** Two different models were fit to the same time series. The first 100 time periods were used for the training set and the last 12 periods were treated as a holdout set. Assume that both models make sense practically and fit the data pretty well. Below are the RMSE values for each of the models:

	Training Set	Validation Set
Model A	543	690
Model B	669	675

a. Which model appears more useful for explaining the different components of this time series? Why?

b. Which model appears to be more useful for forecasting purposes? Why?

16.3   **Forecasting Department Store Sales.** The file DepartmentStoreSales.xls contains data on the quarterly sales for a department store over a six-year period (data courtesy of Chris Albright).

   a. Create a well-formatted time plot of the data.

   b. Which of the four components (level, trend, seasonality, noise) seem to be present in this series?

16.4   **Shipments of Household Appliances.** The file ApplianceShipments.xls contains the series of quarterly shipments (in million $) of US household appliances between 1985–1989 (data courtesy of Ken Black).

   a. Create a well-formatted time plot of the data.

   b. Which of the four components (level, trend, seasonality, noise) seem to be present in this series?

16.5   **Canadian Manufacturing Workers Workhours.** The time plot in Figure 16.6 describes the average annual number of weekly hours spent by Canadian manufacturing workers (data are available in CanadianWorkHours.xls—thanks to Ken Black for the data).

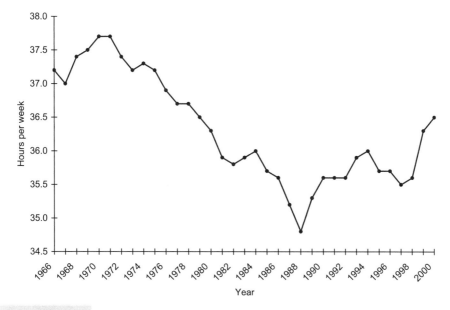

FIGURE 16.6   AVERAGE ANNUAL WEEKLY HOURS SPENT BY CANADIAN MANUFACTURING WORKERS

   a. Reproduce the time plot.

   b. Which of the four components (level, trend, seasonality, noise) seem to be present in this series?

16.6 **Souvenir Sales:** The file `SouvenirSales.xls` contains monthly sales for a souvenir shop at a beach resort town in Queensland, Australia, between 1995–2001 (source: Hyndman, R.J. Time Series Data Library, http://data.is/TSDLdemo. Accessed on 07/25/15). Back in 2001, the store wanted to use the data to forecast sales for the next 12 months (year 2002). They hired an analyst to generate forecasts. The analyst first partitioned the data into training and validation sets, with the validation set containing the last 12 months of data (year 2001). She then fit a regression model to sales, using the training set.

   a. Create a well-formatted time plot of the data.

   b. Change the scale on the *x*-axis, or on the *y*-axis, or on both to log-scale in order to achieve a linear relationship. Select the time plot that seems most linear.

   c. Comparing the two time plots, what can be said about the type of trend in the data?

   d. Why were the data partitioned? Partition the data into the training and validation set as explained above.

16.7 **Shampoo Sales.** The file `ShampooSales.xls` contains data on the monthly sales of a certain shampoo over a three-year period (source: Hyndman, R.J. Time Series Data Library, http://data.is/TSDLdemo. Accessed on 07/25/15).

   a. Create a well-formatted time plot of the data.

   b. Which of the four components (level, trend, seasonality, noise) seems to be present in this series?

   c. Do you expect to see seasonality in sales of shampoo? Why?

   d. If the goal is forecasting sales in future months, which of the following steps should be taken?

- Partition the data into training and validation sets.
- Tweak the model parameters to obtain good fit to the validation data.
- Look at MAPE and RMSE values for the training set.
- Look at MAPE and RMSE values for the validation set.

## CHAPTER 17

Regression-Based Forecasting

A popular forecasting tool is based on multiple linear regression models, using suitable predictors to capture trend and/or seasonality. In this chapter we show how a linear regression model can be set up to capture a time series with a trend and/or seasonality. The model, which is estimated from the data, can produce future forecasts from the relevant predictor information that is inserted into the estimated regression equation. We describe different types of common trends (linear, exponential, polynomial), as well as two types of seasonality (additive and multiplicative). Then we show how a regression model can be used to quantify the correlation between neighboring values in a time series (called autocorrelation). This type of model, called an autoregressive model, is useful for improving forecast precision by making use of the information contained in the autocorrelation (beyond trend and seasonality). It is also useful for evaluating the predictability of a series, by evaluating whether the series is a "random walk." The various steps of fitting linear regression and autoregressive models, using them to generate forecasts, and assessing their predictive accuracy, are illustrated using the Amtrak ridership series.

17.1 A MODEL WITH TREND[1]

Linear Trend

To create a linear regression model that captures a time series with a global linear trend, the output variable (Y) is set as the time series measurement or

[1]This and subsequent sections in this chapter © 2016 Statistics.com and Galit Shmueli. Used by permission.

Data Mining for Business Analytics: Concepts, Techniques, and Applications with XLMiner®, Third Edition. Galit Shmueli, Peter C. Bruce, and Nitin R. Patel.
© 2016 John Wiley & Sons Inc. Published 2016 by John Wiley & Sons, Inc.

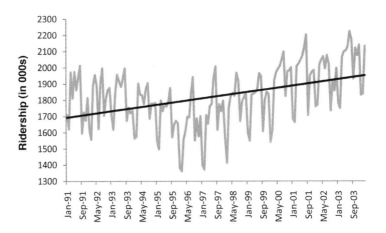

FIGURE 17.1   **A LINEAR TREND FIT TO AMTRAK RIDERSHIP**

some function of it, and the predictor (X) is set as a time index. Let us consider a simple example: fitting a linear trend to the Amtrak ridership data. This type of trend is shown in Figure 17.1. From the time plot it is obvious that the global trend is not linear. However, we use this example to illustrate how a linear trend is fitted, and later we consider more appropriate models for this series.

To obtain a linear relationship between Ridership and Time, we set the output variable Y as the Amtrak Ridership and create a new variable that is a time index $t = 1, 2, 3, \ldots$. This time index is then used as a single predictor in the regression model:

$$Y_t = \beta_0 + \beta_1 t + \epsilon,$$

where Y_t is the Ridership at time point t and ϵ is the standard noise term in a linear regression. Thus we are modeling three of the four time series components: level (β_0), trend (β_1), and noise (ϵ). Seasonality is not modeled. A snapshot of the two corresponding columns $(Y$ and $t)$ in Excel are shown in Table 17.1.

After partitioning the data into training and validation sets, the next step is to fit a linear regression model to the training set, with t as the single predictor. Applying this to the Amtrak ridership data (with a validation set consisting of the last 12 months) results in the estimated model shown in Figure 17.2. The actual and fitted values and the residuals are shown in the two lower panels in time plots. Note that examining only the estimated coefficients and their statistical significance can be very misleading! In this example they would indicate that the linear fit is reasonable, although it is obvious from the time plots that the trend is not linear. The difference in the magnitude of the validation average error is also indicative of an inadequate trend shape. But an

TABLE 17.1	OUTPUT VARIABLE (MIDDLE) AND PREDICTOR VARIABLE (RIGHT) USED TO FIT A LINEAR TREND	
Month	Ridership	t
Jan-91	1709	1
Feb-91	1621	2
Mar-91	1973	3
Apr-91	1812	4
May-91	1975	5
Jun-91	1862	6
Jul-91	1940	7
Aug-91	2013	8
Sep-91	1596	9
Oct-91	1725	10
Nov-91	1676	11
Dec-91	1814	12
Jan-92	1615	13
Feb-92	1557	14

inadequate trend shape is easiest to detect by examining the time plot of the residuals.

Exponential Trend

Several alternative trend shapes are useful and easy to fit via a linear regression model. Recall Excel's *Trendline* and other plotting options that help assess the type of trend in the data. One such shape is an exponential trend. An exponential trend implies a multiplicative increase/decrease of the series over time ($Y_t = ce^{\beta_1 t + \varepsilon}$). To fit an exponential trend, simply replace the output variable Y with $\log Y$ (where $\log$ is the natural logarithm), and fit a linear regression ($\log Y_t = \beta_0 + \beta_1 t + \varepsilon$). In the Amtrak example we would fit a linear regression of $\log$(Ridership) on the index variable t. Exponential trends are popular in sales data, where they reflect percentage growth.

Note: As in the general case of linear regression, when comparing the predictive accuracy of models that have a different output variable, such as a linear trend model (with Y) and an exponential trend model (with $\log Y$), it is essential to compare forecasts or forecasts errors on the same scale. An exponential trend model will produce forecasts of $\log Y$, and the forecast errors reported by the software will therefore be $\log Y - \widehat{\log Y}$. To obtain forecasts in the original units, create a new column that takes an exponent of the model forecasts. Then use this column to create an additional column of forecast errors, by subtracting the original Y. An example is shown in Figures 17.3 and 17.4, where an exponential

The Regression Model

Input variables	Coefficient	Std. error	p-Value	SS
Constant term	1713.028809	27.08552361	0	477456500
t	1.2053107	0.31751993	0.00021544	384546.3125

Training Data Scoring–Summary Report

Total sum of squared errors	RMS error	Average error
3869551.676	162.2451256	−3.84852E-05

Validation Data Scoring–Summary Report

Total sum of squared errors	RMS error	Average error
529326.616	210.0251207	168.8524156

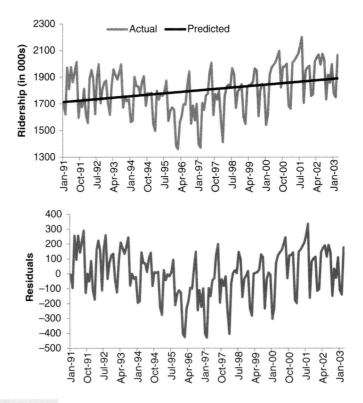

FIGURE 17.2   FITTED REGRESSION MODEL WITH LINEAR TREND. REGRESSION OUTPUT (TOP), AND TIME PLOTS OF ACTUAL AND FITTED SERIES (MIDDLE) AND RESIDUALS (BOTTOM)

trend is fit to the Amtrak ridership data. Note that the performance measures for the training and validation data are not comparable to those from the linear trend model shown in Figure 17.2. Instead, we manually compute two new columns

Regression Model

Input variables	Coefficient	Std. Error	p-Value	SS
Constant term	7.44398642	0.01547452	0	8251.513672
t	0.00065125	0.00018141	0.00045169	0.11226512

Training Data Scoring–Summary Report

Total sum of squared errors	RMS error	Average error
1.263050414	0.092694011	−5.27755E-08

Validation Data Scoring–Summary Report

Total sum of squared errors	RMS error	Average error
0.139731707	0.107908799	0.08800547

FIGURE 17.3   OUTPUT FROM REGRESSION MODEL WITH EXPONENTIAL TREND, FIT TO TRAINING DATA

in Figure 17.4, one that gives forecasts of ridership (in thousands) and the other which gives the forecast errors in terms of ridership. To compare RMS Error or Average Error, we would now proceed to the new forecast errors and compute their standard deviation (for RMS Error) or their average (for Average Error). These would then be comparable to the numbers in Figure 17.2.

Predicted Value	Actual Value	Residual	t	Predicted Ridership	Forecast Errors
7.54037142	7.649168201	0.108796781	148	1882.529104	216.3698957
7.54102267	7.652028465	0.111005795	149	1883.755501	221.1554993
7.54167392	7.663722787	0.122048867	150	1884.982696	244.688304
7.54232517	7.706769897	0.164444727	151	1886.210691	337.1383092
7.54297642	7.684489647	0.141513227	152	1887.439486	286.9205144
7.54362767	7.566003514	0.022375844	153	1888.669081	42.73691909
7.54427892	7.659864524	0.115585604	154	1889.899477	231.5705227
7.54493017	7.638224256	0.093294086	155	1891.130675	184.9233248
7.54558142	7.668877413	0.123295993	156	1892.362675	248.3143249
7.54623267	7.51289495	-0.03333772	157	1893.595478	-62.08747771
7.54688392	7.516436567	-0.030447353	158	1894.829083	-56.82308341
7.54753517	7.665024957	0.117489787	159	1896.063493	236.3825072

FIGURE 17.4   ADJUSTING FORECASTS OF LOG(RIDERSHIP) TO THE ORIGINAL SCALE (FIFTH COLUMN), AND COMPUTING FORECAST ERRORS IN THE ORIGINAL SCALE (RIGHT COLUMN)

Polynomial Trend

Another nonlinear trend shape that is easy to fit via linear regression is a polynomial trend, and in particular, a quadratic relationship of the form $Y_t = \beta_0 + \beta_1 t + \beta_2 t^2 + \varepsilon$. This is done by creating an additional predictor t^2 (the square of t), and fitting a multiple linear regression with the two predictors t and t^2. For the Amtrak ridership data, we have already seen a U-shaped trend in the data. We therefore fit a quadratic model, concluding from the plots of model fit and residuals (Figure 17.5) that this shape adequately captures the trend. The residuals are now devoid of trend and exhibit only seasonality.

FIGURE 17.5 REGRESSION MODEL WITH QUADRATIC TREND FIT TO THE TRAINING DATA (TOP) AND MODEL RESIDUALS (BOTTOM)

In general, any type of trend shape can be fit as long as it has a mathematical representation. However, the underlying assumption is that this shape is applicable throughout the period of data that we have and also during the period that we are going to forecast. Do not choose an overly complex shape. Although it will fit the training data well, it will in fact be overfitting them. To avoid overfitting, always examine performance on the validation set and refrain from choosing overly complex trend patterns.

17.2   A MODEL WITH SEASONALITY

A seasonal pattern in a time series means that observations that fall in some seasons have consistently higher or lower values than those that fall in other seasons. Examples are day-of-week patterns, monthly patterns, and quarterly patterns. The Amtrak ridership monthly time series, as can be seen in the time plot, exhibits strong monthly seasonality (with highest traffic during summer months). Seasonality is modeled in a regression model by creating a new categorical variable that denotes the season for each observation. This categorical variable is then turned into dummies, which in turn are included as predictors in the regression model. To illustrate this, we created a new "Month" column for the Amtrak ridership data, as shown in Table 17.2.

TABLE 17.2	NEW CATEGORICAL VARIABLE (RIGHT) TO BE USED (VIA DUMMIES) AS PREDICTOR(S) IN A LINEAR REGRESSION MODEL

Month	Ridership	Season
Jan-91	1709	Jan
Feb-91	1621	Feb
Mar-91	1973	Mar
Apr-91	1812	Apr
May-91	1975	May
Jun-91	1862	Jun
Jul-91	1940	Jul
Aug-91	2013	Aug
Sep-91	1596	Sep
Oct-91	1725	Oct
Nov-91	1676	Nov
Dec-91	1814	Dec
Jan-92	1615	Jan
Feb-92	1557	Feb
Mar-92	1891	Mar
Apr-92	1956	Apr
May-92	1885	May

In order to include the season categorical variable as a predictor in a regression model for Y (e.g., Ridership), we turn it into dummies. For m seasons we create $m - 1$ dummies, which are binary variables that take on the value 1 if the record falls in that particular season and 0 otherwise.[2] We then partition the data into training and validation sets (see Section 16.5) and fit the regression model to the training data. The top panels of Figure 17.6 show the output of a linear regression fit to Ridership (Y) on 11 month dummies (using the training data). The fitted series and the residuals from this model are shown in the lower panels. The model appears to capture the seasonality in the data. However, since we have not included a trend component in the model (as shown in Section 17.1), the fitted values do not capture the existing trend. Therefore the residuals, which are the difference between the actual and the fitted values, clearly display the remaining U-shaped trend.

When seasonality is added as described above (create categorical seasonal variable, create dummies from it, then regress on Y), it captures *additive seasonality*. This means that the average value of Y in a certain season is a fixed amount more or less than that in another season. For example, in the Amtrak ridership, the coefficient for August (139.39) indicates that the average number of passengers in August is higher by 140 thousand passengers than the average in April (the reference category). In using regression models, we can also capture *multiplicative seasonality*, where values in a certain season are on average higher or lower by a percentage amount compared to another season. To fit multiplicative seasonality, we use the same model as above, except that we use $\log Y$ as the output variable.

17.3 A MODEL WITH TREND AND SEASONALITY

Last, we can create models that capture both trend and seasonality by including predictors of both types. For example, from our exploration of the Amtrak Ridership data, it appears that a quadratic trend and monthly seasonality are both warranted. We therefore fit a model to the training data with 13 predictors: 11 dummies for month, and t and t^2 for trend. The output and fit from this final model are shown in Figure 17.7. If we are satisfied with this model after evaluating its predictive performance on the validation data and comparing it against alternatives, we would re-fit it to the entire unpartitioned series. This re-fitted model can then be used to generate k-step-ahead forecasts (denoted by F_{t+k}) by plugging in the appropriate month and index terms.

---

[2] We use only $m - 1$ dummies because information about the $m - 1$ seasons is sufficient. If all $m - 1$ variables are zero, then the season must be the mth season. Including the mth variable would cause redundant information and multicollinearity errors.

Regression Model

Input variables	Coefficient	Std. Error	p-Value	SS
Constant term	1855.235962	33.95079803	0	477456500
season_Aug	139.3903351	48.01367569	0.00431675	483721.3125
season_Dec	−19.82307816	48.01367569	0.68036187	33314.77734
season_Feb	−288.9631348	47.08128357	0	665331.9375
season_Jan	−251.2854462	47.08128357	0.00000034	598841.0625
season_Jul	94.34428406	48.01367569	0.05147372	187691.7656
season_Jun	−10.11090946	48.01367569	0.83352947	11869.09277
season_Mar	11.57308865	47.08128357	0.80620199	48930.94922
season_May	31.24033737	48.01367569	0.51637506	114420.9141
season_Nov	−63.96651077	48.01367569	0.18502063	3121.062012
season_Oct	−54.12883377	48.01367569	0.26158884	14579.31641
season_Sep	−193.6371613	48.01367569	0.00009163	224972.1094

Training Data Scoring–Summary Report

Total sum of squared errors	RMS error	Average error
1867303.623	112.7064583	−5.14163E-05

Validation Data Scoring–Summary Report

Total sum of squared errors	RMS error	Average error
841206.3548	264.765046	262.1077072

FIGURE 17.6 FITTED REGRESSION MODEL WITH SEASONALITY. REGRESSION OUTPUT (TOP), PLOTS OF FITTED AND ACTUAL SERIES (MIDDLE), AND MODEL RESIDUALS (BOTTOM)

Regression Model

Input variables	Coefficient	Std. error	p-Value	SS
Constant term	1932.998779	27.85863113	0	477456500
season_Aug	135.1726227	30.52143288	0.00001955	483721.3125
season_Dec	−29.65872955	30.53801155	0.33320817	33314.77734
season_Feb	−306.3078308	29.94875526	0	665331.9375
season_Jan	−267.444458	29.94642067	0	598841.0625
season_Jul	91.31225586	30.5189991	0.00330446	187691.7656
season_Jun	−12.04474545	30.51724434	0.69370645	11869.09277
season_Mar	−7.04482555	29.95207596	0.81441271	48930.94922
season_May	30.31717491	30.51618195	0.32228076	114420.9141
season_Nov	−72.26641083	30.53282547	0.01938256	3121.062012
season_Oct	−60.98049164	30.52834129	0.04781064	14579.31641
season_Sep	−199.1280975	30.52454758	0	224972.1094
t	−5.246521	0.58674908	0	398979.7188
t^2	0.0437566	0.00384071	0	725213.9375

Training Data Scoring–Summary Report

Total sum of squared errors	RMS error	Average error
743110.0191	71.0997201	−6.05149E−05

Validation Data Scoring–Summary Report

Total sum of squared errors	RMS error	Average error
30722.61731	50.59859789	−34.11397564

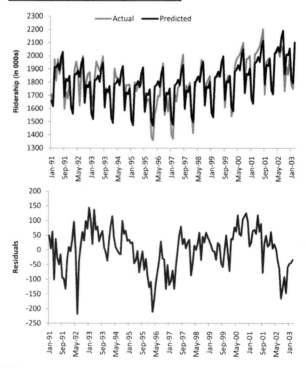

FIGURE 17.7 REGRESSION MODEL WITH MONTHLY (ADDITIVE) SEASONALITY AND QUADRATIC TREND, FIT TO AMTRAK RIDERSHIP DATA. REGRESSION OUTPUT (TOP), PLOTS OF FITTED AND ACTUAL SERIES (MIDDLE), AND MODEL RESIDUALS (BOTTOM)

17.4   AUTOCORRELATION AND ARIMA MODELS

When we use linear regression for time series forecasting, we are able to account for patterns such as trend and seasonality. However, ordinary regression models do not account for dependence between observations, which in cross-sectional data is assumed to be absent. Yet, in the time series context, observations in neighboring periods tend to be correlated. Such correlation, called *autocorrelation*, is informative and can help in improving forecasts. If we know that a high value tends to be followed by high values (positive autocorrelation), then we can use that to adjust forecasts. We will now discuss how to compute the autocorrelation of a series, and how best to utilize the information for improving forecasts.

Computing Autocorrelation

Correlation between values of a time series in neighboring periods is called *autocorrelation* because it describes a relationship between the series and itself. To compute autocorrelation, we compute the correlation between the series and a lagged version of the series. A *lagged series* is a "copy" of the original series that is moved forward one or more time periods. A lagged series with lag-1 is the original series moved forward one time period, a lagged series with lag-2 is the original series moved forward two time periods, and so on. Table 17.3 shows the first 24 months of the Amtrak ridership series, the lag-1 series and the lag-2 series.

Next, to compute the lag-1 autocorrelation, which measures the linear relationship between values in consecutive time periods, we compute the correlation between the original series and the lag-1 series (e.g., via the Excel function CORREL) to be 0.08. Note that although the original series shown above has 24 time periods, the lag-1 autocorrelation will only be based on 23 pairs (because the lag-1 series does not have a value for Jan-91). Similarly, the lag-2 autocorrelation, measuring the relationship between values that are two time periods apart, is the correlation between the original series and the lag-2 series (yielding -0.15).

We can use XLMiner's *ACF (autocorrelations)* utility within the *Time Series* menu to directly compute the autocorrelation of a series at different lags. For example, the output for the 24-month ridership is shown in Figure 17.8. To display a bar chart of the autocorrelation at different lags, check the *Plot ACF* option.

The typical autocorrelation behaviors that are useful to explore are the following three:

TABLE 17.3   FIRST 24 MONTHS OF AMTRAK RIDERSHIP SERIES

Month	Ridership	Lag-1 Series	Lag-2 Series
Jan-91	1709		
Feb-91	1621	1709	
Mar-91	1973	1621	1709
Apr-91	1812	1973	1621
May-91	1975	1812	1973
Jun-91	1862	1975	1812
Jul-91	1940	1862	1975
Aug-91	2013	1940	1862
Sep-91	1596	2013	1940
Oct-91	1725	1596	2013
Nov-91	1676	1725	1596
Dec-91	1814	1676	1725
Jan-92	1615	1814	1676
Feb-92	1557	1615	1814
Mar-92	1891	1557	1615
Apr-92	1956	1891	1557
May-92	1885	1956	1891
Jun-92	1623	1885	1956
Jul-92	1903	1623	1885
Aug-92	1997	1903	1623
Sep-92	1704	1997	1903
Oct-92	1810	1704	1997
Nov-92	1862	1810	1704
Dec-92	1875	1862	1810

Strong autocorrelation (positive or negative) at a lag k larger than 1 and its multiples $(2k, 3k, ...)$ typically reflects a cyclical pattern. For example, strong positive lag-12 autocorrelation in monthly data will reflect an annual seasonality (where values during a given month each year are positively correlated).

Positive lag-1 autocorrelation (called "stickiness") describes a series where consecutive values move generally in the same direction. In the presence of a strong linear trend, we would expect to see a strong and positive lag-1 autocorrelation.

Negative lag-1 autocorrelation reflects swings in the series, where high values are immediately followed by low values, and vice versa.

Examining the autocorrelation of a series can therefore help detect seasonality patterns. In Figure 17.8, for example, we see that the strongest autocorrelation is at lag 6 and is negative. This indicates a bi-annual pattern in ridership, with 6-month switches from high to low ridership. A look at the time plot confirms the high-summer low-winter pattern.

FIGURE 17.8 **XLMINER OUTPUT SHOWING AUTOCORRELATION AT LAGS 1 TO 12 FOR THE 24 MONTHS OF AMTRAK RIDERSHIP**

Besides looking at the autocorrelation of the raw series, it is useful to look at the autocorrelation of *residual series*. For example, after fitting a regression model (or using any other forecasting method), we can examine the autocorrelation of the series of residuals. If we have adequately modeled the seasonal pattern, then the residual series should show no autocorrelation at the season's lag. Figure 17.9 displays the autocorrelations for the residuals from the regression model with seasonality and quadratic trend shown in Figure 17.7. It is clear that the 6-month (and 12-month) cyclical behavior no longer dominates the series of residuals, indicating that the regression model captured them adequately. However, we can also see a strong positive autocorrelation from lag 1 on, indicating a positive relationship between neighboring residuals. This is valuable information that can be used to improve forecasts.

Improving Forecasts by Integrating Autocorrelation Information

In general, there are two approaches to taking advantage of autocorrelation. One is by directly building the autocorrelation into the regression model, and the other is by constructing a second-level forecasting model on the residual series.

Among regression-type models that directly account for autocorrelation are *autoregressive* (AR) models, or the more general class of models called ARIMA

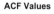

XLMiner : Time Series—ACF (Autocorrelations)

Inputs

Data	
# Records in input data	147
Selected variable	Residual

Parameters/Options	
Max Lag	12

ACF Values

Lags	ACF
0	1
1	0.64821321
2	0.51890093
3	0.40798336
4	0.31966141
5	0.26237851
6	0.21345751
7	0.22334783
8	0.22640951
9	0.19724335
10	0.14933859
11	0.17307311
12	0.12726976

FIGURE 17.9   **AMTRAK DATA: XLMINER OUTPUT SHOWING AUTOCORRELATION OF RESIDUAL SERIES**

models (autoregressive integrated moving average models). Autoregressive models are similar to linear regression models, except that the predictors are the past values of the series. For example, an autoregressive model of order 2, denoted AR(2), can be written as

$$Y_t = \beta_0 + \beta_1 Y_{t-1} + \beta_2 Y_{t-2} + \varepsilon. \tag{17.1}$$

Estimating such models is roughly equivalent to fitting a linear regression model with the series as the output, and the two lagged series (at lag 1 and 2 in this example) as the predictors. However, it is better to use designated ARIMA estimation methods (e.g., those available in XLMiner's Time Series > ARIMA menu), over ordinary linear regression estimation, to produce more accurate results.[3] Moving from AR to ARIMA models creates a larger set of more flexible forecasting models, but also requires much more statistical expertise. Even with the simpler AR models, fitting them to raw time series that contain patterns such as trends and seasonality requires the user to perform several initial data transformations and to choose the order of the model. These are

---

[3] ARIMA model estimation differs from ordinary regression estimation by accounting for the dependence between observations.

not straightforward tasks. Because ARIMA modeling is less robust and requires more experience and statistical expertise than other methods, the use of such models for forecasting raw series is generally less popular in practical forecasting. We therefore direct the interested reader to classic time series textbooks (e.g., see Chapter 4 in Chatfield, 2003).

However, we do discuss one particular use of AR models that is straightforward to apply in the context of forecasting and that can provide a significant improvement to short-term forecasts. This relates to the second approach for utilizing autocorrelation, which requires constructing a second-level forecasting model for the residuals, as follows:

1. Generate a k-step-ahead forecast of the series (F_{t+k}), using any forecasting method.

2. Generate a k-step-ahead forecast of the forecast error (residual) (E_{t+k}), using an AR (or other) model.

3. Improve the initial k-step-ahead forecast of the series by adjusting it according to its forecasted error: *Improved* $F^*_{t+k} = F_{t+k} + E_{t+k}$.

In particular, we can fit low-order AR models to series of residuals (or *forecast errors*) that can then be used to forecast future forecast errors. By fitting the series of residuals, rather than the raw series, we avoid the need for initial data transformations (because the residual series is not expected to contain any trends or cyclical behavior besides autocorrelation).

To fit an AR model to the series of residuals, we first examine the autocorrelations of the residual series. We then choose the order of the AR model according to the lags in which autocorrelation appears. Often, when autocorrelation exists at lag 1 and higher, it is sufficient to fit an AR(1) model of the form

$$E_t = \beta_0 + \beta_1 E_{t-1} + \varepsilon, \tag{17.2}$$

where E_t denotes the residual (or *forecast error*) at time t. For example, although the autocorrelations in Figure 17.9 appear large from lags 1 to 10 or so, it is likely that an AR(1) would capture all of these relationships. The reason is that if immediate neighboring values are correlated, then the relationship propagates to values that are two periods away, then three periods away, and so on. The result of fitting an AR(1) model to the Amtrak ridership residual series is shown in Figure 17.10. The AR(1) coefficient (0.65) is close to the lag-1 autocorrelation that we found earlier (Figure 17.9). The forecasted residual for April 2003, given at the bottom, is computed by plugging in the most recent residual from March 2003 (equal to -33.786) into the AR(1) model: $0 + (0.647)(-33.786) = -21.866$. The negative value tells us that the regression

XLMiner : ARIMA Model

Output Navigator			
Inputs	ARIMA Model	Residuals	ACF Plot
Elapsed Time	Forecast	Var Covar	

Inputs

Data	
# Records in input data	147
Input data	Data!$A$2:$C$148
Selected variable	Residual

Parameters/Options	
Do Not Fit Constant Term	No
AR	1
MA	0
Ordinary Difference	0
Show Forecasting Output	Yes
# Forecasts	1
Confidence Level	95
Show Residual Output	Yes

ARIMA Model

ARIMA	Coeff	Std. error	p-Value
Const. term	−0.00002147	4.86492062	0.99999648
AR1	0.64688748	0.06221873	0

Mean	−0.000061
-2LogL	1590.729736
S	54.047494
#Iterations	8

Lag	12	24	36	48
p-Value	0.577618	0.48778015	0.60872775	0.52462631
ChiSq	9.48063087	22.54175186	32.10258102	45.74367905
df	11	23	35	47

Forecast

Month	Forecast
Apr-03	−21.85559082

Time Plot of Actual Residual vs. Forecasted Residual (Training Data)

Actual ——Forecast

FIGURE 17.10   FITTING AN AR(1) MODEL TO THE RESIDUAL SERIES model will produce a ridership forecast for April 2003 that is too high, and that we should adjust it down by subtracting 21.866 thousand riders. In this example the regression model (with quadratic trend and seasonality) produced a forecast of 2115 thousand riders, and the improved two-stage model (regression + AR(1) correction) corrected it by reducing it to 2093 thousand riders. The actual value for April 2003 turned out to be 2099 thousand riders—much closer to the improved forecast.

From the plot of the actual vs. forecasted residual series, we can see that the AR(1) model fits the residual series quite well. However, the plot is based on the training data (until March 2003). To evaluate predictive performance of the two-level model (regression + AR(1)), we would have to examine performance (e.g., via MAPE or RMSE metrics) on the validation data in a fashion similar to the calculation that we performed for April 2003 above.

> To fit an AR (autoregression) model in XLMiner, use ARIMA in the Time Series menu. In the "Nonseasonal Parameters," set *Autoregressive (p)* to the required order, and *Moving Average (q)* to 0. The *Advanced* menu will allow you to request forecasts and to display fitted values and residuals.

Last, to examine whether we have indeed accounted for the autocorrelation in the series and that no more information remains in the series, we examine the autocorrelations of the series of residuals-of-residuals (the residuals obtained after the AR(1) was applied to the regression residuals). This can be seen in Figure 17.11. It is clear that no more autocorrelation remains, and that the addition of the AR(1) model has captured the autocorrelation information adequately.

We mentioned earlier that improving forecasts via an additional AR layer is useful for short-term forecasting. The reason is that an AR model of order k will usually only provide useful forecasts for the next k periods, and after that forecasts will rely on earlier forecasts rather than on actual data. For example, to forecast the residual of May 2003 when the time of prediction is March 2003, we would need the residual for April 2003. However, because that value is not available, it would be replaced by its forecast. Hence the forecast for May 2003 would be based on the forecast for April 2003.

Evaluating Predictability

Before attempting to forecast a time series, it is important to determine whether it is predictable, in the sense that its past can be used to predict its future beyond the naive forecast. One useful way to assess predictability is to test whether the series is a *random walk*. A random walk is a series in which changes from one time

XLMiner : Time Series—ACF (Autocorrelations)

Inputs

Data	
# Records in input data	147
Selected variable	Residuals

Parameters/Options	
Max Lag	12

ACF Values

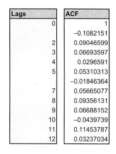

Lags	ACF
0	1
1	−0.1082151
2	0.09046599
3	0.06693597
4	0.0296591
5	0.05310313
6	−0.01846364
7	0.05665077
8	0.09356131
9	0.06688152
10	−0.0439739
11	0.11453787
12	0.03237034

FIGURE 17.11 AUTOCORRELATIONS OF RESIDUALS-OF-RESIDUALS SERIES

period to the next are random. According to the Efficient Market Hypothesis in economics, asset prices are random walks, and therefore predicting stock prices is a game of chance.[4]

A random walk is a special case of an AR(1) model, where the slope coefficient is equal to 1:

$$Y_t = \beta_0 + Y_{t-1} + \varepsilon_t. \tag{17.3}$$

We can also write this as

$$Y_t - Y_{t-1} = \beta_0 + \varepsilon_t. \tag{17.4}$$

We see from the last equation that the difference between the values at periods $t - 1$ and t is random, hence the term "random walk." Forecasts from such a model are basically equal to the most recent observed value (the naive forecast), reflecting the lack of any other information.

To test whether a series is a random walk, we fit an AR(1) model and test the hypothesis that the slope coefficient is equal to 1 ($H_0 : \beta_1 = 1$ vs. $H_1 : \beta_1 \neq 1$).

---

[4] There is some controversy surrounding the Efficient Market Hypothesis, with claims that there is slight autocorrelation in asset prices, which does make them predictable to some extent. However, transaction costs and bid–ask spreads tend to offset any prediction benefits.

If the null hypothesis is rejected (reflected by a small *p*-value), then the series is not a random walk and we can attempt to predict it.

As an example, consider the AR(1) model shown in Figure 17.10. The slope coefficient (0.647) is more than 3 standard errors away from 1, indicating that this is not a random walk. In contrast, consider the AR(1) model fitted to the series of S&P500 monthly closing prices between May 1995 and August 2003 (available in SP500.xls, shown in Figure 17.12). Here the slope coefficient is 0.985, with a standard error of 0.015. The coefficient is sufficiently close to 1 (around one standard error away), indicating that this is a random walk. Forecasting this series using any of the methods described earlier (aside from the naive forecast) is therefore futile.

XLMiner : ARIMA Model

Output Navigator			
Inputs	ARIMA Model	Residuals	ACF Plot
Elapsed Time	Forecast	Var Covar	

Inputs

Data	
# Records in input data	100
Input data	Data!$A$2:$C$101
Selected variable	Close

Parameters/Options	
Do Not Fit Constant Term	No
AR	1
MA	0
Ordinary Difference	0

ARIMA Model

ARIMA	Coeff	StErr	p-Value
Const. term	15.62566853	3.68750787	0.00002261
AR1	0.98479182	0.01436355	0

FIGURE 17.12 AR(1) MODEL FITTED TO S&P500 MONTHLY CLOSING PRICES (MAY 1995–AUG 2003). (NOTE: IGNORE THE P-VALUES REPORTED IN THE OUTPUT; THEY TEST A DIFFERENT HYPOTHESIS THAN THE ONE DISCUSSED IN THE TEXT.)

PROBLEMS

17.1   **Impact of September 11 on Air Travel in the United States.** The Research and Innovative Technology Administration's Bureau of Transportation Statistics conducted a study to evaluate the impact of the September 11, 2001 terrorist attack on US transportation. The 2006 study report and the data can be found at http://goo.gl/w2lJPV. The goal of the study was stated as follows:

> The purpose of this study is to provide a greater understanding of the passenger travel behavior patterns of persons making long distance trips before and after 9/11.

The report analyzes monthly passenger movement data between January 1990 and May 2004. Data on three monthly time series are given in file `Sept11Travel.xls` for this period: (1) Actual airline revenue passenger miles (Air), (2) Rail passenger miles (Rail), and (3) Vehicle miles traveled (Car).

In order to assess the impact of September 11, BTS took the following approach: using data before September 11, they forecasted future data (under the assumption of no terrorist attack). Then they compared the forecasted series with the actual data to assess the impact of the event. Our first step, therefore, is to split each of the time series into two parts: pre– and post–September 11. We now concentrate only on the earlier time series.

a.   Plot the pre-event AIR time series.

   i.   What time series components appear from the plot?

b.   Figure 17.13 shows a time plot of the **seasonally adjusted** pre–September–11 AIR series.

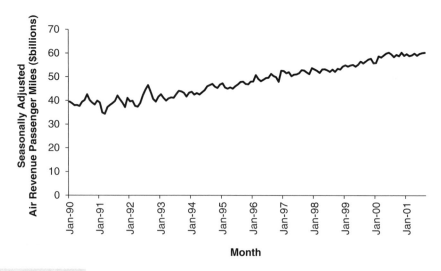

FIGURE 17.13   SEASONALLY ADJUSTED PRE–SEPTEMBER–11 AIR SERIES

Which of the following methods would be adequate for forecasting the series shown in the figure?

- Linear regression model with dummies
- Linear regression model with trend
- Linear regression model with dummies and trend

c. Specify a linear regression model for the AIR series that would produce a seasonally adjusted series similar to the one shown in Figure 17.13, with multiplicative seasonality. What is the output variable? What are the predictors?

d. Run the regression model from (c). Remember to create dummy variables for the months (XLMiner will create 12 dummies; use only 11 and drop the April dummy) and to use only pre-event data.

  i. What can we learn from the statistical insignificance of the coefficients for October and September?

  ii. The actual value of AIR (air revenue passenger miles) in January 1990 was 35.153577 billion. What is the residual for this month, using the regression model? Report the residual in terms of air revenue passenger miles.

e. Create an ACF (autocorrelation) plot of the regression residuals.

  i. What does the ACF plot tell us about the regression model's forecasts?

  ii. How can this information be used to improve the model?

f. Fit linear regression models to Air, Rail, and Auto with additive seasonality and an appropriate trend. For Air and Rail, fit a linear trend. For Rail, use a quadratic trend. Remember to use only pre-event data. Once the models are estimated, use them to forecast each of the three post-event series.

  i. For each series (Air, Rail, Auto), plot the complete pre-event and post-event actual series overlaid with the predicted series.

  ii. What can be said about the effect of the September 11 terrorist attack on the three modes of transportation? Discuss the magnitude of the effect, its time span, and any other relevant aspects.

17.2 Analysis of Canadian Manufacturing Workers Workhours. The time plot in Figure 17.14 describes the average annual number of weekly hours spent by Canadian manufacturing workers (data are available in CanadianWorkHours.xls, data courtesy of Ken Black).

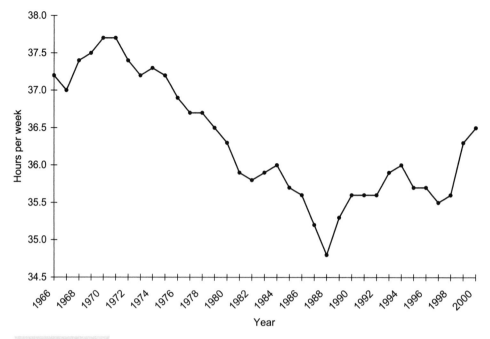

FIGURE 17.14 AVERAGE ANNUAL WEEKLY HOURS SPENT BY CANADIAN
MANUFACTURING WORKERS

a. Which of the following regression-based models would fit the series best (choose one)?

- Linear trend model
- Linear trend model with seasonality
- Quadratic trend model
- Quadratic trend model with seasonality

b. If we computed the autocorrelation of this series, would the lag-1 autocorrelation exhibit negative, positive, or no autocorrelation? How can you see this from the plot?

c. Compute the autocorrelation of the series and produce an ACF plot. Verify your answer to the previous question.

17.3 Toys "R" Us Revenues. Figure 17.15 is a time plot of the quarterly revenues of Toys "R" Us between 1992–1995 (thanks to Chris Albright for suggesting the use of these data, which are available in ToysRUsRevenues.xls).

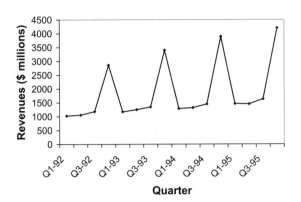

QUARTERLY REVENUES OF TOYS "R" US, 1992–1995

a. Fit a regression model with a linear trend and seasonal dummies. Use the entire series (excluding the last two quarters) as the training set.

b. A partial output of the regression model is shown in Figure 17.16. Use this output to answer the following questions:

i. Which two statistics (and their values) measure how well this model fits the training data?

ii. Which two statistics (and their values) measure the predictive accuracy of this model?

iii. After adjusting for trend, what is the average difference between sales in Q3 and sales in Q1?

iv. After adjusting for seasonality, which quarter (Q_1, Q_2, Q_3, or Q_4) has the highest average sales?

Regression Model

Input variables	Coefficient	Std. Error	p-value	SS
Constant term	906.749939	115.3461227	0.00002541	41669100
Trend	47.1071434	11.25662899	0.00235907	825673.5625
Quarter_2	−15.10719299	119.6596069	0.9023084	1335472
Quarter_3	89.16661835	128.6739807	0.50581801	1357627
Quarter_4	2101.726074	129.1654205	0.00000001	7514922

Residual df	9
Multiple R-squared	0.977372001
Std. dev. estimate	168.4737854
Residual SS	255450.7656

Training Data Scoring–Summary Report

Total sum of squared errors	RMS error	Average error
255450.7619	135.0795432	0.000106558

Validation Data Scoring–Summary Report

Total sum of squared errors	RMS error	Average error
196792.9676	313.6821382	183.1429921

OUTPUT FOR REGRESSION MODEL FITTED TO TOYS "R" US TIME SERIES

17.4   **Walmart Stock.** Figure 17.17 shows the series of Walmart daily closing prices between February 2001 and February 2002 (thanks to Chris Albright for suggesting the use of these data, which are publicly available, e.g., at http://finance.yahoo.comand are in the file WalMartStock.xls).

FIGURE 17.17   DAILY CLOSE PRICE OF WALMART STOCK, FEB 2001–2002

a. Fit an AR(1) model to the close price series. Report the coefficient table.

b. Which of the following is/are relevant for testing whether this stock is a random walk?

- The autocorrelations of the close prices series
- The AR(1) slope coefficient
- The AR(1) constant coefficient

c. Does the AR model indicate that this is a random walk? Explain how you reached your conclusion.

d. What are the implications of finding that a time series is a random walk? Choose the correct statement(s) below.

- It is impossible to obtainforecasts that are more accurate than naive forecasts forthe series.
- The series is random.
- The changes in the series from one period to the next are random.

17.5 **Department Store Sales.** The time plot in Figure 17.18 describes actual quarterly sales for a department store over a 6-year period (data are available in `DepartmentStore-Sales.xls`, data courtesy of Chris Albright).

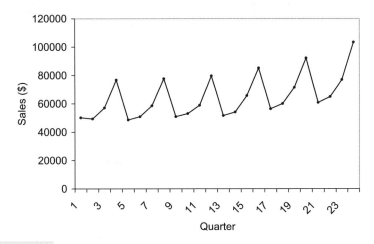

FIGURE 17.18   DEPARTMENT STORE QUARTERLY SALES SERIES

   a. The forecaster decided that there is an exponential trend in the series. In order to fit a regression-based model that accounts for this trend, which of the following operations must be performed?

- Take log of quarter index.
- Take log of sales.
- Take an exponent of sales.
- Take an exponent of quarter index.

   b. Fit a regression model with an exponential trend and seasonality, using the first 20 quarters as the training data (remember to first partition the series into training and validation series).

   c. A partial output is shown in Figure 17.19. From the output, after adjusting for trend, are Q2 average sales higher, lower, or approximately equal to the average Q1 sales?

Regression Model

Input variables	Coefficient	Std. error	p-Value	SS
Constant term	10.74894524	0.01872449	0	2429.415771
Quarter	0.01108785	0.0012952	0.00000033	0.18121047
Qtr_2	0.02495589	0.02076364	0.24803306	0.11009274
Qtr_3	0.165343	0.02088447	0.00000094	0.00970232
Qtr_4	0.43374524	0.02108433	0	0.45436361

Residual df	15
Multiple R-squared	0.979125117
Std. dev. estimate	0.03276626
Residual SS	0.01610442

FIGURE 17.19   OUTPUT FROM REGRESSION MODEL FIT TO DEPARTMENT STORE SALES TRAINING SERIES

   d. Use this model to forecast sales in quarters 21 and 22.

e. The plots in Figure 17.20 describe the fit (top) and forecast errors (bottom) from this regression model.

    i. Recreate these plots.

    ii. Based on these plots, what can you say about your forecasts for quarters 21 and 22? Are they likely to over-forecast, under-forecast or be reasonably close to the real sales values?

FIGURE 17.20   FIT OF REGRESSION MODEL FOR DEPARTMENT STORE SALES

f. From the forecast errors plot, which of the following statements appear true?

- Seasonality is not captured well.
- The regression model fits the data well.
- The trend in the data is not captured well by the model.

g. Which of the following solutions is adequate *and* a parsimonious solution for improving model fit?

- Fit a quadratic trend model to the residuals (with Quarter and Quarter^2).
- Fit an AR model to the residuals.
- Fit a quadratic trend model to Sales (with Quarter and Quarter^2).

17.6   **Souvenir Sales.** Figure 17.21 shows a time plot of monthly sales for a souvenir shop at a beach resort town in Queensland, Australia, between 1995–2001 (data are available in `SouvenirSales.xls`, source: Hyndman, R.J. Time Series Data Library, http://data.is/TSDLdemo, accessed on 07/25/15). The series is presented twice, in Australian dollars and in log-scale. Back in 2001, the store wanted to use the data to forecast sales for the next 12 months (year 2002). They hired an analyst to generate forecasts. The analyst first partitioned the data into training and validation sets, with the validation set containing the last 12 months of data (year 2001). She then fit a regression model to sales, using the training set.

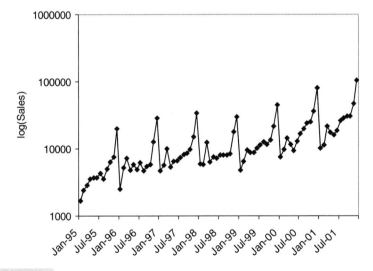

FIGURE 17.21   MONTHLY SALES AT AUSTRALIAN SOUVENIR SHOP IN DOLLARS (TOP) AND IN LOG-SCALE (BOTTOM)

a. Based on the two time plots, which predictors should be included in the regression model? What is the total number of predictors in the model?

b. Run a regression model with Sales (in Australian dollars) as the output variable, and with a linear trend and monthly predictors. Remember to fit only the training data. Call this model A.

    i. Examine the estimated coefficients. Which month tends to have the highest average sales during the year? Why is this reasonable?

    ii. The estimated trend coefficient is 245.36. What does this mean?

c. Run a regression model with log(Sales) as the output variable, and with a linear trend and monthly predictors. Remember to fit only the training data. Call this model B.

    i. Fitting a model to log(Sales) with a linear trend is equivalent to fitting a model to Sales (in dollars) with what type of trend?

    ii. The estimated trend coefficient is 0.02. What does this mean?

    iii. Use this model to forecast the sales in February 2002. What is the extra step needed?

d. Compare the two regression models (A and B) in terms of forecast performance. Which model is preferable for forecasting? Mention at least two reasons based on the information in the outputs.

e. Continuing with model B (with log(Sales) as output), create an ACF plot until lag 15 for the forecast errors. Now fit an AR model with lag 2 [ARIMA(2, 0, 0)] to the forecast errors.

    i. Examining the ACF plot and the estimated coefficients of the AR(2) model (and their statistical significance), what can we learn about the forecasts that result from model B?

    ii. Use the autocorrelation information to compute an improved forecast for January 2002, using model B and the AR(2) model above.

f. How would you model these data differently if the goal was to understand the different components of sales in the souvenir shop between 1995–2001? Mention two differences.

17.7 **Shipments of Household Appliances.** The time plot in Figure 17.22 shows the series of quarterly shipments (in million dollars) of US household appliances between 1985–1989 (data are available in `ApplianceShipments.xls`, data courtesy of Ken Black).

    If we compute the autocorrelation of the series, which lag (>0) is most likely to have the largest coefficient (in absolute value)? Create an ACF plot and compare with your answer.

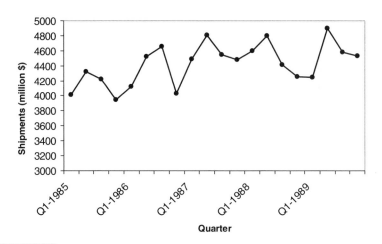

FIGURE 17.22   QUARTERLY SHIPMENTS OF US HOUSEHOLD APPLIANCES OVER 5 YEARS

17.8   Australian Wine Sales. Figure 17.23 shows time plots of monthly sales of six types of Australian wines (red, rose, sweet white, dry white, sparkling, and fortified) for 1980–1994 (data are available in `AustralianWines.xls`, source: Hyndman, R.J. Time Series Data Library, http://data.is/TSDLdemo, accessed on 07/25/15). The units are thousands of liters. You are hired to obtain short term forecasts (2 to 3 months ahead) for each of the six series, and this task will be repeated every month.

a. Which forecasting method would you choose if you had to choose the same method for all series? Why?

b. Fortified wine has the largest market share of the above six types of wine. You are asked to focus on fortified wine sales alone, and produce as accurate as possible forecasts for the next 2 months.

- Start by partitioning the data using the period until December 1993 as the training set.

- Fit a regression model to sales with a linear trend and seasonality.

  i. Create the "actual vs. forecast" plot. What can you say about the model fit?

  ii. Use the regression model to forecast sales in January and February 1994.

c. Create an ACF plot for the residuals from the model above until lag 12. Examining this plot, which of the following statements are reasonable?

- Decembers (month 12) are not captured well by the model.

- There is a strong correlation between sales on the same calendar month.

- The model does not capture the seasonality well.

- We should try to fit an autoregressive model with lag 12 to the residuals.

FIGURE 17.23   MONTHLY SALES OF SIX TYPES OF AUSTRALIAN WINES BETWEEN 1980–1994

## CHAPTER 18

Smoothing Methods

In this chapter we describe a set of popular and flexible methods for forecasting time series that rely on *smoothing*. Smoothing is based on averaging over multiple periods in order to reduce the noise. We start with two simple smoothers, the *moving average* and *simple exponential smoother*, which are suitable for forecasting series that contain no trend or seasonality. In both cases forecasts are averages of previous values of the series (the length of the series history that is considered and the weights that are used in the averaging differ between the methods). We also show how a moving average can be used, with a slight adaptation, for data visualization. We then proceed to describe smoothing methods that are suitable for forecasting series with a trend and/or seasonality. Smoothing methods are data-driven, and are able to adapt to changes in the series over time. Although highly automated, the user must specify *smoothing constants*, which determine how fast the method adapts to new data. We discuss the choice of such constants, and their meaning. The different methods are illustrated using the Amtrak ridership series.

18.1 INTRODUCTION[1]

A second class of methods for time series forecasting is *smoothing methods*. Unlike regression models, which rely on an underlying theoretical model for the components of a time series (e.g., linear model or multiplicative seasonality),

---

[1]This and subsequent sections in this chapter © 2016 Statistics.com and Galit Shmueli. Used by permission.

Data Mining for Business Analytics: Concepts, Techniques, and Applications with XLMiner®, Third Edition.
Galit Shmueli, Peter C. Bruce, and Nitin R. Patel.
© 2016 John Wiley & Sons Inc. Published 2016 by John Wiley & Sons, Inc.

smoothing methods are data-driven, in the sense that they estimate time series components directly from the data without a predetermined structure. Data-driven methods are especially useful in series where patterns change over time. Smoothing methods "smooth" out the noise in a series in an attempt to uncover the patterns. Smoothing is done by averaging the series over multiple periods, where different smoothers differ by the number of periods averaged, how the average is computed, how many times averaging is performed, and so on. We now describe two types of smoothing methods that are popular in business applications due to their simplicity and adaptability. These are the moving average method and exponential smoothing.

18.2 MOVING AVERAGE

The moving average is a simple smoother: it consists of averaging across a window of consecutive observations, thereby generating a series of averages. A moving average with window width w means averaging across each set of w consecutive values, where w is determined by the user.

In general, there are two types of moving averages: a *centered moving average* and a *trailing moving average*. Centered moving averages are powerful for visualizing trends because the averaging operation can suppress seasonality and noise, thereby making the trend more visible. In contrast, trailing moving averages are useful for forecasting. The difference between the two is in terms of the window's location on the time series.

Centered Moving Average for Visualization

In a centered moving average, the value of the moving average at time t (MA_t) is computed by centering the window around time t and averaging across the w values within the window:

$$MA_t = \left(Y_{t-(w-1)/2} + \cdots + Y_{t-1} + Y_t + Y_{t+1} + \cdots + Y_{t+(w-1)/2} \right) / w.$$

For example, with a window of width $w = 5$, the moving average at time point $t = 3$ means averaging the values of the series at time points 1,2,3,4,5; at time point $t = 4$ the moving average is the average of the series at time points 2,3,4,5,6, and so on.[2] This is illustrated in the top panel of Figure 18.1.

Choosing the window width in a seasonal series is straightforward: because the goal is to suppress seasonality for better visualizing the trend, the default choice should be the length of a seasonal cycle. Returning to the Amtrak ridership data, the annual seasonality indicates a choice of $w = 12$. Figure 18.2 shows a centered moving average line overlaid on the original series. We can

---

[2] For an even window width, for example $w = 4$, obtaining the moving average at time point $t = 3$ requires averaging across two windows: across time points $1, 2, 3, 4$; across time points $2, 3, 4, 5$; and finally the average of the two averages is the final moving average.

Centered window (w = 5)

t − 2   t − 1   t   t + 1   t + 2

Trailing window (w = 5)

t − 4   t − 3   t − 2   t − 1   t

FIGURE 18.1 SCHEMATIC OF CENTERED MOVING AVERAGE (TOP) AND TRAILING MOVING AVERAGE (BOTTOM), BOTH WITH WINDOW WIDTH W = 5

FIGURE 18.2 CENTERED MOVING AVERAGE WITH WINDOW W = 12, OVERLAID ON AMTRAK RIDERSHIP SERIES. THIS HELPS VISUALIZE TRENDS

see a global U-shape, but unlike the regression model that fits a strict U-shape, the moving average shows some deviation, such as the slight dip during the last year. This plot is obtained by using Excel's *Trendline* (as explained in Chapter 16), and c hoosing the *Moving Average* option, where you can set the length of the window (see Figure 18.3).

Trailing Moving Average for Forecasting

Centered moving averages are computed by averaging across data in the past and the future of a given time point. In that sense, they cannot be used for forecasting because at the time of forecasting, the future is typically unknown. Hence, for purposes of forecasting, we use *trailing moving averages*, where the window of width w is set on the most recent available w values of the series. The k-step-ahead forecast F_{t+k} ($k = 1,2,3,\ldots$) is then the average of these w values (see also bottom plot in Figure 18.1):

$$F_{t+k} = \left(Y_t + Y_{t-1} + \ldots + Y_{t-w+1} \right) / w.$$

FIGURE 18.3   EXCEL'S *TRENDLINE* MENU.

For example, in the Amtrak ridership series, to forecast ridership in February 1992 or later months, given information until January 1992 and using a moving average with window width $w = 12$, we would take the average ridership during the most recent 12 months (February 1991 to January 1992).

Computing a trailing moving average can be done via XLMiner's *Moving Average* menu (within *Time Series > Smoothing*). This will yield forecasts and forecast errors for the training set, and if checked, for the validation set as well. The default window width (called *Interval* in the *Weights* box) is $w = 2$, which should be modified by the user.

Using XLMiner, we illustrate a 12-month moving average forecaster for the Amtrak ridership. We partitioned once again the Amtrak ridership series, leaving the last 12 months as the validation set. Applying a moving average forecaster with window $w = 12$, we obtained the output partially shown in Figure 18.4. Note that for the first 11 records of the training set, there is no forecast because there are less than 12 past values to average. Also note that the forecasts for all months in the validation set are identical (1942.73), which is because the method assumes that information is known only until March 2003. In this example, it is clear that the moving average forecaster is inadequate for generating monthly forecasts because it does not capture the seasonality in the data. Hence seasons with high ridership are under-forecasted, and seasons with low ridership are over-forecasted. A similar issue arises when forecasting a series with a trend: the moving average "lags behind," thereby under-forecasting in the presence of an increasing trend, and over-forecasting in the presence of a decreasing trend.

In general, the moving average should be used for forecasting *only in series that lack seasonality and trend*. Such a limitation might seem impractical. However, there are a few popular methods for removing trends (de-trending) and removing seasonality (de-seasonalizing) from a series, such as regression models. The moving average can then be used to forecast such de-trended and de-seasonalized series, and then the trend and seasonality can be added back to the forecast. For example, consider the regression model shown in Figure 17.7 in Chapter 17, which yields residuals devoid of seasonality and trend (bottom plot). We can apply a moving average forecaster to that series of residuals (also called forecast errors), thereby creating a forecast for the next *forecast error*. For example, to forecast ridership in April 2003 (the first period in the validation set), assuming that we have information until March 2003, we use the regression model in Figure 17.7 to generate a forecast for April 2003 (which yields 2115 thousand riders). We then use a 12-month moving average (using the period April 2002 to March 2003) to forecast the *forecast error* for April 2003, which yields −66.33 (manually, or using XLMiner, as shown in Figure 18.5). The negative value implies that the regression model's forecast for April 2003 is too high, and therefore we should adjust it by reducing approximately 66 thousand riders from the regression model's forecast of 2,115,000 riders.

Choosing Window Width (w)

With moving average forecasting or visualization, the only choice that the user must make is the width of the window (w). As with other methods such as k-nearest-neighbors, the choice of the smoothing parameter is a balance between

XLMiner : Time Series — Moving Average Smoothing

Output Navigator

| Inputs | Train. Error Measures | Valid. Error Measu | Fitted Md | Forecast |

Inputs

Data

Workbook	Amtrak-Smoothing.xlsx
Worksheet	Data_PartitionTS
Range	$BS20:$CS179
Selected Variable	Ridership
#Records in Training Data	147
#Records in Validation Data	12

Parameters/Options

Interval	12
Forecast	Yes
#Forecasts	12

Training Error Measures

Mean Absolute Percentage Error (MAPE)	6.85518801
Mean Absolute Deviation (MAD)	119.866987
Mean Square Error (MSE)	21069.3208
Tracking Signal Error (TSE)	5.8559945
Cumulative Forecast Error (CFE)	701.940417
Mean Forecast Error (MFE)	5.19955864

Validation Error Measures

Mean Absolute Percentage Error (MAPE)	7.71191897
Mean Absolute Deviation (MAD)	162.040708
Mean Square Error (MSE)	30531.4915
Tracking Signal Error (TSE)	9.19460311
Cumulative Forecast Error (CFE)	1489.9
Mean Forecast Error (MFE)	124.158333

Fitted Model

Month	Actual	F'recast	Residuals
Jan-91	1708.92	•	
Feb-91	1620.59	•	
Mar-91	1972.72	•	
Apr-91	1811.67	•	
May-91	1974.96	•	
Jun-91	1862.36	•	
Jul-91	1939.86	•	
Aug-91	2013.26	•	
Sep-91	1595.66	•	
Oct-91	1724.92	•	
Nov-91	1675.67	•	
Dec-91	1813.86	•	
Jan-92	1614.83	1809.537	-194.7095
Feb-92	1557.09	1801.696	-244.6077
Mar-92	1891.02	1796.404	94.818833
Apr-92	1955.98	1789.613	166.36783

Time Plot of Actual Vs. Forecast (Training Data)

Forecast

Month	Actual	Forecast	Error	LCI	UCI
Apr-03	2098.9	1942.74	156.161	1659.2	2226.27
May-03	2104.91	1942.74	162.173	1659.2	2226.28
Jun-03	2129.67	1942.74	186.933	1659.19	2226.29
Jul-03	2223.35	1942.74	280.611	1659.18	2226.29
Aug-03	2174.36	1942.74	231.622	1659.18	2226.3
Sep-03	1931.41	1942.74	-11.3321	1659.17	2226.31
Oct-03	2121.47	1942.74	178.732	1659.16	2226.31
Nov-03	2076.05	1942.74	133.316	1659.16	2226.32
Dec-03	2140.68	1942.74	197.939	1659.15	2226.33
Jan-04	1831.51	1942.74	-111.23	1659.14	2226.33
Feb-04	1838.01	1942.74	-104.732	1659.14	2226.34
Mar-04	2132.45	1942.74	189.708	1659.13	2226.35

Time Plot of Actual vs. Forecast (Validation)

FIGURE 18.4 PARTIAL OUTPUT FOR MOVING AVERAGE FORECASTER WITH $W = 12$ APPLIED TO AMTRAK RIDERSHIP SERIES.

XLMiner : Time Series—Moving Average Smoothing

Output Navigator			
Inputs	Train. Error Measure	Fitted Model	Forecast

Inputs

Data	
Workbook	Amtrak-Smoothing.xlsx
Worksheet	RegResiduals
Range	$A$1:$C$148
Selected Variable	Regression Residual
# Records in Input Data	147

Parameters/Options	
Interval	12
Forecast	Yes
#Forecasts	1

Training Error Measures

Mean Absolute Percentage Error (MAPE)	212.094
Mean Absolute Deviation (MAD)	51.4816
Mean Square Error (MSE)	4098.27
Tracking Signal Error (TSE)	-2.16418
Cumulative Forecast Error (CFE)	-111.415
Mean Forecast Error (MFE)	-0.8253

Forecast

Month	Forecast	LCI	UCI
Apr-03	-66.3322	-191.3953611	58.731

Fitted Model

Month	Actual	Forecast	Residuals
Apr-02	19.604	40.2643	-20.6603
May-02	2.9528	36.1846	-33.2318
Jun-02	-35.9678	31.7798	-67.7476
Jul-02	-65.2079	18.9207	-84.1286
Aug-02	-164.728	8.23974	-172.968
Sep-02	-129.881	-12.9137	-116.968
Oct-02	-92.5494	-17.2763	-75.2731
Nov-02	-146.914	-26.7603	-120.153
Dec-02	-58.8259	-43.1088	-15.7171
Jan-03	-46.7583	-48.9705	2.21226
Feb-03	-43.9256	-54.3547	10.4291
Mar-03	-33.7857	-63.1189	29.3331

FIGURE 18.5   APPLYING MOVING AVERAGE TO THE RESIDUALS FROM THE REGRESSION MODEL (WHICH LACK TREND AND SEASONALITY), TO FORECAST THE APRIL 2003 RESIDUAL.

under-smoothing and over-smoothing. For visualization (using a centered window), wider windows will expose more global trends, while narrow windows will reveal local trends. Hence examining several window widths is useful for exploring trends of differing local/global nature. For forecasting (using a trailing window), the choice should incorporate domain knowledge in terms of relevance of past observations and how fast the series changes. Empirical predictive evaluation can also be done by experimenting with different values of w and comparing performance. However, care should be taken not to overfit!

18.3 SIMPLE EXPONENTIAL SMOOTHING

A popular forecasting method in business is exponential smoothing. Its popularity derives from its flexibility, ease of automation, cheap computation, and good performance. Simple exponential smoothing is similar to forecasting with a moving average, except that instead of taking a simple average over the w most recent values, we take a *weighted average* of *all* past values, such that the weights decrease exponentially into the past. The idea is to give more weight to recent information, yet not to completely ignore older information.

Like the moving average, simple exponential smoothing should only be used for forecasting *series that have no trend or seasonality*. As mentioned earlier, such series can be obtained by removing trend and/or seasonality from raw series, and then applying exponential smoothing to the series of residuals (which are assumed to contain no trend or seasonality).

The exponential smoother generates a forecast at time $t + 1$ (F_{t+1}) as follows:

$$F_{t+1} = \alpha Y_t + \alpha(1 - \alpha)Y_{t-1} + \alpha(1 - \alpha)^2 Y_{t-2} + \dots, \tag{18.1}$$

where α is a constant between 0 and 1 called the *smoothing parameter*. This formulation displays the exponential smoother as a weighted average of all past observations, with exponentially decaying weights.

It turns out that we can write the exponential forecaster in another way, which is very useful in practice:

$$F_{t+1} = F_t + \alpha E_t, \tag{18.2}$$

where E_t is the forecast error at time t. This formulation presents the exponential forecaster as an "active learner": It looks at the previous forecast (F_t) and how far it was from the actual value (E_t), and then corrects the next forecast based on that information. If in one period the forecast was too high, the next period is adjusted down. The amount of correction depends on the value of the smoothing parameter α. The formulation in (18.2) is also advantageous in terms of data storage and computation time: it means that we need to store and use only the forecast and forecast error from the most recent period, rather than the entire series. In applications where real-time forecasting is done, or many series are being forecasted in parallel and continuously, such savings are critical.

Note that forecasting further into the future yields the same forecast as a one-step-ahead forecast. Because the series is assumed to lack trend and seasonality, forecasts into the future rely only on information until the time of prediction. Hence the k-step-ahead forecast is equal to

$$F_{t+k} = F_{t+1}.$$

Choosing Smoothing Parameter α

The smoothing parameter α, which is set by the user, determines the rate of learning. A value close to 1 indicates fast learning (i.e., only the most recent observations have influence on forecasts), whereas a value close to 0 indicates slow learning (past observations have a large influence on forecasts). This can be seen by plugging 0 or 1 into equation (18.1) or (18.2). Hence the choice of α depends on the required amount of smoothing, and on how relevant the history is for generating forecasts. Default values that have been shown to work well are around 0.1 and 0.2. Some trial and error can also help in the choice of α: examine the time plot of the actual and predicted series, as well as the predictive accuracy (e.g., MAPE or RMSE of the validation set). Finding the α value that optimizes predictive accuracy on the validation set can be used to determine the degree of local vs. global nature of the trend (e.g., by using "optimize" in XLMiner's "Weights" box). However, beware of choosing the "best α" for forecasting purposes, as this will most likely lead to model overfitting and low predictive accuracy on future data.

In XLMiner, forecasting using simple exponential smoothing is done via the Exponential menu (within *Time Series > Smoothing*). This will yield forecasts and forecast errors for both the training and validation sets. You can use the default value of $\alpha = 0.2$, set it to another value, or choose to find the optimal α in terms of minimizing RMSE of the validation data.

To illustrate forecasting with simple exponential smoothing, we return to the residuals from the regression model, which are assumed to contain no trend or seasonality. To forecast the residual on April 2003, we apply exponential smoothing to the entire period until March 2003, and use the default $\alpha = 0.2$ value. The output is shown in Figure 18.6. We see that the forecast for the residual is -58.220 (in thousands of riders), implying that we should adjust the regression's forecast by reducing 58,220 riders from that forecast.

Relation between Moving Average and Simple Exponential Smoothing

In both smoothing methods the user must specify a single parameter: in moving averages, the window width (w) must be set; in exponential smoothing, the smoothing parameter (α) must be set. In both cases, the parameter determines the importance of fresh information over older information. In fact, the two

XLMiner : Time Series—Exponential Smoothing

Output Navigator			
Inputs	Train. Error Measure	Fitted Model	Forecast

Inputs

Data	
Workbook	Amtrak-Smoothing.xlsx
Worksheet	RegResiduals
Range	$A$1:$C$148
Selected Variable	Regression Residual
#Records in Input Data	147

Parameters/Options	
Optimization Selected	No
Alpha (Level)	0.2
Forecast	Yes
#Forecasts	1

Training Error Measures

Mean Absolute Percentage Error (MAPE)	201.519
Mean Absolute Deviation (MAD)	46.8629
Mean Square Error (MSE)	3565.13
Tracking Signal Error (TSE)	-11.3934
Cumulative Forecast Error (CFE)	-533.929
Mean Forecast Error (MFE)	-3.65705

Forecast

Month	Forecast	LCI	UCI
Apr-03	-58.2203	-174.8485142	58.4079

Fitted Model

Month	Actual	Forecast	Residuals
Jan-91	48.5654	*	*
Feb-91	4.21307	48.5654	-44.3524
Mar-91	62.1068	39.695	22.4118
Apr-91	-101.048	44.1773	-145.225
May-91	36.7867	15.1323	21.6544
Jun-91	-28.6941	19.4632	-48.1573
Jul-91	-49.8695	9.83173	-59.7012
Aug-91	-15.7357	-2.10851	-13.6271
Sep-91	-94.5393	-4.83394	-89.7053
Oct-91	-99.0047	-22.775	-76.2297
Nov-91	-132.648	-38.021	-94.6272
Dec-91	-32.8197	-56.9464	24.1267
Jan-92	10.0826	-52.1211	62.2037
Feb-92	-4.72795	-39.6803	34.9524
Mar-92	34.1216	-32.6899	66.8115
Apr-92	95.7249	-19.3276	115.052

FIGURE 18.6   PARTIAL OUTPUT FOR SIMPLE EXPONENTIAL SMOOTHING FORECASTER WITH $\alpha = 0.2$, APPLIED TO THE SERIES OF RESIDUALS FROM THE REGRESSION MODEL (WHICH LACK TREND AND SEASONALITY). THE FORECAST FOR APRIL 2003 RESIDUAL IS AT THE BOTTOM.

smoothers are approximately equal if the window width of the moving average is equal to $w = 2/\alpha - 1$.

18.4 ADVANCED EXPONENTIAL SMOOTHING

As mentioned earlier, both the moving average and simple exponential smoothing should only be used for forecasting series with no trend or seasonality, that is, series that have only a level and noise. One solution for forecasting series with trend and/or seasonality is first to remove those components (e.g., via regression models). Another solution is to use a more sophisticated version of exponential smoothing, which can capture trend and/or seasonality.

Series with a Trend

For series that contain a trend, we can use "double exponential smoothing." Unlike in regression models, the trend shape is not assumed to be global, but rather, it can change over time. In double exponential smoothing, the local trend is estimated from the data and is updated as more data arrive. Similar to simple exponential smoothing, the level of the series is also estimated from the data and is updated as more data arrive. The k-step-ahead forecast is given by combining the level estimate at time t (L_t) and the trend estimate at time t (T_t):

$$F_{t+k} = L_t + kT_t \qquad (18.3)$$

Note that in the presence of a trend, one-, two-, three-step-ahead (etc.) forecasts are no longer identical. The level and trend are updated through a pair of updating equations:

$$L_t = \alpha Y_t + (1 - \alpha)(L_{t-1} + T_{t-1}), \qquad (18.4)$$

$$T_t = \beta \left(L_t - L_{t-1}\right) + (1 - \beta)T_{t-1}. \qquad (18.5)$$

The first equation means that the level at time t is a weighted average of the actual value at time t and the level in the previous period, adjusted for trend (in the presence of a trend, moving from one period to the next requires factoring in the trend). The second equation means that the trend at time t is a weighted average of the trend in the previous period and the more recent information on the change in level.[3] Here there are two smoothing parameters, α and β, that determine the rate of learning. As in simple exponential smoothing, they are both constants in the range [0, 1], set by the user, with higher values leading to faster learning (more weight to most recent information).

Series with a Trend and Seasonality

For series that contain both trend and seasonality, the "Holt–Winters exponential smoothing" method can be used. This is a further extension of double exponential smoothing, where the k-step-ahead forecast also takes into account the seasonality at period $t + k$. Assuming seasonality with M seasons (e.g., for weekly seasonality $M = 7$), the forecast is given by

$$F_{t+k} = \left(L_t + kT_t\right) S_{t+k-M}. \qquad (18.6)$$

---

[3] There are various ways to estimate the initial values L_1 and T_1, but the differences among these ways usually disappear after a few periods.

(Note that by the time of forecasting t, the series must have included at least one full cycle of seasons in order to produce forecasts using this formula, i.e., $t > M$.) Being an adaptive method, Holt–Winters exponential smoothing allows the level, trend, and seasonality patterns to change over time. These three components are estimated and updated as more information arrives. The three updating equations are given by

$$L_t = \alpha Y_t / S_{t-M} + (1 - \alpha)(L_{t-1} + T_{t-1}), \tag{18.7}$$

$$T_t = \beta\left(L_t - L_{t-1}\right) + (1 - \beta)T_{t-1}, \tag{18.8}$$

$$S_t = \gamma Y_t / L_t + (1 - \gamma)S_{t-M}. \tag{18.9}$$

The first equation is similar to that in double exponential smoothing, except that it uses the seasonally adjusted value at time t rather than the raw value. This is done by dividing Y_t by its seasonal index, as estimated in the last cycle. The second equation is identical to double exponential smoothing. The third equation means that the seasonal index is updated by taking a weighted average of the seasonal index from the previous cycle and the current trend-adjusted value. Note that this formulation describes a multiplicative seasonal relationship, where values on different seasons differ by percentage amounts. There is also an additive seasonality version of Holt–Winters exponential smoothing, where seasons differ by a constant amount (also available in XLMiner; see more in Shmueli, 2012).

To illustrate forecasting a series with the Holt–Winters method, we again consider the raw Amtrak ridership data. As we observed earlier, the data contain both a trend and monthly seasonality. Figure 18.7 shows part of XLMiner's output. We see the values of the three smoothing parameters (left at their defaults), and the chosen 12-month cycle.

Series with Seasonality (No Trend)

Finally, for series that contain seasonality but no trend, we can use a Holt–Winters exponential smoothing formulation that lacks a trend term, by deleting the trend term in the forecasting equation and updating equations (in XLMiner this is called Holt–Winters No Trend).

Inputs

Data

Data	
Workbook	Amtrak-Smoothin
Worksheet	Data_PartitionTS
Range	$B$20:$C$179
Selected Variable	Ridership
# Records in Training Data	147
# Records in Validation Data	12

Parameters/Options

Parameters/Options	
Optimize Weights	No
Alpha (Level)	0.2
Beta (Trend)	0.15
Gamma (Seasonality)	0.05
Season length	12
Number of seasons	12
Forecast	Yes
#Forecasts	12

Fitted Model

Month	Actual	Forecast	Residuals
Jan-91	1708.92	1656.21	52.7023
Feb-91	1620.59	1612.22	8.36838
Mar-91	1972.72	1928.36	44.352
Apr-91	1811.67	1929.43	-117.762
May-91	1974.96	1918.03	56.9986
Jun-91	1862.36	1813.33	49.0289
Jul-91	1939.86	1953.9	-14.0384
Aug-91	2013.26	2022.6	-9.33473
Sep-91	1595.66	1702.71	-107.055
Oct-91	1724.92	1783.25	-58.3214
Nov-91	1675.67	1760.59	-84.9196
Dec-91	1813.86	1780.8	33.0589
Jan-92	1614.83	1627.49	-12.659
Feb-92	1557.09	1564.27	-7.18351
Mar-92	1891.22	1861.15	30.076
Apr-92	1955.98	1845.51	110.471

Training Error Measures

Mean Absolute Percentage Error (MAPE)	3.2051
Mean Absolute Deviation (MAD)	56.4561
Mean Square Error (MSE)	5016.4146
Tracking Signal Error (TSE)	2.9756
Cumulative Forecast Error (CFE)	167.9896
Mean Forecast Error (MFE)	1.1428

Validation Error Measures

Mean Absolute Percentage Error (MAPE)	5.3831
Mean Absolute Deviation (MAD)	111.0447
Mean Square Error (MSE)	14985.9661
Tracking Signal Error (TSE)	12.0000
Cumulative Forecast Error (CFE)	1332.5364
Mean Forecast Error (MFE)	111.0447

Forecast

Month	Actual	Forecast	Error	LCI	UCI
Apr-03	2098.9	2036.57	62.3276	1896.72	2176.43
May-03	2104.91	2061.73	43.1785	1925.5	2197.97
Jun-03	2129.67	1971.37	158.297	1808.57	2134.18
Jul-03	2223.35	2092.87	130.478	1930.1	2255.64
Aug-03	2174.36	2152.57	21.7908	1987.79	2317.35
Sep-03	1931.41	1804.36	127.044	1646.8	1961.92
Oct-03	2121.47	1933.85	187.618	1766.58	2101.13
Nov-03	2076.05	1922.26	153.79	1750.22	2094.31
Dec-03	2140.68	1968.26	172.412	1824.05	2112.48
Jan-04	1831.51	1760.98	70.5324	1606.41	1915.54
Feb-04	1838.01	1715.4	122.607	1561.76	1869.04
Mar-04	2132.45	2049.99	82.4601	1892.67	2207.3

FIGURE 18.7 PARTIAL OUTPUT FOR HOLT–WINTERS EXPONENTIAL SMOOTHING APPLIED TO AMTRAK RIDERSHIP SERIES.

PROBLEMS

18.1 **Impact of September 11 on Air Travel in the United States.** The Research and Innovative Technology Administration's Bureau of Transportation Statistics conducted a study to evaluate the impact of the September 11, 2001, terrorist attack on US transportation. The 2006 study report and the data can be found at http://goo.gl/w2lJPV. The goal of the study was stated as follows:

> The purpose of this study is to provide a greater understanding of the passenger travel behavior patterns of persons making long distance trips before and after 9/11.

The report analyzes monthly passenger movement data between January 1990 and May 2004. Data on three monthly time series are given in file `Sept11Travel.xls` for this period: (1) actual airline revenue passenger miles (Air), (2) rail passenger miles (Rail), and (3) vehicle miles traveled (Car).

In order to assess the impact of September 11, BTS took the following approach: using data before September 11, they forecasted future data (under the assumption of no terrorist attack). Then they compared the forecasted series with the actual data to assess the impact of the event. Our first step therefore is to split each of the time series into two parts: pre– and post–September 11. We now concentrate only on the earlier time series.

a. Create a time plot for the pre-event AIR time series. What time series components appear from the plot?

b. Figure 18.8 shows a time plot of the **seasonally adjusted** pre September 11 AIR series. Which of the following smoothing methods would be adequate for forecasting this series?

- Moving average (with what window width?)
- Simple exponential smoothing
- Holt exponential smoothing
- Holt–Winters exponential smoothing

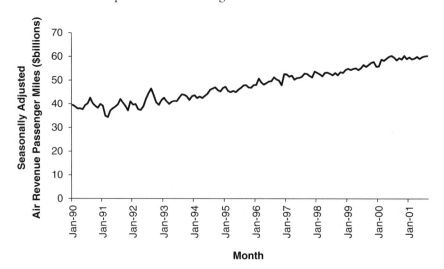

FIGURE 18.8 SEASONALLY ADJUSTED PRE–SEPTEMBER-11 AIR SERIES.

18.2   Relation between Moving Average and Exponential Smoothing. Assume that we apply a moving average to a series, using a very short window span. If we wanted to achieve an equivalent result using simple exponential smoothing, what value should the smoothing coefficient take?

18.3   Forecasting with a Moving Average. For a given time series of sales, the training set consists of 50 months. The first five months' data are shown in Table 18.1:

TABLE 18.1

Month	Sales
Sept 98	27
Oct 98	31
Nov 98	58
Dec 98	63
Jan 99	59

a. Compute the sales forecast for January 1999 based on a moving average with $w = 4$.

b. Compute the forecast error for the above forecast.

18.4   Optimizing Holt–Winters Exponential Smoothing. The output in Figure 18.9 from applying Holt–Winters exponential smoothing to data, using "optimal" smoothing constants.

Winters' exponential smoothing	
Smoothing constant(s)	
Level	1.000
Trend	0.000
Seasonality	0.246

FIGURE 18.9   OPTIMIZED SMOOTHING CONSTANTS.

a. The value of zero that is obtained for the trend smoothing constant means that (choose one of the following):

- There is no trend.
- The trend is estimated only from the first two periods.
- The trend is updated throughout the data.
- The trend is statistically insignificant.

b. What is the danger of using the optimal smoothing constant values?

18.5 **Department Store Sales.** The time plot in Figure 18.10 describes actual quarterly sales for a department store over a 6-year period (data are available in `DepartmentStore-Sales.xls`, data courtesy of Chris Albright).

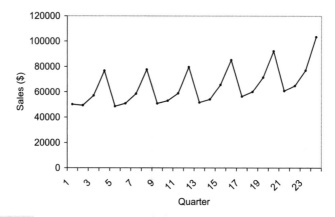

FIGURE 18.10   DEPARTMENT STORE QUARTERLY SALES SERIES.

a. Which of the following methods would **not** be suitable for forecasting this series?
- Moving average of raw series
- Moving average of deseasonalized series
- Simple exponential smoothing of the raw series
- Double exponential smoothing of the raw series
- Holt–Winters exponential smoothing of the raw series
- Regression model fit to the raw series
- Random walk model fit to the raw series

b. The forecaster was tasked to generate forecasts for 4 quarters ahead. He therefore partitioned the data such that the last 4 quarters were designated as the validation period. The forecaster approached the forecasting task by using multiplicative Holt–Winters exponential smoothing. The smoothing parameters used were $\alpha = 0.2, \beta = 0.15, \gamma = 0.05$.

  i. Run this method on the data.

  ii. The forecasts for the validation set are given in Figure 18.11. Compute the MAPE values for the forecasts of quarters 21 and 22 for each of the two models (regression and exponential smoothing).

Quarter	Actual	Forecast	Error	LCI	UCI
21	60800	60847.97	–47.9659	55603.43	66092.51
22	64900	62487.87	2412.127	57101.99	67873.76
23	76997	72125.26	4871.74	65908.72	78341.8
24	103337	95491.26	7845.741	87260.78	103721.7

FIGURE 18.11   FORECASTS FOR VALIDATION PERIOD USING EXPONENTIAL SMOOTHING.

c. The fit and residuals from the exponential smoothing and regression models are compared in Figure 18.12. Using all the information thus far, which model is more suitable for forecasting quarters 21 and 22?

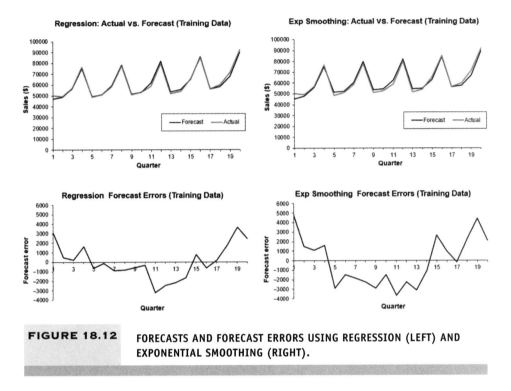

FIGURE 18.12 FORECASTS AND FORECAST ERRORS USING REGRESSION (LEFT) AND EXPONENTIAL SMOOTHING (RIGHT).

18.6 **Shipments of Household Appliances.** The time plot below shows the series of quarterly shipments (in million dollars) of US household appliances between 1985-1989 (data are available in `ApplianceShipments.xls`, data courtesy of Ken Black).

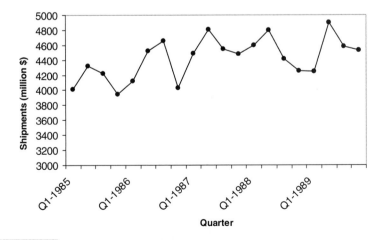

FIGURE 18.13 QUARTERLY SHIPMENTS OF US HOUSEHOLD APPLIANCES OVER 5 YEARS.

   a. Which of the following methods would be suitable for forecasting this series if applied to the raw data?

      ● Moving average

      ● Simple exponential smoothing

      ● Double exponential smoothing

      ● Holt–Winters exponential smoothing

   b. Apply a moving average with window span $w = 4$ to the data. Use all but the last year as the training set. Create a time plot of the moving average series.

      i. What does the MA(4) chart reveal?

      ii. Use the MA(4) model to forecast appliance sales in Q1-1990.

      iii. Use the MA(4) model to forecast appliance sales in Q1-1991.

      iv. Is the forecast for Q1-1990 most likely to underestimate, overestimate, or accurately estimate the actual sales on Q1-1990? Explain.

      v. Management feels most comfortable with moving averages. The analyst therefore plans to use this method for forecasting future quarters. What else should be considered before using the MA(4) to forecast future quarterly shipments of household appliances?

   c. We now focus on forecasting beyond 1989. In the following, continue to use all but the last year as the training set, and the last four quarters as the validation set. First, fit a regression model to sales with a linear trend and quarterly seasonality to the training data. Next, apply Holt–Winters exponential smoothing (with the default smoothing coefficient values) to the training data. Choose an adequate "season length."

      vi. Compute the MAPE for the validation data using the regression model.

      vii. Compute the MAPE for the validation data using Holt–Winters exponential smoothing.

      viii. Which model would you prefer to use for forecasting Q1-1990? Give three reasons.

      ix. If we optimize the smoothing parameters in the Holt–Winters method, is it likely to get values that are close to zero? Why or why not?

18.7 **Shampoo Sales.** The time plot in Figure 18.14 describes monthly sales of a certain shampoo over a three year period (data are available in `ShampooSales.xls`, source: Hyndman, R.J. Time Series Data Library, http://data.is/TSDLdemo, accessed on 07/25/15).

   Which of the following methods would be suitable for forecasting this series if applied to the raw data?

   ● Moving average

   ● Simple exponential smoothing

   ● Double exponential smoothing

   ● Holt–Winters exponential smoothing

18.8 **Natural Gas Sales.** Below is a time plot of quarterly natural gas sales (in billions of BTU) of a certain company, over a period of four years (data courtesy of George McCabe). The company's analyst is asked to use a moving average to forecast sales in Winter 2005.

   a. Reproduce the time plot with the overlaying MA(4) line (use Excel's "trendline").

   b. What can we learn about the series from the MA line?

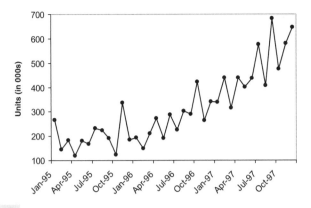

FIGURE 18.14   **MONTHLY SALES OF A CERTAIN SHAMPOO.**

FIGURE 18.15   **QUARTERLY SALES OF NATURAL GAS OVER 4 YEARS.**

   c. Run a moving average forecaster with adequate season length. Are forecasts generated by this method expected to over-forecast, under-forecast, or
accurately forecast actual sales? Why?

18.9 **Australian Wine Sales.** Figure 18.16 shows time plots of monthly sales of six
types of Australian wines (red, rose, sweet white, dry white, sparkling, and fortified) for 1980–1994 (data are available in `AustralianWines.xls`, source:
Hyndman, R.J. Time Series Data Library, http://data.is/TSDLdemo, accessed
on 07/25/15). The units are thousands of liters. You are hired to obtain short-
term forecasts (2–3 months ahead) for each of the six series, and this task will be
repeated every month.

a. Which forecasting method would you choose if you had to choose the same method for all series? Why?

b. Fortified wine has the largest market share of the six types of wine. You are asked to focus on fortified wine sales alone, and produce as accurate as possible forecasts for the next two months.

- Start by partitioning the data using the period until December 1993 as the training set.

- Apply Holt–Winters exponential smoothing to sales with an appropriate season length (use the default values for the smoothing constants).

c. Create an ACF plot for the residuals from the Holt–Winters exponential smoothing until lag 12.

i. Examining this plot, which of the following statements are reasonable?

- Decembers (month 12) are not captured well by the model.

- There is a strong correlation between sales on the same calendar month.

- The model does not capture the seasonality well.

- We should try to fit an autoregressive model with lag 12 to the residuals.

- We should first deseasonalize the data and then apply Holt–Winters exponential smoothing.

ii. How can you handle the effect above without adding another layer to your model?

FIGURE 18.16   MONTHLY SALES OF SIX TYPES OF AUSTRALIAN WINES BETWEEN 1980 AND 1994.

## PART VII

Data Analytics

## CHAPTER 19

Social Network Analytics[1]

In this chapter we will examine the basic ways to visualize and describe social networks, measure linkages, and analyze the network with both supervised and unsupervised techniques. The methods we will use long pre-date the Internet, but gained widespread use with the explosion in social media data. Twitter, for example, makes its feed available for public analysis, and some other social media firms make some of their data available to programmers and developers via an application programming interface (API). Two powerful and free tools for network analysis and visualization are NodeXL (an Excel add-on) and Gephi. NodeXL is available for Windows at http://nodexl.codeplex.com, and Gephi at gephi.org.

19.1 INTRODUCTION[2]

The use of social media began its rapid growth in the early 2000s with the advent of Friendster and MySpace, and, in 2004, Facebook. LinkedIn, catering to professionals, soon followed as did Twitter, Tumblr, Instagram, Yelp, TripAdvisor, and others. These information-based companies quickly began generating a deluge of data—especially data on links among people (friends, followers, connections, etc.).

---

[1]The organization of ideas in this chapter owes much to Jennifer Golbeck and her *Analyzing the Social Web*. The contribution of Marc Smith, developer and shepherd of NodeXL, is also acknowledged.

[2]This and subsequent sections in this chapter © 2016 Statistics.com and Galit Shmueli. Used by permission.

Data Mining for Business Analytics: Concepts, Techniques, and Applications with XLMiner®, Third Edition.
Galit Shmueli, Peter C. Bruce, and Nitin R. Patel.
© 2016 John Wiley & Sons Inc. Published 2016 by John Wiley & Sons, Inc.

For some companies, like Facebook, Twitter, and LinkedIn, nearly the entire value of the company lies in the analytic and predictive value of these data from their social networks. As of this writing (August 2015), Facebook was worth more than double General Motors and Ford combined. Other companies, like Amazon and Pandora, use social network data as important components of predictive engines aimed at selling products and services.

Social networks are basically entities (e.g., people) and the connections among them. Let's look at the building blocks for describing, depicting, and analyzing networks. The basic elements of a network are:

- Nodes (also called vertices or vertexes)
- Edges (connections or links between nodes)

A very simple LinkedIn network might be depicted as shown in Figure 19.1. This network has six nodes depicting members, with edges connecting some, but not all, of the pairs of nodes.

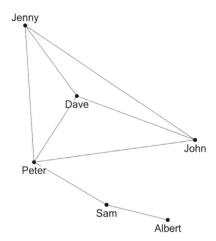

FIGURE 19.1   TINY HYPOTHETICAL LINKEDIN NETWORK. THE EDGES REPRESENT CONNECTIONS AMONG THE MEMBERS.

19.2 DIRECTED VS. UNDIRECTED NETWORKS

In the graph shown in Figure 19.1, edges are bi-directional or undirected, meaning that if John is connected to Peter, then Peter must also be connected to John, and there is no difference in the nature of these connections. You can see from this graph that there is a group of well-connected members (Peter, John, Dave, and Jenny), plus two less-connected members (Sam and Albert).

Connections might also be directional, or directed. For example, in Twitter, Dave might follow Peter, but Peter might not follow Dave. A simple Twitter network (using the same members and connections) might be depicted using edges with arrows as shown in Figure 19.2.

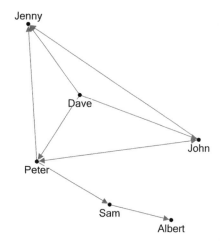

FIGURE 19.2   TINY HYPOTHETICAL TWITTER NETWORK WITH DIRECTED EDGES
(ARROWS) SHOWING WHO FOLLOWS WHOM

Edges can also be weighted to reflect attributes of the connection. For example, the thickness of the edge might represent the level of email traffic between two members in a network, or the bandwidth capacity between two nodes in a digital network (as illustrated in Figure 19.3). The length of an edge can also be used to represent attributes like physical distance between two points on a map.

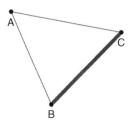

FIGURE 19.3   EDGE WEIGHTS REPRESENTED BY LINE THICKNESS (E.G., BANDWIDTH
CAPACITY) BETWEEN NODES IN A DIGITAL NETWORK

19.3   VISUALIZING AND ANALYZING NETWORKS

You have probably seen graphs used as a tool for visualizing and exploring networks; they are used widely in the news media. Jason Buch and Guillermo

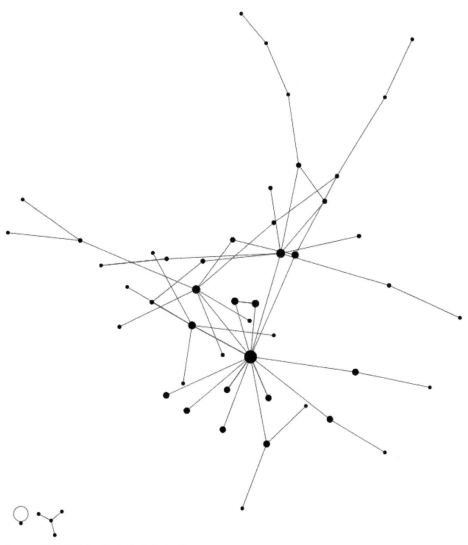

Created with NodeXL (http://nodexl.codeplex.com)

FIGURE 19.4   DRUG MONEY LAUNDERING NETWORK IN SAN ANTONIO, TX.

Contreras, reporters for the *San Antonio Express News*,[3] pored over law enforcement records and produced the network diagram shown in Figure 19.4 to understand and illustrate the connections used to launder drug money.[4] You can see that there is a well-connected central node; it is the address of an expensive residence in the gated Dominion community in San Antonio owned by accused launderers Mauricio and Alejandro Sanchez Garza. There are several entities in

---

[3] *San Antonio Express News*, May 23, 2012, accessed June 16, 2014.
[4] The original visualization can be seen at www.google.com/fusiontables/DataSource?snapid=S457047pVkn, accessed June 13, 2014.

the lower left connected only to themselves and one singleton. Nodes are sized according to how central they are to the network (specifically, in proportion to their *eigenvector centrality*, which is discussed in Section 19.4).

Graph Layout

It is important to note that *x-y* coordinates usually carry no meaning in network graphs; the meaning is conveyed in other elements such as node size, edge width, labels, and directional arrows. Consequently the same network may be depicted by two very different looking graphs. For example, Figure 19.5 presents two different layouts of the hypothetical LinkedIn network.

As a result visualization tools face innumerable choices in graph layout. The first step in making a choice is to establish what principles should govern the layout. Dunne and Shneiderman (2009, cited in Golbeck, 2013) list these four graph readability principles:

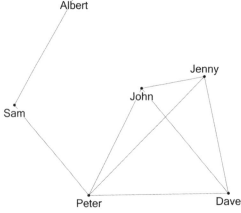

FIGURE 19.5 **TWO DIFFERENT LAYOUTS OF THE SMALL LINKEDIN NETWORK PRESENTED EARLIER.**

1. Every node should be visible.

2. For every node, you should be able to count its degree (explained below).

3. For every link, you should be able to follow it from source to destination.

4. Clusters and outliers should be identifiable.

These general principles are then translated into readability metrics by which graphs can be judged. Two simple layouts are *circular* (all nodes lie in a circle) and *grid* (all nodes lie at the intersection of grid lines in a rectangular grid).

You can probably think of alternate layouts that more clearly reveal structures such as clusters and singletons, and so can computers using a variety of algorithms. These algorithms typically use a combination of fixed arbitrary starting structures, random tweaking, analogues to physical properties (e.g., springs connecting nodes), and a sequence of iteration and measurement against the readability principles. A detailed discussion of layout algorithms is beyond the scope of this chapter; see Golbeck (2013, chapter 4) for an introduction to layout issues, including algorithms, the use of size, shape and color, scaling issues, and labeling.

Adjacency List

A network graph such as the one in Figure 19.4 is always tied to a data table called an *adjacency list*, or an *edge list*. Table 19.1 shows an excerpt from the data table used to generate Figure 19.4.

6451 Babcock Road	Q & M LLC
Q & M LLC	10 Kings Heath
Maurico Sanchez	Q & M LLC
Hilda Riebeling	Q & M LLC
Ponte Vedra Apartments	Q & M LLC
O S F STEAK HOUSE, LLC	Mauricio Sanchez
Arturo Madrigal	O S F STEAK HOUSE
HARBARD BAR, LLC	Arturo Madrigal
10223 Sahara Street	O S F STEAK HOUSE
HARBARD BAR, LLC	Maurico Sanchez
9510 Tioga Drive, Suite 206	Mauricio Sanchez
FDA FIBER, INC	Arturo Madrigal
10223 Sahara Street	O S F STEAK HOUSE
A G Q FULL SERVICE, LLC	Alvaro Garcia de Quevedo
19510 Gran Roble	Arturo Madrigal
Lorenza Madrigal Cristan	19519 Gran Roble
Laredo National Bank	19519 Gran Roble

TABLE 19.1   ADJACENCY LIST EXCERPT CORRESPONDING TO THE DRUG LAUNDERING NETWORK IN FIGURE 19.4

In a typical network visualization tool, you can select a row from the data table and see its node and connections highlighted in the network graph. Likewise, in the graph, you can click on a node and see it highlighted in the data table.

All the entities in both columns are nodes, and each row represents a link between the two nodes. If the network is directional, the link is usually structured from the left column to the right column.

Adjacency Matrix

The same relationships can be presented in a matrix. The adjacency matrix for the small directed graph for Twitter in Figure 19.2 is shown in Table 19.2.

Each cell in the matrix indicates an edge, with the originating node in the left header column and the destination node in the top row of headers. Reading the first row, we see that Dave is following three people—Peter, Jenny, and John.

	Dave	Peter	Jenny	Sam	John	Albert
Dave	0	1	1	0	1	0
Peter	0	0	1	1	1	0
Jenny	0	0	0	0	0	0
Sam	0	0	0	0	0	1
John	0	1	1	0	0	0
Albert	0	0	0	0	0	0

TABLE 19.2   ADJACENCY MATRIX EXCERPT CORRESPONDING TO THE TWITTER DATA IN FIGURE 19.2

Using Network Data in Classification and Prediction

In our discussion of classification and prediction, as well as clustering and data reduction, we were dealing mostly with highly structured data in the form of a matrix - columns were variables (features), and rows were records. We saw how to use XLMiner to sample from relational databases to bring data into the form of a flat matrix.

Highly structured data like this can be used for network analysis, but network data often starts out in a more unstructured or semi-structured format. Twitter provides a public feed of a portion of its voluminous stream of tweets, which has captured researchers' attention and accelerated interest in the application of network analytics to social media data. Network analysis can take this unstructured data and turn it into structured data with usable metrics.

We now turn our attention to those metrics. These metrics can be used not only to describe the attributes of networks, but as inputs to more traditional data mining methods.

19.4 SOCIAL DATA METRICS AND TAXONOMY

Several popular network metrics are used in network analysis. Before introducing them, we introduce some basic network terminology used for constructing the metrics.

Edge weight measures the strength of the relationship between the two connected nodes. For example, in an email network, there might be an edge weight that reflects the number of emails exchanged between two individuals linked by that edge.

Path and **path length** are important for measuring distance between nodes. A path is the route of nodes needed to go from node A to node B; path length is the number of edges in that route. Typically, these terms refer to the shortest route. In a weighted graph, the shortest path does not necessarily reflect the path with the fewest edges, but rather the path with the least weight. For example, if the weights reflect a cost factor, the shortest path would reflect the minimal cost.

Connected network A network is *connected* if each node in the network has a path, of any length, to all other nodes. A network may be unconnected in its entirety but consist of segments that are connected within themselves. In the money laundering visualization (Figure 19.4), the network as a whole is unconnected—the nodes are not all connected to one another. You can see one large connected segment, and in the lower left, a small connected segment and a singleton.

A clique is a network in which each node is directly connected by an edge to every other node. The connections must all be single edges—a connection via a multi-node path does not count.

A singleton is an unconnected node. It might arise when an individual signs up for a social network service (e.g., to read reviews) and does not participate in any networking activities.

Degree measures the number of connections to a node. In Figure 19.1, the Albert node is of degree 1, Sam of degree 2, and Jenny of degree 3.

Node-Level Centrality Metrics

Often we may be interested in the importance or influence of a particular individual or node, which is reflected in how central that node is in the network. One way to calculate this is by degree—how many edges are connected to the node. Nodes with many connections are more central.

Another metric for a node's centrality is *closeness*—how close the node is to the other nodes in the network. This is measured by finding the shortest path from that node to all the other nodes, then taking the average path length.

Still another metric is *betweenness*—the extent to which a given node lies on the shortest path between pairs of nodes. The calculation starts with the given node, say, node A, and two other nodes, say B and C, out of perhaps many nodes in a network. The shortest paths between B and C are listed, and the proportion of paths that include A is noted. This proportion is noted also for all other nodal pairs, and betweenness is the average proportion.

An aphorism relevant for social media networks is "it's not what you know, but who you know." A more accurate rendition would qualify it further—"it's who you know and who they know." A link to a member that has many other connections can be more valuable than a link to a member with few connections. A metric that measures this connectivity aspect is *eigenvector centrality*, which factors in both the number of links from a node and the number of onward connections from those links. The mathematics of the calculation are not discussed here, but the result always lies between 0 (not central) and 1 (maximum centrality).

Centrality can be depicted on a network graph by the size of a node—the more central the node, the larger it is.

Egocentric Network

It is often important to gain information that comes only from the analysis of individuals and their connections. For example, an executive recruiting firm may be interested in individuals with certain job titles and the people those individuals are connected to.

An egocentric network is the network of connections centered around an individual node. A degree 1 egocentric network consists of all the edges connected to the individual node, plus their nodes. A degree 1.5 egocentric network is the network of those nodes, plus all the edges among them. A degree 2 egocentric network is the network of all those nodes and edges, plus the edges and nodes connected to them.

The degree 1 and degree 1.5 egocentric networks for Peter in the LinkedIn graph are shown in Figure 19.6. The degree 2 egocentric network for Peter is the entire graph shown in Figure 19.1.

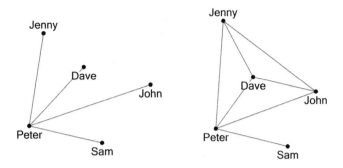

FIGURE 19.6   THE DEGREE 1 (LEFT) AND DEGREE 1.5 (RIGHT) EGOCENTRIC NETWORKS FOR PETER, FROM THE LINKEDIN GRAPH IN FIGURE 19.1.

Network Metrics

To this point we have discussed metrics and terms that apply to nodes and edges. We can also measure attributes of the network as a whole:

Degree distribution describes the range of *connectedness* of the nodes—how many nodes have (for example) 5 connections, how many have 4 connections, how many have 3, and so on. In the tiny LinkedIn network (Figure 19.1), we see that Peter has four connections, Dave, John, and Jenny each have three, Sam has two, and Albert has one. A table of this degree distribution is shown in Table 19.3.

TABLE 19.3   DEGREE DISTRIBUTION OF THE TINY LINKEDIN NETWORK

Node	Connections
Peter	4
Dave	3
John	3
Jenny	3
Sam	2
Albert	1

Density Another way to describe the overall connectedness of a graph is *density*, which focuses on the edges, not the nodes. The metric looks at the ratio of the actual number of edges to the maximum number of potential edges (i.e., if every node were connected to every other node) in a network with a fixed number of nodes. For a directed network with n nodes, there can be a maximum of $n(n - 1)$ edges. For an undirected network, the number is $n(n - 1)/2$. More formally, density calculations for directed and undirected networks are as follows:

$$\text{density (directed)} = \frac{e}{n(n - 1)}, \tag{19.1}$$

$$\text{density (undirected)} = \frac{e}{n(n - 1)/2}, \tag{19.2}$$

where e = number of edges and n = number of nodes. This metric ranges between just above zero (not dense at all) and one (as dense as possible). Figures 19.7 and 19.8 illustrate a sparse and dense network, respectively.

FIGURE 19.7 A RELATIVELY SPARSE NETWORK.

19.5 USING NETWORK METRICS IN PREDICTION AND CLASSIFICATION

Network attributes can be used along with other predictors in standard classification and prediction procedures. The most common applications involve the concept of *matching*. Online dating services, for example, will predict for their members which other members might be potentially compatible. Their algorithms typically involve calculation of a distance measure between a member seeking a relationship and candidate matches. It might also go beyond the mem-

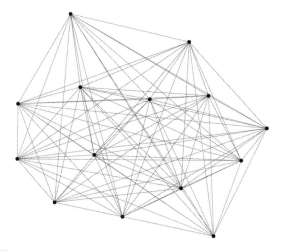

FIGURE 19.8   **A RELATIVELY DENSE NETWORK.**

bers' self-reported features and incorporate information about links between the member and candidate matches. A link might represent the action "viewed candidate profile."

Link Prediction

Social networks such as Facebook and LinkedIn use network information to recommend new connections. The translation of this goal into an analytics problem is:

> If presented with a network, can you predict the next link to form?

Prediction algorithms list all possible node pairs, then assign a score to each pair that reflects the similarity of the two nodes. The pair that scores as most similar (closest) is the next link predicted to form, if it does not already exist. See Chapter 15 for a discussion of such distance measures. Some variables used in calculating similarity measures are the same as those based on non–network information (e.g., years of education, age, sex, location). Other metrics used in link prediction apply specifically to network data:

- shortest path
- number of common neighbors
- edge weight

Link prediction is also used in targeting intelligence surveillance. "Collecting everything" may be technically, politically, or legally unfeasible, and an agency must therefore identify a priori a smaller set of individuals requiring surveillance.

The agency will often start with known targets, then use link prediction to identify additional targets and prioritize collection efforts.

Entity Resolution

Governments use network analysis to track terrorist networks, and a key part of that effort is identification of individuals. The same individual may appear multiple times from different data sources, and the agencies want to know, for example, whether individual A identified by French surveillance in Tunisia is the same as individual AA identified by Israeli intelligence in Lebanon and individual AAA identified by US intelligence in Pakistan.

One way to evaluate whether an individual appears in multiple databases is to measure distances and use them in a similar fashion to nearest-neighbors or clustering. In Chapter 15, we looked in particular at Euclidean distance, and discussed this metric not in terms of the network an individual belongs to but, rather, in terms of the profile (predictor values) of the individual. When basing entity resolution on these variables, it is useful to bring domain knowledge into the picture to weight the importance of each variable. For example, two variables in an individual's record might be street address and zip code. A match on street address is more definitive than a match on zip code, so we would probably want to give street address more weight in the scoring algorithm. For a more detailed discussion of automated weight calculation and assignment, see Golbeck (2013, p. 137).

In addition to measuring distance based on individual profiles, we can bring network attributes into the picture. Consider the simple networks for each individual in Figure 19.9, showing connections to known individuals: Based on network connections, you would conclude that A and AA are likely the same person, while AAA is probably a different person. The metrics that can formalize this search, and be used in automated fashion where human-intermediated visualization is not practical, are the same ones that are used in link prediction.

Entity resolution is also used extensively in customer record management and search. For example, a customer might contact a company inquiring about a product or service, triggering the creation of a customer record. The same customer might later inquire again, or purchase something. The customer database system should flag the second interaction as belonging to the first customer. Ideally, the customer will enter his or her information exactly the same way in each interaction, facilitating the match, but this does not necessarily happen. Failing an exact match, other customers may be proposed as matches based on proximity.

Another area where entity resolution is used is fraud detection. For example, a large telecom used link resolution to detect customers who "disappeared" after accumulating debt but then reappeared by opening a new account. The

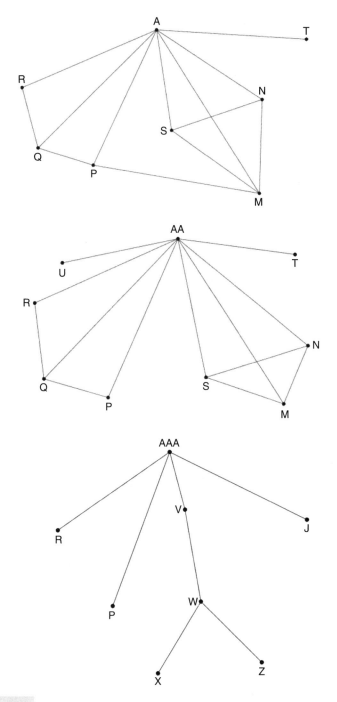

FIGURE 19.9   THE NETWORKS FOR SUSPECT A (TOP), SUSPECT AA (MIDDLE), AND SUSPECT AAA (BOTTOM).

network of phone calls to and from such people tends to remain stable, assisting the company to identify them.

In traditional business operations, the match may be based on variables such as name, address, and postal code. In social media products, matches may also be calculated on the basis of network similarity.

Collaborative Filtering

We saw in Chapter 14 that collaborative filtering uses similarity metrics to identify similar individuals, and thereby develop recommendations for a particular individual. Companies that have a social media component to their business can use information about network connections to augment other data in measuring similarity.

For example, in a company where internet advertising revenue is important, a key question is what ads to show consumers.

Consider the following small illustration for a company whose business is centered around online users: User A has just logged on, and is to be compared to users B-D. Table 19.4 shows some demographic and user data for each of the users.

TABLE 19.4   FOUR MEASUREMENTS FOR USERS A, B, C, AND D

user	months as cust.	age	spending	education
A	7	23	0	3
B	3	45	0	2
C	5	29	100	3
D	11	59	0	3

Our initial step is to compute distances based on these values, to determine which user is closest to user A. First, we convert the raw data to normalized values to place all the measurements on the same scale (for education, 1 = high school, 2 = college, 3 = post college degree). Normalizing means subtracting the mean and dividing by the standard deviation. The normalized data are shown in Table 19.5.

TABLE 19.5   NORMALIZED MEASUREMENTS FOR USERS A, B, C, AND D

user	months as cust.	age	spending	education
A	0.17	−1.14	−0.58	0.58
B	−1.18	0.43	−0.58	−1.73
C	−0.51	−0.71	1.73	0.58
D	1.52	1.42	−0.58	0.58

Next we calculate the Euclidean distance between A and each of the other users (see Table 19.6). Based on these calculations, which only take into account the demographic and user data, user C is the closest one to the new user A.

TABLE 19.6   EUCLIDEAN DISTANCE BETWEEN EACH PAIR OF USERS

pair	months as cust.	age	spending	education	Euclidean dist.
A-B	1.83	2.44	0	5.33	3.1
A-C	0.46	0.18	5.33	0	2.44
A-D	1.83	6.55	0	0	2.89

Let's now bring in network metrics, and suppose that the inter-user distances in terms of shortest path are A to B = 2, A to C = 4, and A to D = 3 (Table 19.7).

Finally, we can combine this network metric with the other user measurement distances calculated earlier. There is no need to calculate and normalize differences, as we did with the other user measurements, since the shortest-path metric itself already measures distance between records. We therefore take a weighted average of the network and non-network metrics, using weights that reflect the relative importance we want to attach to each. Using equal weights (Table 19.8), user B is scored as the closest to A, and could be recommended as a link for user A (in a social environment), or could be used as a source for product and service recommendations. With different weights it is possible to obtain different results. Choosing weights is not a scientific process; rather, it depends on the business judgment about the relative importance of the network metrics vs. the non-network user measurements.

TABLE 19.7   NETWORK METRICS

pair	shortest path
A-B	2
A-C	4
A-D	3

TABLE 19.8   COMBINING THE NETWORK AND NON-NETWORK METRICS

pair	shortest path	weight	non-network	weight	mean
A-B	2	0.5	3.1	0.5	2.55
A-C	4	0.5	2.44	0.5	3.22
A-D	3	0.5	2.89	0.5	2.95

In addition to recommending social media connections and providing data for recommendations, network analysis has been used for identifying clusters of similar individuals based on network data (e.g., for marketing purposes), identifying influential individuals (e.g., to target with promotions or outreach), and understanding—in some cases, controlling—the propagation of disease and information.

19.6 ADVANTAGES AND DISADVANTAGES

The key value of social network data to businesses is the information it provides about individuals and their needs, desires, and tastes. This information can be used to improve target advertising—and perfectly targeted advertising is the Holy Grail of the advertising business. People often disdain or ignore traditional "broad spectrum" advertising, whereas they pay close attention when presented with information about something specific that they are interested in. The power of network data is that it allows capturing information on individual needs and tastes without measuring or collecting that data directly. Moreover network data are often user-provided rather than actively collected.

To see the power of social network data, one need look no further than the data-based social media giants of the 21st century—Facebook, LinkedIn, Twitter, Yelp, and others. They have built enormous value based almost exclusively on the information contained in their social data—they manufacture no products and sell no services (in the traditional sense) to their users. The main value they generate is the ability to generate real-time information at the level of the individual to better target ads.

Other companies saw this value, and have tried to add a social component to what they produce. Google added a variety of social and sharing facilities with *Google+*. Amazon has experimented with adding a social component to its recommendation system.

It is important to distinguish between social network *engagement* and the use of social network *analytics*. Many organizations have social media policies and use social media as part of their communications and advertising strategies; however, this does not mean they have access to detailed social network data that they can use for analysis. Abrams Research, an Internet marketing agency, cited the edgy online retailer Zappos in 2009 for the best use of social media, but Zappos' orientation was engagement—use of social media for product support and customer service—rather than analytics.

Reliance on social network analytics comes with hazards and challenges. One major hazard is the dynamic, faddish, and somewhat ephemeral nature of social media use. In part this is because social media involvement is relatively new, and the landscape is changing with the arrival of new players and new technologies. In part it stems from the essential nature of social media. People use social media not to provide essential needs, like food and shelter, but as a voluntary avocation to provide entertainment and interaction with others. Tastes change, and what's in and out of fashion shifts rapidly in these spheres.

Facebook was a pioneer, and its initial appeal was to the college crowd and later to young adults. Eight years later, nearly half its users were age 45 or older, significantly higher than in the rest of the social media industry. These users spend more time per visit and are wealthier than college students, so Facebook may be benefiting from this demographic trend. However, this rapid shift shows how fast the essentials of their business model can shift.

Other challenges lie in the public and personal nature of the data involved. Although individuals nearly always engage with social media voluntarily, this does not mean they do so responsibly or with knowledge of the potential consequences. Information posted on Facebook became evidence in 33 percent of US divorce cases, and numerous cases have been cited of individuals being fired because of information they posted on Facebook.

One enterprising programmer created the website pleaserobme.com (since removed) that listed the real-time location of people who had "checked in" via FourSquare to a cafe or restaurant, thus letting the world know that they were not at home. FourSquare was not making this information easily available to hackers, but the check-ins were auto-posted to Twitter, where they could be harvested via an API (application programming interface). Thus information collected for the purpose of enriching the "targetability" of advertising to users ended up creating a liability for both FourSquare and Twitter. Twitter's liability came about indirectly, and it would have been easy for Twitter to overlook this risk in setting up their connection with FourSquare.

In summary, despite the potential for abuse and exposure to risk, the value provided by social network analytics seems large enough to ensure that the analytics methods outlined in this chapter will continue to be in demand.

PROBLEMS

19.1 **Describing a Network.** Consider an undirected network for individuals A, B, C, D, and E. A is connected to B and C. B is connected to A and C. C is connected to A, B, and D. D is connected to C and E. E is connected to D.

   a. Produce a network graph for this network.

   b. What node(s) would need to be removed from the graph for the remaining nodes to constitute a clique?

   c. What is the degree for node A?

   d. Which node(s) have the lowest degree?

   e. Tabulate the degree distribution for this network.

   f. Is this network connected?

   g. Calculate the density of the network.

19.2 **Network Density and Size.** Imagine that two new nodes are added to the undirected network in the previous exercise.

   a. By what percentage has the number of nodes increased?

   b. By what percentage has the number of possible edges increased?

   c. Suppose the new node has a typical (median) number of connections. What will happen to network density?

   d. Comment on comparing densities in networks of different sizes.

   e. Now add those two nodes, connect them however you wish, and tabulate the degree distribution of the resulting network.

19.3 **Link Prediction.** Consider the network shown in Figure 19.10.

   a. Using the number of common neighbors score, predict the next link to form (i.e., suggest which new link has the best chance of success).

   b. Using the shortest path score, identify the link that is least likely to form.

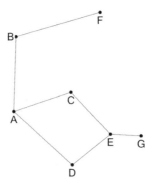

FIGURE 19.10   **NETWORK FOR LINK PREDICTION EXERCISE.**

19.4 **Startup Market Exploration.** Social media has given a boost to the artisan and craft trades. Many people embrace pastimes such as beer brewing, bread making, and pottery making. Social media provides both a mechanism for a diasapora community to form online, and a medium for instructions and advice to spread quickly. It also provides a vehicle for new ideas, products, and services to go viral. This has opened up opportunities for entrepreneurs and established companies to provide equipment and supplies to the artisanal community, and become small to medium scale producers themselves. Before the advent of social media, the high cost of advertising and marketing confined most artisans to the ranks of dedicated hobbyists. Now the availability of social media as a marketing tool brings commercialization and expansion within reach of hundreds of thousands of would-be entrepreneurs, and their financial backers. And because social media has become the communication vehicle of choice for many in the artisan community, it is an important channel for established supply companies whose marketing budgets are not similarly constrained.

In the following problems, use the graph tool in NodeXL, along with inspection of the data, to learn something about the network. This is an exploratory task, with no correct or incorrect answers.

Note: At publication date, NodeXL Pro was required for YouTube import, and Gephi, an alternate multi-platform tool, did not have a Twitter/YouTube import facility.

a. Use NodeXL's import facility to import data from the YouTube video and user networks. Pick an artisan topic as suggested above (examples in the area of food making alone abound—e.g. pastries, cheese, smoked meats, backyard chickens, etc.) Limit the number of videos initially to 100 or fewer, and experiment with the options for how to define edges. Produce graphs of the networks you examine, and explore their structure.

   i. Can you identify influential videos?

   ii. Can you identify influential users?

   iii. Imagine that you are a marketing consultant called to assist with the promotion of a product associated with this topic. Can you determine what social media strategies are in use among other marketers to this community?

b. Use NodeXL's import facility to import data from the Twitter search network on a topic of your choice. Limit the number of tweets initially to 100 or fewer, and increase it step-wise as you determine the required search time. *N*ote: With Twitter, especially, choosing a topic that is too popular risks submerging the interesting information in a sea of irrelevant chatter.

   i. Can you identify influential users?

   ii. Looking at the data supporting the graph, can you derive some broad categories
   of tweets, and their relative share in the whole?

   iii. Can you discern and describe marketing campaigns (e.g., the use of Twitter and
   other share buttons)?

c. Explore whether similar explorations are useful for other products and services that you are familiar with, such as consulting, software, and mobile apps.

## CHAPTER 20

Text Mining

In this chapter, we introduce text as a form of data. First, we discuss a spreadsheet representation of text data in which each column is a word, each row is a document, and each cell is a 0 or 1, indicating whether that column's word is present in that row's document. Then, we consider how to move from unstructured documents to this structured matrix. Finally, we illustrate how to integrate this process into the standard data mining procedures that we have already covered.

20.1 INTRODUCTION[1]

Up to this point, we have been dealing with three types of data:

- Numerical
- Binary (yes/no)
- Multicategory

In some common predictive analytics applications, though, data come in text form. An Internet service provider, for example, might want to use an automated algorithm to classify support tickets as urgent or routine, so that the urgent ones can receive immediate human review. A law firm facing a massive discovery process (review of large numbers of documents) would benefit from a document

---

[1]This and subsequent sections in this chapter © 2016 Statistics.com and Galit Shmueli. Used by permission.

Data Mining for Business Analytics: Concepts, Techniques, and Applications with XLMiner®, Third Edition.
Galit Shmueli, Peter C. Bruce, and Nitin R. Patel.
© 2016 John Wiley & Sons Inc. Published 2016 by John Wiley & Sons, Inc.

review algorithm that could classify documents as relevant or irrelevant. In such cases, the input variables (features) are embedded as text in documents.

Text mining methods have gotten a boost from the availability of social media data—the Twitter feed, for example, as well as blogs, online forums, review sites, and news articles. According to the Rexer Analytics 2013 Data Mining Survey, between 25% and 40% of data miners responding to the survey use data mining with text from these sources. Their generally public availability and high profile has provided a huge repository of data on which researchers can hone text mining methods. One area of growth has been the application of text mining methods to notes and transcripts from contact centers and service centers.

20.2   THE SPREADSHEET REPRESENTATION OF TEXT: "BAG-OF-WORDS"

Consider the following two sentences:

- First we consider the spreadsheet model.
- Then we consider another model.

We can represent the words in these two sentences in a spreadsheet (see Table 20.1), where each column is a word and each row is a sentence.

| **TABLE 20.1** | **SPREADSHEET REPRESENTATION OF WORDS IN SENTENCES S1 AND S2** |

	first	we	consider	the	spreadsheet	model	then	another
S1	1	1	1	1	1	1	0	0
S2	0	1	1	0	0	1	1	1

Note that all the words in both sentences are represented in the table and each word has one column. Although the words are listed in some form of order of appearance, order is not important—a "1" in a cell simply means that the word appeared at least once in the sentence for that row; a "0" means it did not appear. This is the *bag-of-words* approach, where the document is treated simply as a collection of words in which order, grammar, and syntax do not matter.

The two sentences have now been transformed into a data format just like those we have seen to this point. In a simple world, this binary matrix could be used in clustering, or, with the appending of a target variable, classification. However, the text mining world is not a simple one. Even confining our analysis to the bag-of-words approach, considerable thinking and preprocessing may be required. Some human review of the documents, beyond simply classifying them for training purposes, may be indispensable.

20.3 BAG-OF-WORDS VS. MEANING EXTRACTION AT DOCUMENT LEVEL

We can distinguish between two undertakings in text mining:

- Labeling a document as belonging to a class, or clustering similar documents
- Extracting more detailed meaning from a document

The first goal requires a sizable collection of documents, or a *corpus*,[2] the ability to extract predictor variables from documents, and for the classification task, lots of pre-labeled documents to train a model. The models that are used, though, are the standard statistical and machine learning predictive models that we have already dealt with for numerical and categorical data.

The second goal might involve a single document, and is much more ambitious. The computer must learn at least some version of the complex "algorithms" that make up human language comprehension: grammar, syntax, punctuation, and so on. In other words, it must undertake the processing of a natural (i.e. noncomputer) language to *understand* documents in that language. Understanding the meaning of one document on its own is a far more formidable task than probabilistically assigning a class to a document based on rules derived from hundreds or thousands of similar documents.

For one thing, text comprehension requires maintenance and consideration of word order. "San Francisco beat Boston in last night's baseball game" is very different from "Boston beat San Francisco in last night's baseball game."

Even identical words in the same order can carry different meanings, depending on the cultural and social context: "Hitchcock shot The Birds in Bodega Bay," to an avid outdoors person indifferent to capitalization and unfamiliar with Alfred Hitchcock's films, might be about bird hunting. Ambiguity resolution is a major challenge in text comprehension—does "bot" mean "bought," or does it refer to robots?

Our focus will remain with the overall focus of the book and the easier goal—probabilistically assigning a class to a document, or clustering similar documents. The second goal—deriving understanding from a single document—is the subject of the field of natural language processing (NLP).

---

[2] The term "corpus" is often used to refer to a large, fixed standard set of documents that can be used by text preprocessing algorithms, often to train algorithms for a specific type of text, or to compare the results of different algorithms. A specific text mining setting may rely on algorithms trained on a corpus specially suited to that task. One early general-purpose standard corpus was the Brown corpus of 500 English language documents of varying types (named Brown because it was compiled at Brown University in the early 1960s).

20.4 PREPROCESSING THE TEXT

The simple example we presented had ordinary words separated by spaces, and a period to denote the end of the each sentence. A fairly simple algorithm could break the sentences up into the word matrix with a few rules about spaces and periods. It should be evident that the rules required to parse data from real world sources will need to be more complex. It should also be evident that the preparation of data for text mining is a more involved undertaking than the preparation of numerical or categorical data for predictive models. For simplicity, we will use as our framework the choices you face in processing text data in XLMiner.

*N*ote: While Excel was not designed as a text processing tool, it has native facilities that can be used for this purpose. XLMiner substantially expands those capabilities. Some projects, though, will inevitably require the flexibility and power of a standard programming language.

Tokenization

Our simple data set was composed entirely of words found in the dictionary. A real set of documents will have more variety—it will contain numbers, alphanumeric strings like date stamps or part numbers, web and email addresses, abbreviations, slang, proper nouns, misspellings, and more.

Tokenization is the process of taking a text and, in an automated fashion, dividing it into separate "tokens" or terms. A token (term) is the basic unit of analysis. A word separated by spaces is a token. 2 + 3 would need to be separated into three tokens, while 23 would remain as one token. Punctuation might also stand as its own token (e.g., the @ symbol). These tokens become the column headers in the data matrix. Each text mining software program will have its own list of delimiters (spaces, commas, colons, etc.) that it uses to divide up the text into tokens. XLMiner uses a wide variety of characters for delimiting, but defines certain sequence types as being exempt from splitting (e.g., email addresses, url's, monetary amounts). In some programs the delimiters themselves are retained as tokens, while in others (XLMiner included) they are immediately removed from the document.

For a sizable corpus, this will also result in a huge number of predictor variables - the English language has over a million words, let alone the non-word terms that will be encountered in typical documents. Anything that can be done in the preprocessing stage to reduce the number of terms will aid in the analysis. The initial focus is on eliminating terms that simply add bulk and noise.

Some of the terms that result from the initial parsing of the corpus may not be useful in prediction and can be eliminated in the preprocessing stage. For example, in a legal discovery case, one corpus of documents might be emails, all of which have company information and some boilerplate as part of the signature. These terms might be added to a *stop list* of terms that are to be automatically eliminated in the preprocessing stage (*stopwords*).

Text Reduction

Most text-processing software (XLMiner included) come with a generic *stopword* list of frequently occurring terms to be removed. If you review the XLMiner stopword list during preprocessing stages, you will see that it contains a large number of terms to be removed (the "a's" contain, for example, "a," "about," "actually," "again," "after," etc.; see Figure 20.2). You can add additional terms or remove existing terms from the generic list.[3]

Another approach is to specify the terms that you want the analysis to focus on, rather than the terms you want it to ignore (the "specified terms only" option in XLMiner). This is useful when there is accurate and informative domain knowledge about the documents. For example, in a forensic review of voluminous company communications, those heading the investigation might supply a list of individual and company names to focus on.

Additional techniques to reduce the volume of text ("vocabulary reduction") and focus on the most meaningful text include:

- *Stemming*, a linguistic method that reduces different variants of words to a common core.
- Frequency filters can be used to eliminate either terms that occur in a great majority of documents or very rare terms. They can also be used to limit the vocabulary to the *n* most frequent terms.
- Synonyms or synonymous phrases may be consolidated.
- Case can be ignored.
- A variety of specific terms in a category can be replaced with the category name. This is called *normalization*. For example, different email addresses or different numbers might all be replaced with "emailtoken" or "numbertoken."

Presence/Absence vs. Frequency

The bag-of-words approach can be implemented in terms of either frequency or presence/absence of terms. The latter might be appropriate in some

---

[3] XLMiner actually has two term-removal facilities. The stopword list identifies terms to be removed when they appear in the document exactly in the form in which they appear in the list. The *exclusion list* removes all the variants of the term that stem from the same core (see "stemming").

circumstances—in a forensic accounting model, for example, the presence or absence of a particular vendor name may be a key predictor variable, without regard to how often it appears in a given document. *Frequency* may be important in other circumstances, however. For example, in processing support tickets a single mention of "IP address" may be non-meaningful—all support tickets might involve a user's IP address as part of the submission. Repetition of the phrase multiple times, however, might provide useful information that IP address is part of the problem (e.g., DNS resolution).

XLMiner has a base option of presence/absence in its text preprocessing. It also has the option of producing a matrix in which term frequency, rather than a 0 or 1, is the entry in each cell.

Term Frequency—Inverse Document Frequency (TF-IDF)

There are additional popular options that factor in both the frequency of a term in a document and the frequency of documents with that term. One such popular option is Term Frequency—Inverse Document Frequency (TF-IDF). For a given document and term, TF-IDF is the product of term frequency *TF* (the number of times a term appears in a document) and inverse document frequency *IDF* (the log of the inverse of the frequency with which documents have that term). There are multiple ways to define both TF and IDF, so there are a variety of possible definitions of TF-IDF. The general idea is that it identifies documents with frequent occurrences of rare terms. TF-IDF yields high values for documents that have a relatively high frequency for terms that are relatively rare overall, and near-zero values for terms that are absent from a document, or present in most documents.

From Terms to Concepts: Latent Semantic Indexing

In Chapter 4 we showed how numerous continuous numeric predictor variables can be reduced to a small number of "principal components" that explain most of the variation in a set of variables. The principal components are linear combinations of the original (typically, correlated) variables, and a subset of them serve as new variables to replace the numerous original variables.

An analogous dimension reduction method—latent semantic indexing—can be applied to text data. The mathematics of the algorithm are beyond the scope of this book, but a good intuitive explanation comes from the XLMiner user guide:

For example: if we inspected our document collection, we might find that each time the term "alternator" appeared in an automobile document, the document also included the terms "battery" and "headlights." Or each time the term "brake" appeared in an automobile document, the terms "pads" and "squeaky" also appeared. However there is no detectable pattern regarding the use of the terms "alternator" and "brake" together. Documents including "alternator" might or might not include "brake" and documents including "brake" might or might not include "alternator." Our four terms, battery, headlights, pads, and squeaky describe two different automobile repair issues: failing brakes and a bad alternator.

Analytic Solver Platform, XLMiner Platform, *Data Mining User Guide*, 2014, Frontline *Systems, p. 245*

So, in this case, latent semantic indexing would reduce the four terms to two concepts:

- Brake failure
- Alternator failure

Extracting Meaning

In this simple example, the concepts to which the terms map (failing brakes, bad alternator) are clear and understandable. In many cases, unfortunately, this will not be true—the concepts to which the terms map will not be obvious. In such cases, latent semantic indexing will greatly enhance the manageability of the text for model-building purposes, and sharpen its predictive power by reducing noise, but it will turn the model into a black-box device for prediction, not so useful for understanding the roles that terms and concepts play. This is OK for our purposes—as noted earlier, we are focusing on text mining to classify or cluster new documents, not to extract meaning.

20.5 IMPLEMENTING DATA MINING METHODS

After the text has gone through the preprocessing stage, it is in a numeric matrix format; then you can apply the various data mining methods discussed earlier in this book. Clustering methods can be used to identify clusters of documents—for example, large numbers of medical reports can be mined to identify clusters of symptoms. Prediction methods could be used with tech support tickets to predict how long it will take to resolve an issue. Perhaps the most popular application of text mining is for classification—also termed labeling—of documents.

Note: XLMiner, while primarily a tool for predictive modeling, visualization, and clustering, does provide some facilities in its text mining module for extraction of meaning from documents. In our illustrations and exercises, we skip over those capabilities.

20.6 Example: Online Discussions on Autos and Electronics

This example (from the XLMiner user guide, with minor modifications) illustrates a classification task—to classify Internet discussion posts as either auto-related or electronics-related. One post looks like this:

> From: smith@logos.asd.sgi.com (Tom Smith) Subject: Ford Explorer 4WD - do I need performance axle?
> We're considering getting a Ford Explorer XLT with 4WD and we have the following questions (All we would do is go skiing – no off-roading):
> 1. With 4WD, do we need the "performance axle" – (limited slip axle). Its purpose is to allow the tires to act independently when the tires are on different terrain.
> 2. Do we need the all-terrain tires (P235/75X15) or will the all-season (P225/70X15) be good enough for us at Lake Tahoe?
> Thanks,
> Tom
>
> –
> ==
> Tom Smith Silicon Graphics smith@asd.sgi.com 2011 N. Shoreline Rd. MS 8U-815 415-962-0494 (fax) Mountain View, CA 94043
> ==

The posts are taken from Internet groups that are devoted to autos and electronics, so are pre-labeled. This one, clearly, is auto-related. A related organizational scenario might involve messages received by a medical office that must be classified as medical or non-medical (the messages in such a real scenario would probably have to be labeled by humans as part of the preprocessing).

The posts are in the form of small files whose names are numbers—all the auto posts start with "1" and all the electronics posts start with "5." In the following, we describe the main steps from preprocessing to building a classification model using a sample of 985 posts.

Importing and Labeling the Records

Importing and preprocessing text data in XLMiner involves more steps and decisions than opening the other datasets used in this book. Here is a schematic of the whole classification process in XLMiner:

1. Import data via the *Sample* menu. In this example format, each document is a file, and the files are to be imported from a folder (in XLMiner either use the "write file contents" option at import, or, if you write file paths, you can use the option "Text variables contain file paths" at a later stage).

Another possible import format is one in which all the documents are in a single text file, one document per row.

2. From the *Text* menu, use "Pre-processing" and "Representation" for the various preprocessing steps discussed here.

3. At some point after tokenization, add or restore labels.

4. From the *Text* menu, use "Output options" to produce the matrix that is needed to build a predictive model. You will need to add the target variable.

5. Using the matrix thus produced, use the *Classify* menu to select and configure various classification models.

IMPORTANT: This is just a schematic, covering only the major phases of analysis, using XLMiner v.15.0.2. The XLMiner user guide section on Text Mining should be consulted in detail, as the process is more complex than with mining numeric data, and the user interface is not a sufficient guide by itself.

Note that prior to preprocessing, each document is treated as a single row in Excel, or a single variable. It has not yet been split up into tokens. How will we handle the document labels? If they are part of the document itself, they will be subsumed, and perhaps lost, as part of the preprocessing.

It is best to add, or restore, the labels after the automated preprocessing that XLMiner does. In this example, the labels correspond to the first character in the file names—1's for autos and 5's for electronics. We can add labels later, as long as we know which documents (rows in Excel) come from which source.

Tokenization

The first preprocessing step, tokenization, is typically automated. XLMiner will split the document into tokens or terms (words, or non-word character strings). For example, the term "4WD" is isolated as a term because, in two instances, it has spaces on either side and, in one instance, it has a space on one side and a comma (another delimiter) on the other. Some text processing algorithms treat the delimiters themselves as terms (tokens), others not. Most will treat the full email address as a single term, and not split it up.

In this example, you can see that after tokenization, there can be a large number of single character tokens that will probably not contribute useful information to the classification task. Some programs (XLMiner included) eliminate delimiters during tokenization, which removes most of the punctuation delimiters.

The results of tokenization are cumulative from document to document, in the sense that a master term matrix is built up in which a new column is added each time a new term (not previously encountered in the current document or prior documents) is found.

FIGURE 20.1   PREPROCESSING OPTIONS IN XLMINER.

Text Processing and Reduction

The next step is *stemming*, or the consolidation of multiple forms of a word into a single core. For example, "road" and "Rd." might stem to "road." After stemming, XLMiner's default is set to discard tokens that stem to two or fewer characters. The user can modify this.

Other options include removal of words (tokens) on a stop list (see Figure 20.2) and normalization. Normalization is a general term that does not have a single, specific definition; we need to either consult documentation or specify the rules for normalization. For XLMiner, normalization of case means converting everything to lower case. Normalization of certain terms and characters in XLMiner means replacing all unique occurrences in a category (e.g., email addresses) with a single user-specified token that represents that category (see Figure 20.3.)

FIGURE 20.2 THE STOPWORD LIST BUILT INTO XLMINER.

FIGURE 20.3 NORMALIZATION OPTIONS IN XLMINER.

Producing a Concept Matrix

The tokenization and preprocessing will produce a matrix (terms = columns, documents = records). Instead of a term-document matrix, we will ask XLMiner (in the "Representation" menu) to produce the TF–IDF matrix option, described earlier. This measure will incorporate both the frequency of a term and the frequency with which documents containing that term appear in the overall corpus.

The resulting matrix is probably too large to efficiently analyze in a predictive model, so we will have XLMiner use latent semantic indexing to extract a reduced set of columns, termed "concepts." For manageability we will limit the number of concepts to 20 (see Figure 20.4).

We will use this reduced set of concepts simply to facilitate the processing of a predictive model. We will not attempt to interpret the concepts for meaning (although XLMiner has tools to do so, this lies more properly in the realm of natural language processing than text mining).

	A	B	C	D	E	F	G
1	Class	Doc-name	Concept 1	Concept 2	Concept 3	Concept 4	Concept 5
2	0	52434	-0.71609	-0.2249	0.210686	-0.08325	-0.12258
3	0	52446	-0.66838	-0.07859	0.166447	0.097483	0.091504
4	0	52464	-0.7416	-0.20084	0.305943	-0.26083	-0.31168
5	0	52717	-0.39777	0.296412	-0.15754	0.171159	-0.17366
6	0	52718	-0.41584	-0.16891	0.499957	-0.1801	-0.17858
7	0	52719	-0.35528	-0.2398	0.161951	0.03308	-0.21332
8	0	52721	-0.36316	-0.01288	0.395562	-0.39182	-0.21871
9	0	52722	-0.56536	0.019728	0.284984	-0.02301	-0.30336
10	0	52723	-0.60292	-0.48113	0.387998	-0.06242	-0.00975
11	0	52724	-0.56819	-0.07977	0.393881	-0.06076	-0.56577
12	0	52725	-0.62091	0.041464	0.45776	0.0722	-0.31399
13	0	52726	-0.41962	0.000604	0.341336	-0.12855	-0.54949

FIGURE 20.4   A SMALL PORTION OF THE CONCEPT MATRIX FOR THE AUTOS-ELECTRONICS EXAMPLE; THE FULL MATRIX HAS 20 CONCEPTS AND 985 DOCUMENTS. THE CLASS LABEL MUST BE ADDED MANUALLY AFTER THE MATRIX IS PRODUCED.

Labeling the Documents

Now we have a compact numerical matrix to work with—985 rows and just 20 columns. At this point we can add the labels—we will use "1" for autos and "0" for electronics. Recall that document names beginning with "5" are electronics and those beginning with "1" are autos. We need to add a new column for the document class, and looking at the document names we see that the first 490 rows should be classified as "0" and the remaining 495 rows "1."

Fitting a Model

At this point we have transformed the original text data into a familiar form needed for predictive modeling—a single target variable (1 = autos, 0 = electronics), and 20 predictor variables (the concepts).

We can now partition the data (60% training, 40% validation), and try applying several classification models. Here are the results of a neural net model, with "class" as the target and the 20 concept variables as the predictors, using the default values for XLMiner (4/2015 release).

The confusion matrix (Figure 20.5, top) shows reasonably high accuracy in separating the two classes of documents—an error rate of 9.39%. The decile lift chart (Figure 20.5, bottom) confirms the high separability of the classes. For a two class dataset with a nearly 50/50 split between the classes, the maximum lift per decile is 2, and the lift shown here is just under 2 for the first 40% of the cases, and close to 0 for the last 40%. In a decision-making process, human review could be concentrated on the middle ranked 20% where the classification error is most likely to occur.

We can also try other models to see how they compare; this is left as an exercise.

Prediction

The most prevalent application of text mining is classification ("labeling"), but it can also be used for prediction of numerical values. For example, maintenance or support tickets could be used to predict length or cost of repair. The only step that would be different in the process above would be that the label applied after the preprocessing would be a numeric value rather than a class.

20.7 SUMMARY

In this chapter we drew a distinction between text processing for the purpose of extracting meaning from a single document (natural language processing—NLP) and classifying or labeling numerous documents in probabilistic fashion (text mining). We concentrated on the latter, and examined the preprocessing steps that need to occur before text can be mined. Those steps are more varied and involved than those used in preparing numerical data. The ultimate goal is to produce a matrix in which columns are terms and rows are documents. The nature of language is such that the number of terms is excessive for effective model building, so the preprocessing steps include vocabulary reduction. A final

Validation Data Scoring—Summary Report

Cutoff probability value for success (UPDATABLE)	0.5

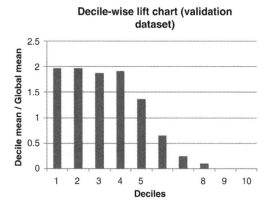

Confusion Matrix

Actual Class	Predicted Class	
	1	0
1	187	13
0	24	170

Error Report

Class	# Cases	# Errors	% Error
1	200	13	6.50
0	194	24	12.37
Overall	394	37	9.39

Decile-wise lift chart (validation dataset)

FIGURE 20.5   VALIDATION DATA CONFUSION MATRIX (TOP) AND LIFT DECILE CHART (BOTTOM), FOR AUTOS-ELECTRONICS DOCUMENT CLASSIFICATION.

major reduction takes place if we use, instead of the terms, a limited set of concepts that represents most of the variation in the documents, in the same way that principal components capture most of the variation in quantitative data. Finally, we end up with a quantitative matrix in which the cells represent the frequency or presence of terms and the rows represent documents. To this we append document labels (classes), and then we are ready to use this matrix for classifying documents using classification methods.

PROBLEMS

20.1 **Tokenization.** Consider the following text version of a post to an online learning forum in a statistics course:

```
Thanks John!<br /><br /><font size="3">
&quot;Illustrations and demos will be
provided for students to work through on
their own&quot;</font>.
Do we need that to finish project? If yes,
where to find the illustration and demos?
Thanks for your help.\<img title="smile"
alt="smile" src="\url{http://lms.statistics.
com/pix/smartpix.php/statistics_com_1/s/smil
ey.gif}" \><br /> <br />
```

  a. Identify 10 non-word tokens in the passage.

  b. Suppose that this passage constitutes a document to be classified, but you are not certain of the business goal of the classification task. Identify material (at least 20% of the terms) that, in your judgment, could be discarded fairly safely without knowing that goal.

  c. Suppose that the classification task is to predict whether this post requires the attention of the instructor, or whether a teaching assistant might suffice. Identify the 20% of the terms that you think might be most helpful in that task.

  d. What aspect of the passage is most problematic from the standpoint of simply using a bag-of-words approach, as opposed to an approach in which meaning is extracted?

20.2 **Classifying Internet Discussion Posts.** In this problem you will use the data and scenario described in this chapter's example, in which the task is to develop a model to classify documents as either auto-related or electronics-related.

  a. From the folder autos-electronics, import the files into XLMiner using the *Sample* menu and the option to read the file contents into individual rows during import. Which row ID's correspond to the autos class? To the electronics class?

  b. Proceed in XLMiner to the preprocessing of the text, accepting defaults except as noted. Explain what would be different if you unchecked the "perform stemming" box (but leave it checked).

  c. Continue to the Representation options in XLMiner, and change the maximum number of concepts to 20. Explain what is different about the Term Frequency matrix, as opposed to the TF-IDF matrix.

  d. Continue to the Output Options section, without changing defaults.

    i. Explain very briefly how the different matrix options differ.

    ii. Restate the goal of the text mining project, and why we are not using the Concept Extraction options.

   e. Going back to your notes about which rows correspond to which class of documents, add class identifications to the concept document matrix. Using this matrix, fit a predictive model (different from the model presented in the chapter illustration) to classify documents (rows) as autos or electronics. Compare its performance to that of the model presented in the chapter illustration.

20.3 **Classifying Classified Ads Submitted Online.** Consider the case of a website that caters to the needs of a specific farming community, and carries classified ads intended for that community. Anyone, including robots, can post an ad via a web interface, and the site owners have problems with ads that are fraudulent, spam, or simply not relevant to the community. They have provided a file with 4143 ads, each ad in a row, and each ad labeled as either −1 (not relevant) or 1 (relevant). An additional 0/1 column has been added with −1 mapping to 0, and 1 mapping to 1, to be used as the target variable (this preprocessing step is not so easy to accomplish in XLMiner). The goal is to develop a predictive model that can classify ads automatically.

- Open the file farm-ads.xls, and briefly review some of the relevant and irrelevant ads to get a flavor for their contents. Since all the documents have been consolidated in a single file, one document per row, the option "Text variables contain file paths" should NOT be checked in XLMiner.

- Preprocess the data in XLMiner, using the defaults, except limit the number of concepts to 20.

   a. Examine the term-document matrix.

      i. Is it sparse or dense?

      ii. Find two non-zero entries and briefly interpret their meaning, in words (you do not need to derive their calculation).

      iii. Had you chosen the presence/absence option for the term-document matrix, what would those two non-zero entries be?

   b. Briefly explain the difference between the term-document matrix and the concept-document matrix. Relate the latter to what you learned in the principal components chapter (Chapter 20).

   c. To start the process of building a predictive model, add a final column of 0's and 1's to the concept-document matrix, to label the −1's as 0's, and the 1's as 1's. To do this, make a note in the original data of where (which row) the −1's stop and the 1's begin. Then consult the first column in the concept-document matrix, which contains this row (document) number. You should also add a name for this first column, say "docnum," so that all columns have names.

   d. Using logistic regression, partition the data (60% training, 40% validation), and develop a model to classify the documents as 1's or 0's. Comment on its efficacy.

   e. Why use the concept-document matrix, and not the term-document matrix, to provide the predictor variables?

## PART VIII

Cases

## CHAPTER 21

Cases

21.1 CHARLES BOOK CLUB[1]

`CharlesBookClub.xls` is the dataset for this case study.

The Book Industry

Approximately 50,000 new titles, including new editions, are published each year in the United States, giving rise to a $25 billion industry in 2001. In terms of percentage of sales, this industry may be segmented as follows:

16%	Textbooks
16%	Trade books sold in bookstores
21%	Technical, scientific, and professional books
10%	Book clubs and other mail-order books
17%	Mass-market paperbound books
20%	All other books

Book retailing in the United States in the 1970s was characterized by the growth of bookstore chains located in shopping malls. The 1980s saw increased purchases in bookstores stimulated through the widespread practice of discounting. By the 1990s, the superstore concept of book retailing gained acceptance and contributed to double-digit growth of the book industry. Conveniently

---

[1]The Charles Book Club case was derived, with the assistance of Ms. Vinni Bhandari, from *The Bookbinders Club, a Case Study in Database Marketing*, prepared by Nissan Levin and Jacob Zahavi, Tel Aviv University; used with permission.

Data Mining for Business Analytics: Concepts, Techniques, and Applications with XLMiner®, Third Edition.
Galit Shmueli, Peter C. Bruce, and Nitin R. Patel.
© 2016 John Wiley & Sons Inc. Published 2016 by John Wiley & Sons, Inc.

situated near large shopping centers, superstores maintain large inventories of 30,000 to 80,000 titles and employ well-informed sales personnel. Book retailing changed fundamentally with the arrival of Amazon, which started out as an online bookseller and, as of 2015, is the world's largest online retailer of any kind. Amazon's margins were small and the convenience factor high, putting intense competitive pressure on all other book retailers. Borders, one of the two major superstore chains, discontinued operations in 2011.

Subscription-based book clubs offer an alternative model that has persisted, though it too has suffered from the dominance of Amazon.

Historically, book clubs offered their readers different types of membership programs. Two common membership programs are the continuity and negative option programs, which are both extended contractual relationships between the club and its members. Under a *continuity program*, a reader signs up by accepting an offer of several books for just a few dollars (plus shipping and handling) and an agreement to receive a shipment of one or two books each month thereafter at more-standard pricing. The continuity program is most common in the children's book market, where parents are willing to delegate the rights to the book club to make a selection, and much of the club's prestige depends on the quality of its selections.

In a *negative option program*, readers get to select how many and which additional books they would like to receive. However, the club's selection of the month is delivered to them automatically unless they specifically mark "no" on their order form by a deadline date. Negative option programs sometimes result in customer dissatisfaction and always give rise to significant mailing and processing costs.

In an attempt to combat these trends, some book clubs have begun to offer books on a *positive option basis*, but only to specific segments of their customer base that are likely to be receptive to specific offers. Rather than expanding the volume and coverage of mailings, some book clubs are beginning to use database-marketing techniques to target customers more accurately. Information contained in their databases is used to identify who is most likely to be interested in a specific offer. This information enables clubs to design special programs carefully tailored to meet their customer segments' varying needs.

Database Marketing at Charles

The Club   The Charles Book Club (CBC) was established in December 1986 on the premise that a book club could differentiate itself through a deep understanding of its customer base and by delivering uniquely tailored offerings. CBC focused on selling specialty books by direct marketing through a variety of channels, including media advertising (TV, magazines, newspapers) and mailing.

CBC is strictly a distributor and does not publish any of the books that it sells. In line with its commitment to understanding its customer base, CBC built and maintained a detailed database about its club members. Upon enrollment, readers were required to fill out an insert and mail it to CBC. Through this process, CBC created an active database of 500,000 readers; most were acquired through advertising in specialty magazines.

The Problem   CBC sent mailings to its club members each month containing the latest offerings. On the surface, CBC appeared very successful: mailing volume was increasing, book selection was diversifying and growing, and their customer database was increasing. However, their bottom-line profits were falling. The decreasing profits led CBC to revisit their original plan of using database marketing to improve mailing yields and to stay profitable.

A Possible Solution   CBC embraced the idea of deriving intelligence from their data to allow them to know their customers better and enable multiple targeted campaigns where each target audience would receive appropriate mailings. CBC's management decided to focus its efforts on the most profitable customers and prospects, and to design targeted marketing strategies to best reach them. The two processes they had in place were:

1. Customer acquisition:
   o New members would be acquired by advertising in specialty magazines, newspapers, and on TV.
   o Direct mailing and telemarketing would contact existing club members.
   o Every new book would be offered to club members before general advertising.
2. Data collection:
   o All customer responses would be recorded and maintained in the database.
   o Any information not being collected that is critical would be requested from the customer.

For each new title, they decided to use a two-step approach:

1. Conduct a market test involving a random sample of 4000 customers from the database to enable analysis of customer responses. The analysis would create and calibrate response models for the current book offering.

2. Based on the response models, compute a score for each customer in the database. Use this score and a cutoff value to extract a target customer list for direct mail promotion.

Targeting promotions was considered to be of prime importance. Other opportunities to create successful marketing campaigns based on customer behavior data (returns, inactivity, complaints, compliments, etc.) would be addressed by CBC at a later stage.

Art History of florence   A new title, *The Art History of florence*, is ready for release. CBC sent a test mailing to a random sample of 4000 customers from its customer base. The customer responses have been collated with past purchase data. The dataset was randomly partitioned into three parts: *Training Data* (1800 customers): initial data to be used to fit models; *Validation Data* (1400 customers): holdout data used to compare the performance of different models; and *Test Data* (800 customers): data to be used only after a final model has been selected to estimate the probable performance of the model when it is deployed. Each row (or case) in the spreadsheet (other than the header) corresponds to one market test customer. Each column is a variable, with the header row giving the name of the variable. The variable names and descriptions are given in Table 21.1.

TABLE 21.1   **LIST OF VARIABLES IN CHARLES BOOK CLUB DATASET**

Variable Name	Description
Seq#	Sequence number in the partition
ID#	Identification number in the full (unpartitioned) market test dataset
Gender	0 = Male 1 = Female
M	Monetary—Total money spent on books
R	Recency—Months since last purchase
F	Frequency—Total number of purchases
FirstPurch	Months since first purchase
ChildBks	Number of purchases from the category child books
YouthBks	Number of purchases from the category youth books
CookBks	Number of purchases from the category cookbooks
DoItYBks	Number of purchases from the category do-it-yourself books
RefBks	Number of purchases from the category reference books (atlases, encyclopedias, dictionaries)
ArtBks	Number of purchases from the category art books
GeoBks	Number of purchases from the category geography books
ItalCook	Number of purchases of book title *Secrets of Italian Cooking*
ItalAtlas	Number of purchases of book title *Historical Atlas of Italy*
ItalArt	Number of purchases of book title *Italian Art*
florence	= 1 if *The Art History of florence* was bought; = 0 if not
Related Purchase	Number of related books purchased

Data Mining Techniques

Various data mining techniques can be used to mine the data collected from the market test. No one technique is universally better than another. The particular context and the particular characteristics of the data are the major factors in determining which techniques perform better in an application. For this assignment we focus on two fundamental techniques: *k*-nearest-neighbors and logistic regression. We compare them with each other as well as with a standard industry practice known as *RFM (recency, frequency, monetary) segmentation.*

RFM Segmentation   The segmentation process in database marketing aims to partition customers in a list of prospects into homogeneous groups (segments) that are similar with respect to buying behavior. The homogeneity criterion we need for segmentation is the propensity to purchase the offering. However, since we cannot measure this attribute, we use variables that are plausible indicators of this propensity.

In the direct marketing business, the most commonly used variables are the *RFM variables*:

R = *recency*, time since last purchase
F = *frequency*, number of previous purchases from the company over a period
M = *monetary*, amount of money spent on the company's products over a period

The assumption is that the more recent the last purchase, the more products bought from the company in the past, and the more money spent in the past buying the company's products, the more likely the customer is to purchase the product offered.

The 1800 observations in the training data and the 1400 observations in the validation data were divided into recency, frequency, and monetary categories as follows:

Recency:

0–2 months (Rcode = 1)
3–6 months (Rcode = 2)
7–12 months (Rcode = 3)
13 months and up (Rcode = 4)

Frequency:

1 book (Fcode = 1)
2 books (Fcode = 2)
3 books and up (Fcode = 3)

Monetary:

$0–$25 (Mcode = 1)
$26–$50 (Mcode = 2)
$51–$100 (Mcode = 3)
$101–$200 (Mcode = 4)
$201 and up (Mcode = 5)

Tables 21.2 and 21.3 display the 1800 customers in the training data cross-tabulated by these categories. Both buyers and nonbuyers are summarized. These tables are available for Excel computations in the RFM spreadsheet in the data file.

Assignment

1. What is the response rate for the training data customers taken as a whole? What is the response rate for each of the $4 \times 5 \times 3 = 60$ combinations of RFM categories? Which combinations have response rates in the training data that are above the overall response in the training data?

2. Suppose that we decide to send promotional mail only to the "above-average" RFM combinations identified in part 1. Compute the response rate in the validation data using these combinations.

3. Rework parts 1 and 2 with three segments:

   Segment 1: RFM combinations that have response rates that exceed twice the overall response rate

   Segment 2: RFM combinations that exceed the overall response rate but do not exceed twice that rate

   Segment 3: remaining RFM combinations

   Draw the lift curve (consisting of three points for these three segments) showing the number of customers in the validation dataset on the x-axis and cumulative number of buyers in the validation dataset on the y-axis.

k-Nearest-Neighbors The k-nearest-neighbors technique can be used to create segments based on product proximity to similar products of the products offered as well as the propensity to purchase (as measured by the RFM variables). For *The Art History of Florence*, a possible segmentation by product proximity could be created using the following variables:

R: recency–months since last purchase

F: frequency–total number of past purchases

M: monetary–total money (in dollars) spent on books

FirstPurch: months since first purchase

TABLE 21.2	RFM COUNTS FOR BUYERS

Rcode = all
Sum of Florence   Mcode

Fcode	1	2	3	4	5	Grand Total
1	2	2	10	7	17	38
2		3	5	9	17	34
3		1	1	15	62	79
Grand Total	2	6	16	31	96	151

Rcode = 1
Sum of Florence   Mcode

Fcode	1	2	3	4	5	Grand Total
1	0	0	0	2	1	3
2		1	0	0	1	2
3		1	0	0	5	6
Grand Total	0	2	0	2	7	11

Rcode = 2
Sum of Florence   Mcode

Fcode	1	2	3	4	5	Grand Total
1	1	0	1	1	5	8
2		0	3	5	5	13
3			0	4	10	14
Grand Total	1	0	4	10	20	35

Rcode = 3
Sum of Florence   Mcode

Fcode	1	2	3	4	5	Grand Total
1	1	0	1	2	5	9
2		1	1	2	4	8
3		0	0	4	31	35
Grand Total	1	1	2	8	40	52

Rcode = 4
Sum of Florence   Mcode

Fcode	1	2	3	4	5	Grand Total
1	0	2	8	2	6	18
2		1	1	2	7	11
3			1	7	16	24
Grand Total	0	3	10	11	29	53

TABLE 21.3   RFM COUNTS FOR ALL CUSTOMERS (BUYERS AND NONBUYERS)

Rcode = all
Count of Florence   Mcode

Fcode	1	2	3	4	5	Grand Total
1	20	40	93	166	219	538
2		32	91	180	247	550
3		2	33	179	498	712
Grand Total	20	74	217	525	964	1800

Rcode = 1
Count of Florence   Mcode

Fcode	1	2	3	4	5	Grand Total
1	2	2	6	10	15	35
2		3	4	12	16	35
3		1	2	11	45	59
Grand Total	2	6	12	33	76	129

Rcode = 2
Count of Florence   Mcode

Fcode	1	2	3	4	5	Grand Total
1	3	5	17	28	26	79
2		2	17	30	31	80
3			3	34	66	103
Grand Total	3	7	37	92	123	262

Rcode = 3
Count of Florence   Mcode

Fcode	1	2	3	4	5	Grand Total
1	7	15	24	51	86	183
2		12	29	55	85	181
3		1	17	53	165	236
Grand Total	7	28	70	159	336	600

Rcode = 4
Count of Florence   Mcode

Fcode	1	2	3	4	5	Grand Total
1	8	18	46	77	92	241
2		15	41	83	115	254
3			11	81	222	314
Grand Total	8	33	98	241	429	809

RelatedPurch: total number of past purchases of related books (i.e., sum of purchases from the art and geography categories and of titles *Secrets of Italian Cooking, Historical Atlas of Italy,* and *Italian Art*)

4. Use the *k*-nearest-neighbor option under the Classify menu choice in XLMiner to classify cases with *k* = 1, *k* = 3, and *k* = 11, using "Florence" as the target variable. Also use XLMiner's automated function to find the best *k*. Use normalized data (note the checkbox "normalize input data" in the dialog box) and all five variables. Create a lift curve for the best *k* model, and report the expected lift for an equal number of customers from the validation dataset.

5. OPTIONAL: Use the *k*-nearest-neighbor option under the Prediction menu choice in XLMiner to compute a lift curve for the validation data for the best *k* calculated above. Use normalized data (note the checkbox "normalize input data" in the dialog box) and all five variables. The *k*-NN prediction algorithm gives a numerical value, which is a weighted average of the values of the Florence variable for the *k*-nearest-neighbors with weights that are inversely proportional to distance. What is the range within which a prediction will fall? How does it compare to the output you get with the *k*-nearest-neighbor classification option?

Logistic Regression The logistic regression model offers a powerful method for modeling response because it yields well-defined purchase probabilities. The model is especially attractive in consumer-choice settings because it can be derived from the random utility theory of consumer behavior.

Use the training set data of 1800 observations to construct three logistic regression models with "Florence" as the target variable and each of the following sets of predictors:

- The full set of 15 predictors in the dataset.
- A subset of predictors that you judge to be the best.
- Only the *R*, *F*, and *M* variables.

6. Score the customers in the validation sample and arrange them in descending order of purchase probabilities.

7. Create a lift chart summarizing the results from the three logistic regression models created above, along with the expected lift for a random selection of an equal number of customers from the validation dataset.

8. If the cutoff criterion for a campaign is a 30% likelihood of a purchase, find the customers in the validation data that would be targeted and count the number of buyers in this set.

21.2 GERMAN CREDIT

GermanCredit.xls is the dataset for this case study.

Background

Money-lending has been around since the advent of money; it is perhaps the world's second-oldest profession. The systematic evaluation of credit risk, though, is a relatively recent arrival, and lending was largely based on reputation and very incomplete data. Thomas Jefferson, the third President of the United States, was in debt throughout his life and unreliable in his debt payments, yet people continued to lend him money. It wasn't until the beginning of the 20th century that the Retail Credit Company was founded to share information about credit. That company is now Equifax, one of the big three credit-scoring agencies (the other two are Transunion and Experion).

Individual and local human judgment are now largely irrelevant to the credit-reporting process. Credit agencies and other big financial institutions extending credit at the retail level collect huge amounts of data to predict whether defaults or other adverse events will occur, based on numerous customer and transaction information.

Data

This case deals with an early stage of the historical transition to predictive modeling, in which humans were employed to label records as either good or poor credit. The German Credit dataset[2] has 30 variables and 1000 records, each record being a prior applicant for credit. Each applicant was rated as "good credit" (700 cases) or "bad credit" (300 cases). Figure 21.1 shows the values

OBS#	CHK_ACCT	DURATION	HISTORY	NEW_CAR	USED_CAR	FURNITURE	RADIO/TV	EDUCATION	RETRAINING	AMOUNT	SAV_ACCT	EMPLOYMENT	INSTALL_RATE	MALE_DIV	MALE_SINGLE	MALE_MAR_WID	CO-APPLICANT	GUARANTOR	PRESENT_RESIDENT	REAL_ESTATE	PROP_UNKN_NONE	AGE	OTHER_INSTALL	RENT	OWN_RES	NUM_CREDITS	JOB	NUM_DEPENDENTS	TELEPHONE	FOREIGN	RESPONSE
1	0	6	4	0	0	0	1	0	0	1169	4	4	4	0	1	0	0	0	4	1	0	67	0	0	1	2	2	1	1	0	1
2	1	48	2	0	0	0	1	0	0	5951	0	2	2	0	0	0	0	0	2	1	0	22	0	0	1	1	2	1	0	0	0
3	3	12	4	0	0	0	0	1	0	2096	0	3	2	0	1	0	0	0	3	1	0	49	0	0	1	1	1	2	0	0	1
4	0	42	2	0	0	1	0	0	0	7882	0	3	2	0	1	0	0	1	4	0	0	45	0	0	0	1	2	2	0	0	1

FIGURE 21.1   **FIRST FOUR RECORDS FROM GERMAN CREDIT DATASET**

---

[2] This dataset is available from ftp.ics.uci.edu/pub/machine-learning-databases/statlog/.

of these variables for the first four records. All the variables are explained in Table 21.6. New applicants for credit can also be evaluated on these 30 predictor variables and classified as a good or a bad credit risk based on the predictor variables.

The consequences of misclassification have been assessed as follows: The costs of a false positive (incorrectly saying that an applicant is a good credit risk) outweigh the benefits of a true positive (correctly saying that an applicant is a good credit risk) by a factor of 5. This is summarized in Table 21.4. The opportunity cost table was derived from the average net profit per loan as shown in Table 21.5.

TABLE 21.4   OPPORTUNITY COST TABLE (DEUTSCHE MARKS)

| | Predicted (Decision) | |
Actual	Good (Accept)	Bad (Reject)
Good	0	100
Bad	500	0

TABLE 21.5   AVERAGE NET PROFIT (DEUTSCHE MARKS)

| | Predicted (Decision) | |
Actual	Good (Accept)	Bad (Reject)
Good	100	0
Bad	−500	0

Because decision makers are used to thinking of their decision in terms of net profits, we use these tables in assessing the performance of the various models.

TABLE 21.6   VARIABLES FOR THE GERMAN CREDIT DATASET

Var.	Variable Name	Description	Variable Type	Code Description
1	OBS#	Observation numbers	Categorical	Sequence number in dataset
2	CHK_ACCT	Checking account status	Categorical	0: < 0 DM
				1: 0–200 DM
				2 : > 200 DM
				3: no checking account
3	DURATION	Duration of credit in months	Numerical	
4	HISTORY	Credit history	Categorical	0: no credits taken
				1: all credits at this bank paid back duly
				2: existing credits paid back duly until now
				3: delay in paying off in the past
				4: critical account
5	NEW_CAR	Purpose of credit	Binary	car (new) 0: No, 1: Yes
6	USED_CAR	Purpose of credit	Binary	car (used) 0: No, 1: Yes
7	FURNITURE	Purpose of credit	Binary	furniture/equipment 0: No, 1: Yes
8	RADIO/TV	Purpose of credit	Binary	radio/television 0: No, 1: Yes
9	EDUCATION	Purpose of credit	Binary	education 0: No, 1: Yes
10	RETRAINING	Purpose of credit	Binary	retraining 0: No, 1: Yes
11	AMOUNT	Credit amount	Numerical	
12	SAV_ACCT	Average balance in savings account	Categorical	0: < 100 DM
				1 : 101–500 DM
				2 : 501–1000 DM
				3 : > 1000 DM
				4 : unknown/ no savings account
13	EMPLOYMENT	Present employment since	Categorical	0 : unemployed
				1: < 1 year
				2: 1–3 years
				3: 4–6 years
				4: ≥ 7 years
14	INSTALL_RATE	Installment rate as % of disposable income	Numerical	
15	MALE_DIV	Applicant is male and divorced	Binary	0: No, 1: Yes

(continued)

TABLE 21.6 *(CONTINUED)*

Var.	Variable Name	Description	Variable Type	Code Description
16	MALE_SINGLE	Applicant is male and single	Binary	0: No, 1: Yes
17	MALE_MAR_WID	Applicant is male and married or a widower	Binary	0: No, 1: Yes
18	CO-APPLICANT	Application has a coapplicant	Binary	0: No, 1: Yes
19	GUARANTOR	Applicant has a guarantor	Binary	0: No, 1: Yes
20	PRESENT_RESIDENT	Present resident since (years)	Categorical	0: $\leq$ 1 year 1: 1–2 years 2: 2–3 years 3: $\geq$ 3 years
21	REAL_ESTATE	Applicant owns real estate	Binary	0: No, 1: Yes
22	PROP_UNKN_NONE	Applicant owns no property (or unknown)	Binary	0: No, 1: Yes
23	AGE	Age in years	Numerical	
24	OTHER_INSTALL	Applicant has other installment plan credit	Binary	0: No, 1: Yes
25	RENT	Applicant rents	Binary	0: No, 1: Yes
26	OWN_RES	Applicant owns residence	Binary	0: No, 1: Yes
27	NUM_CREDITS	Number of existing credits at this bank	Numerical	
28	JOB	Nature of job	Categorical	0 : unemployed/ unskilled— nonresident 1 : unskilled— resident 2 : skilled employee/ official 3 : management/ self-employed/ highly qualified employee/officer
29	NUM_DEPENDENTS	Number of people for whom liable to provide maintenance	Numerical	
30	TELEPHONE	Applicant has phone in his or her name	Binary	0: No, 1: Yes
31	FOREIGN	Foreign worker	Binary	0: No, 1: Yes
32	RESPONSE	Credit rating is good	Binary	0: No, 1: Yes

Note: The original dataset had a number of categorical variables, some of which were transformed into a series of binary variables so that they can be handled appropriately by XLMiner. Several ordered categorical variables were left as is, to be treated by XLMiner as numerical.

Assignment

1. Review the predictor variables and guess what their role in a credit decision might be. Are there any surprises in the data?

2. Divide the data into training and validation partitions, and develop classification models using the following data mining techniques in XLMiner: logistic regression, classification trees, and neural networks.

3. Choose one model from each technique and report the confusion matrix and the cost/gain matrix for the validation data. Which technique has the highest net profit?

4. Let us try and improve our performance. Rather than accept XLMiner's initial classification of all applicants' credit status, use the "predicted probability of success" in logistic regression (where *success* means 1) as a basis for selecting the best credit risks first, followed by poorer-risk applicants.

   - Sort the validation on "predicted probability of success,"
   - For each case, calculate the net profit of extending credit.
   - Add another column for cumulative net profit.

   a. How far into the validation data should you go to get maximum net profit? (Often this is specified as a percentile or rounded to deciles.)

   b. If this logistic regression model is used to score to future applicants, what "probability of success" cutoff should be used in extending credit?

21.3 TAYKO SOFTWARE CATALOGER[3]

Tayko.xls is the dataset for this case study.

Background

Tayko is a software catalog firm that sells games and educational software. It started out as a software manufacturer and later added third-party titles to its offerings. It has recently put together a revised collection of items in a new catalog, which it is preparing to roll out in a mailing.

In addition to its own software titles, Tayko's customer list is a key asset. In an attempt to expand its customer base, it has recently joined a consortium of catalog firms that specialize in computer and software products. The consortium affords members the opportunity to mail catalogs to names drawn from a pooled list of customers. Members supply their own customer lists to the pool, and can "withdraw" an equivalent number of names each quarter. Members are allowed to do predictive modeling on the records in the pool so they can do a better job of selecting names from the pool.

The Mailing Experiment

Tayko has supplied its customer list of 200,000 names to the pool, which totals over 5,000,000 names, so it is now entitled to draw 200,000 names for a mailing. Tayko would like to select the names that have the best chance of performing well, so it conducts a test—it draws 20,000 names from the pool and does a test mailing of the new catalog.

This mailing yielded 1065 purchasers, a response rate of 0.053. Average spending was $103 for each of the purchasers, or $5.46 per catalog mailed. To optimize the performance of the data mining techniques, it was decided to work with a stratified sample that contained equal numbers of purchasers and nonpurchasers. For ease of presentation, the dataset for this case includes just 1000 purchasers and 1000 nonpurchasers, an apparent response rate of 0.5. Therefore, after using the dataset to predict who will be a purchaser, we must adjust the purchase rate back down by multiplying each case's "probability of purchase" by 0.053/0.5, or 0.107.

Data

There are two response variables in this case. *Purchase* indicates whether or not a prospect responded to the test mailing and purchased something. *Spending* indicates, for those who made a purchase, how much they spent. The overall procedure in this case will be to develop two models. One will be used to classify

---

records as *purchase* or *no purchase*. The second will be used for those cases that are classified as *purchase* and will predict the amount they will spend.

Table 21.7 provides a description of the variables available in this case. A partition variable is used because we will be developing two different models in this case and want to preserve the same partition structure for assessing each model.

TABLE 21.7   DESCRIPTION OF VARIABLES FOR TAYKO DATASET

Var.	Variable Name	Description	Variable Type	Code Description
1	US	Is it a U.S. address?	Binary	1: yes 0: no
2–16	Source_*	Source catalog for the record (15 possible sources)	Binary	1: yes 0: no
17	Freq.	Number of transactions in last year at source catalog	Numerical	
18	last_update_days_ago	How many days ago last update was made to customer record	Numerical	
19	1st_update_days_ago	How many days ago first update to customer record was made	Numerical	
20	RFM%	Recency–frequency– monetary percentile, as reported by source catalog (see Section 21.1)	Numerical	
21	Web_order	Customer placed at least one order via Web	Binary	1: yes 0: no
22	Gender=mal	Customer is male	Binary	1: yes 0: no
23	Address_is_res	Address is a residence	Binary	1: yes 0: no
24	Purchase	Person made purchase in test mailing	Binary	1: yes 0: no
25	Spending	Amount (dollars) spent by customer in test mailing	Numerical	
26	Partition	Variable indicating which partition the record will be assigned to	Alphabetical	t: training v: validation s: test

Figure 21.2 shows the first few rows of data (the top shows the sequence number plus the first 14 variables, and the bottom shows the remaining 11 variables for the same rows).

sequence_number	US	source_a	source_c	source_b	source_d	source_e	source_m	source_o	source_h	source_r	source_s	source_t	source_u	source_p	source_x	source_w	Freq	last_update_days_ago	1st_update_days_ago	Web order	Gender = male	Address_is_res	Purchase	Spending	Partition
1	1	0	0	1	0	0	0	0	0	0	0	0	0	0	0	0	2	3662	3662	1	0	1	1	127.87	s
2	1	0	0	0	0	0	1	0	0	0	0	0	0	0	0	0	0	2900	2900	1	1	0	0	0	s
3	1	0	0	0	0	0	0	0	0	1	0	0	0	0	0	0	2	3883	3914	0	0	0	1	127.48	t
4	1	0	1	0	0	0	0	0	0	0	0	0	0	0	0	0	1	829	829	0	1	0	0	0	s
5	1	0	1	0	0	0	0	0	0	0	0	0	0	0	0	0	1	869	869	0	0	0	0	0	t
6	1	0	0	0	0	0	0	0	0	1	0	0	0	0	0	0	1	1995	2002	0	0	1	0	0.06	s
7	1	0	0	0	0	0	0	0	0	0	0	0	0	0	0	1	2	1498	1529	0	0	1	0	0.06	s
8	1	0	0	1	0	0	0	0	0	0	0	0	0	0	0	0	1	3397	3397	0	1	0	0	0.08	t
9	1	1	0	0	0	0	0	0	0	0	0	0	0	0	0	0	4	525	2914	1	1	0	1	488.5	t
10	1	1	0	0	0	0	0	0	0	0	0	0	0	0	0	0	1	3215	3215	0	0	0	1	173.5	v

FIGURE 21.2   FIRST 10 RECORDS FROM TAYKO DATASET

Assignment

1. Each catalog costs approximately $2 to mail (including printing, postage, and mailing costs). Estimate the gross profit that the firm could expect from the remaining 180,000 names if it selects them randomly from the pool.

2. Develop a model for classification of a customer as a purchaser or non-purchaser.

   a. Partition the data into training data on the basis of the partition variable, which has 800 t's, 700 v's, and 500 s's (training data, validation data, and test data, respectively) assigned randomly to cases.

   b. Using the "best-subset" option in logistic regression, implement the full logistic regression model, select the best subset of variables, then implement a regression model with just those variables to classify the data into purchasers and nonpurchasers. (Logistic regression is used because it yields an estimated "probability of purchase," which is required later in the analysis.)

3. Develop a model for predicting spending among the purchasers.

   a. Make a copy of the data sheet (call it data2), sort by the "Purchase" variable, and remove the records where Purchase = 0 (the resulting spreadsheet will contain only purchasers).

   b. Partition this dataset into training and validation partitions on the basis of the partition variable.

   c. Develop models for predicting spending, using:
      i. Multiple linear regression (use best subset selection)
      ii. Regression trees

    d. Choose one model on the basis of its performance with the validation data.

4. Return to the original test data partition. Note that this test data partition includes both purchasers and nonpurchasers. Note also that although it contains the scoring of the chosen classification model, we have not used this partition in our analysis up to this point, so it will give an unbiased estimate of the performance of our models. It is best to make a copy of the test data portion of this sheet to work with, since we will be adding analyses to it. This copy is called *Score Analysis*.

    a. Copy to this sheet the "predicted probability of success" (*success = purchase*) column from the classification of test data.

    b. Score to this data sheet the prediction model chosen.

    c. Arrange the following columns so that they are adjacent:

        i. Predicted probability of purchase (*success*)

        ii. Actual spending (dollars)

        iii. Predicted spending (dollars)

    d. Add a column for "adjusted probability of purchase" by multiplying "predicted probability of purchase" by 0.107. *This is to adjust for oversampling the purchasers* (see above).

    e. Add a column for expected spending [adjusted probability of purchase × predicted spending].

    f. Sort all records on the "expected spending" column.

    g. Calculate cumulative lift [= cumulative "actual spending" divided by the average spending that would result from random selection (each adjusted by 0.107)].

    h. Using this lift curve, estimate the gross profit that would result from mailing to the 180,000 names on the basis of your data mining models.

Note: Although Tayko is a hypothetical company, the data in this case (modified slightly for illustrative purposes) were supplied by a real company that sells software through direct sales. The concept of a catalog consortium is based on the Abacus Catalog Alliance. Details can be found at www.abacusalliance.com/faqs/membership-provide/.

21.4   POLITICAL PERSUASION[4]

`Voter-Persuasion.xlsx` is the dataset for this case study.

Note: Our thanks to Ken Strasma, President of HaystaqDNA and director of targeting for the 2004 Kerry campaign and the 2008 Obama campaign, for the data used in this case, and for sharing the information in the following write-up.

Background

When you think of political persuasion, you may think of the efforts that political campaigns undertake to persuade you that their candidate is better than the other candidate. In truth, campaigns are less about persuading people to change their minds, and more about persuading those who agree with you to actually go out and vote. Predictive analytics now plays a big role in this effort, but in 2004 it was a new arrival in the political toolbox.

Predictive Analytics Arrives in US Politics

In January of 2004, candidates in the US presidential campaign were competing in the Iowa caucuses, part of a lengthy state-by-state primary campaign that culminates in the selection of the Republican and Democratic candidates for president. Among the Democrats, Howard Dean was leading in national polls. The Iowa caucuses, however, are a complex and intensive process attracting only the most committed and interested voters. Those participating are not a representative sample of voters nationwide. Surveys of those planning to take part showed a close race between Dean and three other candidates, including John Kerry. Kerry ended up winning by a surprisingly large margin, and the better than expected performance was due to his campaign's innovative and successful use of predictive analytics to learn more about the likely actions of individual voters. This allowed the campaign to target voters in such a way as to optimize performance in the caucuses. For example, once the model showed sufficient support in a precinct to win that precinct's delegate to the caucus, money and time could be redirected to other precincts where the race was closer.

Political Targeting

Targeting of voters is not new in politics. It has traditionally taken three forms:

- Geographic
- Demographic
- Individual

---
[4] © Statistics.com, Inc. 2015; used with permission.

In geographic targeting, resources are directed to a geographic unit—state, city, county, and so on—on the basis of prior voting patterns or surveys that reveal the political tendency in that geographic unit. This has significant limitations, though. If a county is only, say, 52% in your favor, it may be in the greatest need of attention, but if messaging is directed to everyone in the county, nearly half of it is reaching the wrong people.

In demographic targeting, the messaging is intended for demographic groups—older voters, younger women voters, Hispanic voters, an so on. The limitation of this method is that it is often not easy to implement because messaging is hard to deliver just to single demographic groups.

Traditional individual targeting, the most effective form of targeting, was done on the basis of surveys asking voters how they plan to vote. The big limitation of this method is, of course, the cost. The expense of reaching all voters in a phone or door-to-door survey can be prohibitive.

The use of predictive analytics adds power to the individual targeting method, and reduces cost. A model allows prediction to be rolled out to the entire voter base, not just those surveyed, and brings to bear a wealth of information. Geographic and demographic data remain part of the picture, but they are used at an individual level.

Uplift

In a classical predictive modeling application for marketing, a sample of data is selected and an offer is made (e.g., on the web) or a message is sent (e.g., by mail), and a predictive model is developed to classify individuals as responding or not-responding. The model is then applied to new data, propensities to respond are calculated, individuals are ranked by their propensity to respond, and the marketer can then select those most likely to respond to mailings or offers.

Some key information is missing from this classical approach: how would the individual respond in the absence of the offer or mailing? Might a high propensity customer be inclined to purchase regardless of the offer? Might a person's propensity to buy actually be diminished by the offer? Uplift modeling (see Chapter 13) allows us to estimate the effect of "offer vs. no offer" or "mailing vs. no mailing" at the individual level.

In this case we will apply uplift modeling to actual voter data that were augmented with the results of a hypothetical experiment. The experiment consisted of the following steps:

1. Conduct a pre-survey of the voters to determine their inclination to vote Democratic.

2. Randomly split the voters into two samples—control and treatment.

3. Send a flyer promoting the Democratic candidate to the treatment group.

4. Conduct another survey of the voters to determine their inclination to vote Democratic.

Data

The data in this case are in the file Voter-Persuasion.xlsx. The target variable is MOVED_AD, where a 1 = "opinion moved in favor of the Democratic candidate" and 0 = "opinion did not move in favor of the Democratic candidate." This variable encapsulates the information from the pre- and post-surveys. The important predictor variable is *Flyer*, a binary variable that indicates whether or not a voter received the flyer. In addition there are numerous other predictor variables from these sources:

1. Government voter files

2. Political party data

3. Commercial consumer and demographic data

4. Census neighborhood data

Government voter files are maintained, and made public, to assure the integrity of the voting process. They contain essential data for identification purposes such as name, address, and date of birth. The file used in this case also contains party identification (needed if a state limits participation in party primaries to voters in that party). Parties also staff elections with their own poll-watchers, who record whether an individual votes in an election. These data (termed "derived" in the case data) are maintained and curated by each party, and can be readily matched to the voter data by name. Demographic data at the neighborhood level are available from the census, and can be appended to the voter data by address matching. Consumer and additional demographic data (buying habits, education) can be purchased from marketing firms and appended to the voter data (matching by name and address).

Assignment

The task in this case is to develop an uplift model that predicts the uplift for each voter. Uplift is defined as the increase in propensity to move one's opinion in a Democratic direction. First, review the variables in `Voter-Persuasion.xlsx` and understand which data source they are probably coming from. Then, answer the following questions and perform the tasks indicated:

1. Overall, how well did the flyer do in moving voters in a Democratic direction? (Look at the target variable among those who got the flyer, compared to those who did not.)

2. Explore the data to learn more about the relationships between the predictor variables and MOVED_AD using data visualization. Which of the predictors seem to have good predictive potential? Show supporting charts and/or tables.

3. Partition the data using the partition variable that is in the dataset, make decisions about predictor inclusion, and fit three predictive models accordingly. For each model, give sufficient detail about the method used, its parameters, and the predictors used, so that your results can be replicated.

4. Among your three models, choose the best one in terms of predictive power. Which one is it? Why did you choose it?

5. Using your chosen model, report the propensities for the first three records in the validation set.

6. Create a derived variable that is the opposite of *Flyer*. Call it *Flyer-reversed*. Using your chosen model, re-score the validation data using the *Flyer-reversed* variable as a predictor, instead of *Flyer*. You can do this either by using the stored model utility in XLMiner or by creating a second copy of the validation data with the *Flyer-reversed* variable as a predictor and repeating the model with that copy as "new data" to be scored. Report the propensities for the first three records in the validation set.

7. For each record, uplift is computed based on the following difference:

$$P(success|Flyer = 1) - P(success|Flyer = 0)$$

Compute the uplift for each of the voters in the validation set, and report the uplift for the first three records.

8. If a campaign has the resources to mail the flyer only to 10% of the voters, what uplift cutoff should be used?

21.5 TAXI CANCELLATIONS[5]

`Taxi-cancellation-case.xlsx` is the dataset for this case study.

Business Situation

In late 2013, the taxi company Yourcabs.com in Bangalore, India, was facing a problem with the drivers using their platform—not all drivers were showing up for their scheduled calls. Drivers would cancel their acceptance of a call and, if the cancellation did not occur with adequate notice, the customer would be delayed or even left high and dry.

Bangalore is a key tech center in India, and technology was transforming the taxi industry. Yourcabs.com featured an online booking system (though customers could phone in as well) and presented itself as a taxi booking portal. The Uber ride sharing service would start its Bangalore operations in mid-2014. Yourcabs.com had collected data on its bookings from 2011 to through 2013, and posted a contest on Kaggle, in coordination with the Indian School of Business, to see what it could learn about the problem of cab cancellations.

The data presented for this case are a randomly selected subset of the original data, with 10,000 rows, one row for each booking. There are 17 input variables, including user (customer) ID, vehicle model, whether the booking was made online or via a mobile app, type of travel, type of booking package, geographic information, and the date and time of the scheduled trip. The target variable of interest is the binary indicator of whether a ride was canceled. The overall cancellation rate is between 7% and 8%.

Assignment

1. How can a predictive model based on these data be used by Yourcabs.com?

2. How can a profiling model (identifying predictors that distinguish canceled/uncanceled trips) be used by Yourcabs.com?

3. Explore, prepare, and transform the data to facilitate predictive modeling. Here are some hints:

   o In exploratory modeling it is useful to move fairly soon to at least an initial model without solving *all* data preparation issues. One example is the GPS information—other geographic information is available so you could defer the challenge of how to interpret/use the GPS information.

   o How will you deal with missing data, such as cases where NULL is indicated?

---

[5] © Statistics.com, Inc. and Galit Shmueli 2015; used with permission.

o Think about what useful information might be held within the date and time fields (the booking timestamp and the trip timestamp). The data file contains a worksheet with some hints on how to extract features from the date/time field.

o Think also about the categorical variables, and how to deal with them. Should we turn them all into dummies? Use only some?

4. Fit several predictive models of your choice. Do they provide information on how the predictor variables relate to cancellations?

5. Report the predictive performance of your model in terms of error rates (the confusion matrix). How well does the model perform? Can the model be used in practice?

6. Examine the predictive performance of your model in terms of ranking (lift). How well does the model perform? Can the model be used in practice?

21.6 SEGMENTING CONSUMERS OF BATH SOAP[6]

`BathSoap.xls` is the dataset for this case study.

Business Situation

CRISA is an Asian market research agency that specializes in tracking consumer purchase behavior in consumer goods (both durable and nondurable). In one major research project, CRISA tracks numerous consumer product categories (e.g., "detergents") and, within each category, perhaps dozens of brands. To track purchase behavior, CRISA constituted household panels in over 100 cities and towns in India, covering most of the Indian urban market. The households were carefully selected using stratified sampling to ensure a representative sample; a subset of 600 records is analyzed here. The strata were defined on the basis of socioeconomic status and the market (a collection of cities).

CRISA has both transaction data (each row is a transaction) and household data (each row is a household), and for the household data it maintains the following information:

- Demographics of the households (updated annually)

- Possession of durable goods (car, washing machine, etc., updated annually; an "affluence index" is computed from this information)

- Purchase data of product categories and brands (updated monthly)

---

[6] © Cytel, Inc. and Resampling Stats, Inc. 2006; used with permission.

CRISA has two categories of clients: (1) advertising agencies that subscribe to the database services, obtain updated data every month, and use the data to advise their clients on advertising and promotion strategies; (2) consumer goods manufacturers that monitor their market share using the CRISA database.

Key Problems

CRISA has traditionally segmented markets on the basis of purchaser demographics. They would now like to segment the market based on two key sets of variables more directly related to the purchase process and to brand loyalty:

1. Purchase behavior (volume, frequency, susceptibility to discounts, and brand loyalty)
2. Basis of purchase (price, selling proposition)

Doing so would allow CRISA to gain information about what demographic attributes are associated with different purchase behaviors and degrees of brand loyalty, and thus deploy promotion budgets more effectively. More effective market segmentation would enable CRISA's clients (in this case, a firm called IMRB) to design more cost-effective promotions targeted at appropriate segments. Thus multiple promotions could be launched, each targeted at different market segments at different times of the year. This would result in a more cost-effective allocation of the promotion budget to different market segments. It would also enable IMRB to design more effective customer reward systems and thereby increase brand loyalty.

Data

The data in Table 21.8 profile each household, each row containing the data for one household.

Measuring Brand Loyalty

Several variables in this case measure aspects of brand loyalty. The number of different brands purchased by the customer is one measure of loyalty. However, a consumer who purchases one or two brands in quick succession, and then settles on a third for a long streak, is different from a consumer who constantly switches back and forth among three brands. Therefore how often customers switch from one brand to another is another measure of loyalty. Yet a third perspective on the same issue is the proportion of purchases that go to different brands—a consumer who spends 90% of his or her purchase money on one brand is more loyal than a consumer who spends more equally among several brands.

All three of these components can be measured with the data in the purchase summary worksheet.

TABLE 21.8   **DESCRIPTION OF VARIABLES FOR EACH HOUSEHOLD**

Variable Type	Variable Name	Description
Member ID	Member id	Unique identifier for each household
Demographics	SEC	Socioeconomic class (1 = high, 5 = low)
	FEH	Eating habits(1 = vegetarian, 2 = vegetarian but eat eggs, 3 = nonvegetarian, 0 = not specified)
	MT	Native language (see table in worksheet)
	SEX	Gender of homemaker (1 = male, 2 = female)
	AGE	Age of homemaker
	EDU	Education of homemaker (1 = minimum, 9 = maximum)
	HS	Number of members in household
	CHILD	Presence of children in household (4 categories)
	CS	Television availability (1 = available, 2 = unavailable)
	Affluence Index	Weighted value of durables possessed
Purchase summary over the period	No. of Brands	Number of brands purchased
	Brand Runs	Number of instances of consecutive purchase of brands
	Total Volume	Sum of volume
	No. of Trans	Number of purchase transactions (multiple brands purchased in a month are counted as separate transactions
	Value	Sum of value
	Trans/ Brand Runs	Average transactions per brand run
	Vol/Trans	Average volume per transaction
	Avg. Price	Average price of purchase
Purchase within promotion	Pur Vol	Percent of volume purchased
	No Promo - %	Percent of volume purchased under no promotion
	Pur Vol Promo 6%	Percent of volume purchased under promotion code 6
	Pur Vol OtherPromo %	Percent of volume purchased under other promotions
Brandwise purchase	Br. Cd. **(57, 144)**, **55, 272, 286, 24, 481, 352, 5**, and **999** (others)	Percent of volume purchased of the brand
Price categorywise purchase	Price Cat 1 to 4	Percent of volume purchased under the price category
Selling propositionwise purchase	Proposition Cat 5 to 15	Percent of volume purchased under the product proposition category

Assignment

1. Use *k*-means clustering to identify clusters of households based on:

   a. The variables that describe purchase behavior (including brand loyalty)

   b. The variables that describe the basis for purchase

   c. The variables that describe both purchase behavior and basis of purchase

   > **Note 1:** How should *k* be chosen? Think about how the clusters would be used. It is likely that the marketing efforts would support two to five different promotional approaches.

   > **Note 2:** How should the percentages of total purchases comprised by various brands be treated? Isn't a customer who buys all brand A just as loyal as a customer who buys all brand B? What will be the effect on any distance measure of using the brand share variables as is? Consider using a single derived variable.

2. Select what you think is the best segmentation and comment on the characteristics (demographic, brand loyalty, and basis for purchase) of these clusters. (This information would be used to guide the development of advertising and promotional campaigns.)

3. Develop a model that classifies the data into these segments. Since this information would most likely be used in targeting direct mail promotions, it would be useful to select a market segment that would be defined as a *success* in the classification model.

Appendix

Although not used in the assignment, two additional datasets were used in the derivation of the summary data.

CRISA_Purchase_Data is a transaction database in which each row is a transaction. Multiple rows in this dataset corresponding to a single household were consolidated into a single household row in CRISA_Summary_Data.

The *Durables* sheet in IMRB_Summary_Data contains information used to calculate the affluence index. Each row is a household, and each column represents a durable consumer good. "1" in the column indicates that the durable is possessed by the household; "0" indicates that it is not possessed. This value is multiplied by the weight assigned to the durable item. For example, "5" indicates the weighted value of possessing the durable. The sum of all the weighted values of the durables possessed equals the affluence index.

21.7 DIRECT-MAIL FUNDRAISING

`Fundraising.xls` and `FutureFundraising.xls` are the datasets used for this case study.

Background

Note: Be sure to read the information about oversampling and adjustment in Chapter 5 before starting to work on this case.

A national veterans' organization wishes to develop a predictive model to improve the cost-effectiveness of their direct marketing campaign. The organization, with its in-house database of over 13 million donors, is one of the largest direct-mail fundraisers in the United States. According to their recent mailing records, the overall response rate is 5.1%. Out of those who responded (donated), the average donation is $13.00. Each mailing, which includes a gift of personalized address labels and assortments of cards and envelopes, costs $0.68 to produce and send. Using these facts, we take a sample of this dataset to develop a classification model that can effectively capture donors so that the expected net profit is maximized. Weighted sampling is used, underrepresenting the nonresponders so that the sample has equal numbers of donors and nondonors.

Data

The file `Fundraising.xls` contains 3120 records with 50% donors (TARGET_B = 1) and 50% nondonors (TARGET_B = 0). The amount of donation (TARGET_D) is also included but is not used in this case. The descriptions for the 22 variables (including two target variables) are listed in Table 21.9.

Assignment

Step 1: Partitioning. Partition the dataset into 60% training and 40% validation (set the seed to 12345).

Step 2: Model Building. Follow the following steps to build, evaluate, and choose a model.

1. *Select classification tool and parameters.* Run at least two classification models of your choosing. Be sure NOT to use TARGET_D in your analysis. Describe the two models that you chose, with sufficient detail (method, parameters, variables, etc.) so that it can be replicated.

2. *Classification under asymmetric response and cost.* What is the reasoning behind using weighted sampling to produce a training set with equal numbers of donors and non-donors? Why not use a simple random sample from the original dataset?

TABLE 21.9	DESCRIPTION OF VARIABLES FOR THE FUNDRAISING DATASET

Variable	Description
ZIP	Zip code group (Zip codes were grouped into five groups: 1 = the potential donor belongs to this zip group.) 00000–19999 → zipconvert_1 20000–39999 → zipconvert_2 40000–59999 → zipconvert_3 60000–79999 → zipconvert_4 80000–99999 → zipconvert_5
HOMEOWNER	1 = homeowner, 0 = not a homeowner
NUMCHLD	Number of children
INCOME	Household income
GENDER	0 = male, 1 = female
WEALTH	Wealth rating uses median family income and population statistics from each area to index relative wealth within each state. The segments are denoted 0 to 9, with 9 being the highest wealth group and zero the lowest. Each rating has a different meaning within each state.
HV	Average home value in potential donor's neighborhood in hundreds of dollars
ICmed	Median family income in potential donor's neighborhood in hundreds of dollars
ICavg	Average family income in potential donor's neighborhood in hundreds
IC15	Percent earning less than $15K in potential donor's neighborhood
NUMPROM	Lifetime number of promotions received to date
RAMNTALL	Dollar amount of lifetime gifts to date
MAXRAMNT	Dollar amount of largest gift to date
LASTGIFT	Dollar amount of most recent gift
TOTALMONTHS	Number of months from last donation to July 1998 (the last time the case was updated)
TIMELAG	Number of months between first and second gift
AVGGIFT	Average dollar amount of gifts to date
TARGET_B	Target variable: binary indicator for response 1 = donor, 0 = nondonor
TARGET_D	Target variable: donation amount (in dollars). We will NOT be using this variable for this case.

3. *Calculate net profit.* For each method, calculate the lift of net profit for both the training and validation sets based on the actual response rate (5.1%.) Again, the expected donation, given that they are donors, is $13.00, and the total cost of each mailing is $0.68. (**Hint:** To calculate estimated net profit, we will need to undo the effects of the weighted sampling and calculate the net profit that would reflect the actual response distribution of 5.1% donors and 94.9% nondonors. To do this, divide each row's net profit by the oversampling weights applicable to the actual status of that row. The oversampling weight for actual donors is 50%/5.1% =

9.8. The oversampling weight for actual non–donors is 50%/94.9% = 0.53.)

4. *Draw lift curves.* Draw each model's net profit lift curve for the validation set onto a single graph (net profit on the *y*-axis, proportion of list or number mailed on the *x*-axis). Is there a model that dominates?

5. *Select best model.* From your answer in (4.), what do you think is the "best" model?

Step 3: Testing. The file `FutureFundraising.xls` contains the attributes for future mailing candidates.

6. Using your "best" model from step 2 (number 5), which of these candidates do you predict as donors and non-donors? List them in descending order of the probability of being a donor. Starting at the top of this sorted list, roughly how far down would you go in a mailing campaign?

21.8 CATALOG CROSS-SELLING[7]

`CatalogCrossSell.xls` is the dataset for this case study.

Background

Exeter, Inc. is a catalog firm that sells products in a number of different catalogs that it owns. The catalogs number in the dozens, but fall into nine basic categories:

1. Clothing
2. Housewares
3. Health
4. Automotive
5. Personal electronics

---

[7] © Resampling Stats, Inc. 2006; used with permission.

6. Computers

7. Garden

8. Novelty gift

9. Jewelry

The costs of printing and distributing catalogs are high. By far the biggest cost of operation is the cost of promoting products to people who buy nothing. Having invested so much in the production of artwork and printing of catalogs, Exeter wants to take every opportunity to use them effectively. One such opportunity is in cross-selling—once a customer has "taken the bait" and purchases one product, try to sell them another while you have their attention.

Such cross promotion might take the form of enclosing a catalog in the shipment of a purchased product, together with a discount coupon to induce a purchase from that catalog. Or, it might take the form of a similar coupon sent by email, with a link to the web version of that catalog.

But which catalog should be enclosed in the box or included as a link in the email with the discount coupon? Exeter would like it to be an informed choice—a catalog that has a higher probability of inducing a purchase than simply choosing a catalog at random.

Assignment

Using the dataset `CatalogCrossSell.xls`, perform an association rules analysis, and comment on the results. Your discussion should provide interpretations in English of the meanings of the various output statistics (lift ratio, confidence, support) and include a very rough estimate (precise calculations are not necessary) of the extent to which this will help Exeter make an informed choice about which catalog to cross-promote to a purchaser.

Acknowledgment The data for this case have been adapted from the data in a set of cases provided for educational purposes by the Direct Marketing Education Foundation ("DMEF Academic Data Set Two, Multi Division Catalog Company, Code: 02DMEF"); used with permission.

21.9   PREDICTING BANKRUPTCY

Bankruptcy.xls is the dataset for this case study.

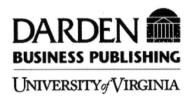

Predicting Corporate Bankruptcy[8]

Just as doctors check blood pressure and pulse rate as vital indicators of the health of a patient, so business analysts scour the financial statements of a corporation to monitor its financial health. Whereas blood pressure, pulse rate, and most medical vital signs are measured through precisely defined procedures, financial variables are recorded under much less specific general principles of accounting. A primary issue in financial analysis, then, is how predictable is the health of a company?

One difficulty in analyzing financial report information is the lack of disclosure of actual cash receipts and disbursements. Users of financial statements have had to rely on proxies for cash flow, perhaps the simplest of which is income (INC) or earnings per share. Attempts to improve INC as a proxy for cash flow include using income plus depreciation (INCDEP), working capital from operations (WCFO), and cash flow from operations (CFFO). CFFO is obtained by adjusting income from operations for all noncash expenditures and revenues and for changes in the current asset and current liabilities accounts.

A further difficulty in interpreting historical financial disclosure information is caused whenever major changes are made in accounting standards. For example, the financial Accounting Standards Board issued several promulgations in the mid-1970s that changed the requirements for reporting accruals pertaining to such things as equity earnings, foreign currency gains and losses, and deferred taxes. One effect of changes of this sort was that earnings figures became less reliable indicators of cash flow.

---

[8] This case was prepared by Professor Mark E. Haskins and Professor Phillip E. Pfeifer. It was written as a basis for class discussion rather than to illustrate effective or ineffective handling of an administrative situation. Copyright 1988 by the University of Virginia Darden School Foundation, Charlottesville, VA. All rights reserved. To order copies, send an email to sales@dardenpublishing.com. No part of this publication may be reproduced, stored in a retrieval system, used in a spreadsheet, or transmitted in any form or by any means—electronic, mechanical, photocopying, recording, or otherwise—without the permission of the Darden School Foundation.

In the light of these difficulties in interpreting accounting information, just what are the important vital signs of corporate health? Is cash flow an important signal? If not, what is? If so, what is the best way to approximate cash flow? How can we predict the impending demise of a company?

To begin to answer some of these important questions, we conducted a study of the financial vital signs of bankrupt and healthy companies. We first identified 66 failed firms from a list provided by Dun and Bradstreet. These firms were in manufacturing or retailing and had financial data available on the Compustat Research tape. Bankruptcy occurred somewhere between 1970 and 1982.

For each of these 66 failed firms, we selected a healthy firm of approximately the same size (as measured by the book value of the firm's assets) from the same industry (3 digit SIC code) as a basis of comparison. This matched sample technique was used to minimize the impact of any extraneous factors (e.g., industry) on the conclusions of the study.

The study was designed to see how well bankruptcy can be predicted two years in advance. A total of 24 financial ratios were computed for each of the 132 firms using data from the Compustat tapes and from Moody's Industrial Manual for the year that was two years prior to the year of bankruptcy. Table 21.10 lists the 24 ratios together with an explanation of the abbreviations used for the fundamental financial variables. All these variables are contained in a firm's annual report with the exception of CFFO. Ratios were used to facilitate comparisons across firms of various sizes.

The first four ratios using CASH in the numerator might be thought of as measures of a firm's cash reservoir with which to pay debts. The three ratios with CURASS in the numerator capture the firm's generation of current assets with which to pay debts. Two ratios, CURDEBT/DEBT and AS-SETS/DEBTS, measure the firm's debt structure. Inventory and receivables turnover are measured by COGS/INV and SALES/REC, and SALES/ASSETS measures the firm's ability to generate sales. The final 12 ratios are asset flow measures.

Assignment

1. What data mining technique(s) would be appropriate in assessing whether there are groups of variables that convey the same information and how important that information is? Conduct such an analysis.

TABLE 21.10	**PREDICTING CORPORATE BANKRUPTCY: FINANCIAL VARIABLES AND RATIOS**		
Abbreviation	**financial Variable**	**Ratio**	**Definition**
ASSETS	Total assets	R_1	CASH/CURDEBT
CASH	Cash	R_2	CASH/SALES
CFFO	Cash flow from operations	R_3	CASH/ASSETS
COGS	Cost of goods sold	R_4	CASH/DEBTS
CURASS	Current assets	R_5	CFFO/SALES
CURDEBT	Current debt	R_6	CFFO/ASSETS
DEBTS	Total debt	R_7	CFFO/DEBTS
INC	Income	R_8	COGS/INV
INCDEP	Income plus depreciation	R_9	CURASS/CURDEBT
INV	Inventory	R_{10}	CURASS/SALES
REC	Receivables	R_{11}	CURASS/ASSETS
SALES	Sales	R_{12}	CURDEBT/DEBTS
WCFO	Working capital from operations	R_{13}	INC/SALES
		R_{14}	INC/ASSETS
		R_{15}	INC/DEBTS
		R_{16}	INCDEP/SALES
		R_{17}	INCDEP/ASSETS
		R_{18}	INCDEP/DEBTS
		R_{19}	SALES/REC
		R_{20}	SALES/ASSETS
		R_{21}	ASSETS/DEBTS
		R_{22}	WCFO/SALES
		R_{23}	WCFO/ASSETS
		R_{24}	WCFO/DEBTS

2. Comment on the distinct goals of profiling the characteristics of bankrupt firms vs. simply predicting (black- box style) whether a firm will go bankrupt and whether both goals, or only one, might be useful. Also comment on the classification methods that would be appropriate in each circumstance.

3. Explore the data to gain a preliminary understanding of which variables might be important in distinguishing bankrupt from nonbankrupt firms. (**Hint:** As part of this analysis, use XLMiner's boxplot option, specifying the bankrupt/not bankrupt variable as the x variable.)

4. For your choice of classifiers, use XLMiner to produce several models to predict whether or not a firm goes bankrupt. Assess model performance on the validation partition.

5. Based on the above, comment on which variables are important in classification, and discuss their effect.

21.10 Time Series Case: Forecasting Public Transportation Demand

`bicup2006.xls` is the dataset for this case study.

Background

Forecasting transportation demand is important for multiple purposes such as staffing, planning, and inventory control. The public transportation system in Santiago de Chile has gone through a major effort of reconstruction. In this context, a business intelligence competition took place in October 2006 that focused on forecasting demand for public transportation. This case is based on the competition, with some modifications.

Problem Description

A public transportation company is expecting an increase in demand for its services and is planning to acquire new buses and to extend its terminals. These investments require a reliable forecast of future demand. To create such forecasts, one can use data on historic demand. The company's data warehouse has data on each 15-minute interval between 6:30am and 22:00pm, on the number of passengers arriving at the terminal. As a forecasting consultant, you have been asked to create a forecasting method that can generate forecasts for the number of passengers arriving at the terminal.

Available Data

Part of the historic information is available in the file `bicup2006.xls`. The file contains the worksheet "Historic Information" with known demand for a 3-week period, separated into 15-minute intervals. The second worksheet ("Future") contains dates and times for a future 3-day period, for which forecasts should be generated (as part of the 2006 competition).

Assignment Goal

Your goal is to create a model/method that produces accurate forecasts. To evaluate your accuracy, partition the given historic data into two periods: a training period (the first two weeks), and a validation period (the last week). Models should be fitted only to the training data and evaluated on the validation data.

Although the competition winning criterion was the lowest Mean Absolute Error (MAE) on the future 3-day data, this is *not* the goal for this assignment. Instead, if we consider a more realistic business context, our goal is to create a model that generates reasonably good forecasts on any time/day of the week.

Consider not only predictive metrics such as MAE, MAPE, and RMSE but also look at actual and forecasted values, overlaid on a time plot, as well as a time plot of the forecast errors.

Assignment

For your final model, present the following summary:

1. Name of the method/combination of methods.
2. A brief description of the method/combination.
3. All estimated equations associated with constructing forecasts from this method.
4. The MAPE and MAE for the training period and the validation period.
5. Forecasts for the future period (March 22–24), in 15-min bins.
6. A single chart showing the fit of the final version of the model to the entire period (including training, validation, and future). Note that this model should be fitted using the combined training + validation data.

Tips and Suggested Steps

1. Use exploratory analysis to identify the components of this time series. Is there a trend? Is there seasonality? If so, how many "seasons" are there? Are there any other visible patterns? Are the patterns global (the same throughout the series) or local?
2. Consider the frequency of the data from a practical and technical point of view. What are some options?
3. Compare the weekdays and weekends. How do they differ? Consider how these differences can be captured by different methods.
4. Examine the series for missing values or unusual values. Think of solutions.
5. Based on the patterns that you found in the data, which models or methods should be considered?
6. Consider how to handle actual counts of zero within the computation of MAPE.

References

Agrawal, R., Imielinski, T., and Swami, A. (1993). Mining associations between sets of items in massive databases, in *Proceedings of the 1993 ACM-SIGMOD International Conference on Management of Data*, pp. 207-216, New York: ACM Press.

Berry, M. J. A., and Linoff, G. S. (1997). *Data Mining Techniques*. New York: Wiley.

Berry, M. J. A., and Linoff, G. S. (2000). *Mastering Data Mining*. New York: Wiley.

Breiman, L., Friedman, J., Olshen, R., and Stone, C. (1984). *Classification and Regression Trees*. Boca Raton, FL: Chapman Hall/CRC (orig. published by Wadsworth).

Chatfield, C. (2003). *The Analysis of Time Series: An Introduction*, 6th ed. Boca Raton, FL:. Chapman Hall/CRC.

Delmaster, R., and Hancock, M. (2001). *Data Mining Explained*. Boston: Digital Press. 148-159.

Efron, B. (1975). The efficiency of logistic regression compared to normal discriminant analysis, *Journal of the American Statistical Association*, vol. 70, number 352, pp. 892-898.

Few, S. (2012). *Show Me the Numbers*, 2nd ed. Oakland, CA: Analytics Press.

Few, S. (2009). *Now You See It*. Analytics Press, Oakland CA, USA.

Golbeck, J. (2013). *Analyzing the Social Web*. Waltham, MA: Morgan Kaufmann.

Han, J., and Kamber, M. (2001). *Data Mining: Concepts and Techniques*. San Diego, CA: Academic Press.

Hand, D., Mannila, H., and Smyth, P. (2001). *Principles of Data Mining*. Cambridge, MA: MIT Press.

Harris, H., Murphy, S., and Vaisman, M. (2013). *Analyzing the Analyzers: An Introspective Survey of Data Scientists and Their Work*, Cambridge, MA: O'Reilly Media.

Hastie, T., Tibshirani, R., and Friedman, J. (2001). *The Elements of Statistical Learning*. New York: Springer.

Hosmer, D. W., and Lemeshow, S. (2000). *Applied Logistic Regression*, 2nd ed. New York: Wiley-Interscience.

Jank, W., and Yahav, I. (2010). E-Loyalty networks in online auctions. *Annals of Applied Statistics*, vol. 4, number 1, pp. 151-178 .

Johnson, W., and Wichern, D. (2002). *Applied Multivariate Statistics*. Upper Saddle River, NJ: Prentice Hall.

Larsen, K. (2005). Generalized naive Bayes classifiers, *SIGKDD Explorations*, vol. 7, number 1, pp. 76-81.

Le, Q. V., Ranzato, M. A., Monga, R., Devin, M., Chen, K., Corrado, G. S., Dean, J., and Ng, A.Y. (2012). Building high-level features using large scale unsupervised learning. In *Proceedings of the Twenty-Ninth International Conference on Machine Learning*. Editors: John Langford and Joelle Pineau. Edinburgh: Omnipress.

Loh, W.-Y., and Shih, Y.-S. (1997), Split selection methods for classification trees, *Statistica Sinica*, vol. 7, pp. 815-840.

Data Mining for Business Analytics: Concepts, Techniques, and Applications with XLMiner®, Third Edition.
Galit Shmueli, Peter C. Bruce, and Nitin R. Patel.
© 2016 John Wiley & Sons Inc. Published 2016 by John Wiley & Sons, Inc.

Lyman, P., and Varian, H. R. (2003). How much information. Retrieved from www.sims.berkeley.edu/how-much-info-2003on Nov. 29, 2005.

Manski, C. F. (1977). The structure of random utility models, *Theory and Decision*, vol. 8, pp. 229-254.

McCullugh, C. E., Paal, B., and Ashdown, S. P. (1998). An optimisation approach to apparel sizing, *Journal of the Operational Research Society,* vol. 49, number 5, pp. 492-499.

Pregibon, D. (1999). 2001: A statistical odyssey. Invited talk at *The Fifth ACM SIGKDD International Conference on Knowledge Discovery and Data Mining*, New York: ACM Press, p. 4.

Shmueli, G. (2012). *Practical Time Series Forecasting: A Hands-On Guide*, 2nd ed. USA: Axelrod-Schnall Publishers.

Siegel, E. (2013). *Predictive Analytics*, New York: Wiley.

Trippi, R., and Turban, E. (eds.) (1996). *Neural Networks in Finance and Investing.* New York: McGraw-Hill.

Veenhoven, R., World Database of Happiness, Erasmus University Rotterdam. Available at http://worlddatabaseofhappiness.eur.nl.

Data Files Used in the Book

1. Accidents.xls
2. Airfares.xls
3. ApplianceShipments.xls
4. AustralianWines.xls
5. Autos-electronics.zip
6. Bankruptcy.xls
7. Banks.xls
8. Banks-adj.xls
9. BathSoap.xls
10. bicup2006.xls
11. BostonHousing.xls
12. CanadianWorkHours.xls
13. CatalogCrossSell.xls
14. Cereals.xlsx
15. CharlesBookClub.xls
16. Cosmetics.xls
17. Cosmetics-small.xls
18. Coursetopics.xlsx
19. DepartmentStoreSales.xls
20. EastWestAirlinesCluster.xls
21. EastWestAirlinesNN.xls
22. eBayAuctions.xls
23. EuropeanJobs.xls
24. Farm-ads.xlsx
25. FlightDelays.xls
26. Fundraising.xls
27. FutureFundraising.xls
28. GermanCredit.xls

Data Mining for Business Analytics: Concepts, Techniques, and Applications with XLMiner®, Third Edition.
Galit Shmueli, Peter C. Bruce, and Nitin R. Patel.
© 2016 John Wiley & Sons Inc. Published 2016 by John Wiley & Sons, Inc.

29. Hair-Care-Product.xls
30. LaptopSales.zip
31. LaptopSalesJanuary2008.xls
32. MortgageDefaulters.xls
33. Pharmaceuticals.xls
34. RidingMowers.xls
35. ShampooSales.xls
36. Sept11Travel.xls
37. SP500.xls
38. SouvenirSales.xls
39. Spambase.xls
40. SystemAdministrators.xls
41. Tayko.xls
42. Textiles.xls
43. ToyotaCorolla.xls
44. ToysRUsRevenues.xls
45. UniversalBank.xls
46. Universities.xls
47. Utilities.xls
48. Voter-Persuasion.xlsx
49. WalMartStock.xls
50. West Roxbury.xls
51. Wine.xls

INDEX

Data Mining for Business Analytics: Concepts, Techniques, and Applications with XLMiner®, Third Edition.
Galit Shmueli, Peter C. Bruce, and Nitin R. Patel.
© 2016 John Wiley & Sons Inc. Published 2016 by John Wiley & Sons, Inc.